Millwright

Level One

Trainee Guide
Fourth Edition

Pearson

NCCER

President and Chief Executive Officer: Boyd Worsham
Vice President – Innovation and Advancement: Jennifer Wilkerson
Chief Learning Officer: Lisa Strite
Millwright Project Manager: Dario VanHorne
Senior Manager of Projects: Chris Wilson
Senior Production Manager: Erin O'Nora
Testing/Assessment Project Manager: Elizabeth Schlaupitz
Testing and Assessment Project Coordinator: Chelsi Causley
Technical Writers: Don Congdon, Troy Staton
Art Manager: Carrie Pelzer
Permissions Specialists: Kay Lewis, Amanda Smith
Desktop Publishing Manager: James McKay
Production Specialists: Gene Page, Eric Caraballoso, Daphney Milian
Digital Content Coordinator: Rachael Downs
Managing Editor: Graham Hack
Editors: Jordan Hutchinson, Karina Kuchta, Hannah Murray, Zi Meng

Pearson

Manager, Program Management: Vanessa Price
Employability Solutions Coordinator: Monica Perez
Senior Content Producer: Alexandrina B. Wolf
Content Producers: Alma Dabral, Aurodeep Mukherjee

Composition: NCCER
Printer/Binder: Lakeside Book Company
Cover Printer: Lakeside Book Company
Text Fonts: Palatino and Univers
Content Technologies: Gnostyx

Cover Image

Pruftechnik, a Fluke Reliability Brand

Credits and acknowledgments for content borrowed from other sources and reproduced, with permission, in this textbook appear at the end of each module.

Pearson

ISBN-13: 978-0-13-765382-9

12 2025

Preface

To the Trainee

In choosing to begin Millwright training, you are seizing the opportunity to enjoy a career that demands complex mechanical knowledge and sharp analytical ability in a variety of industrial work environments. Since its humble beginnings in the construction of wood mills, the millwright trade has expanded to include work in metal and machinery of ever-increasing technology and precision.

Millwrights install, align, and troubleshoot machinery in factories, power plants (particularly the precision machinery required in nuclear power plants), and other industrial sites. They install conveyor systems, connect machinery to power supplies and piping, direct hoisting and setting of machines, and adjust the moving and stationary parts of machines to certain specifications. Millwrights are extremely skilled at mathematics and interpreting blueprints and specs to set machines at perfect measurements, sometimes working with clearances no bigger than thousandths of an inch.

Millwrights are a specialized and elite group, as there are only a little more than 45,000 millwrights employed in the United States today. However, over the next decade, there will be a demand for 3,400 more (US Bureau of Labor Statistics). As the population grows, especially in developing countries, increased demands for energy and travel in particular will require more millwrights working in power plants, refineries, and factories. A trained and experienced millwright enjoys a comfortable salary and the chance for different avenues of professional development. A millwright can progress upwards in the trade, undergoing full apprenticeship, becoming a supervisor, and/or obtaining higher education. Millwrights may also opt for careers in related professions, such as machinist, maintenance mechanic, equipment engineering, or aircraft assembly.

This is the first installment of a four-level curriculum that meets the requirements of a standard millwright apprenticeship program (3–5 years in apprenticeship or 10,000 hours of on-the-job training). If you employ discipline in working with detail, problem-solving, and if you enjoy working independently in a variety of work environments, you will excel in this training program. NCCER wishes you luck as you embark on your chosen career path. With solid millwright training, you are opening the doors to a field scheduled for growth in the next decade.

We invite you to visit the NCCER website at **https://www.nccer.org** for information on the latest product releases and training.

Your feedback is welcome. You may email your comments to **curriculum@nccer.org** or send general comments and inquiries to **info@nccer.org**.

New with *Millwright Level One*

Millwright Level One will help you organize your progression in your journey to master the essential millwrighting skills. Level One is written for students, apprentices, journeymen, and professionals who wish to sharpen their skills in the millwright craft. The text is organized into six modules. The first three modules provide general information about the craft, the tools, and the parts you will come across when you step into a day of work. The following three modules cover the essential practical skills and processes you will employ during a day in the field as a millwright. The accompanying images have been updated to show the most modern tools and equipment in use. We provide you added value by including elective modules, such as Field Sketching, a skill that will allow you to convey technical information confidently and quickly.

NCCER Standardized Curricula

NCCER is a not-for-profit 501(c)(3) education foundation established in 1996 by the world's largest and most progressive construction companies and national construction associations. It was founded to address the severe workforce shortage facing the industry and to develop a standardized training process and curricula. Today, NCCER is supported by hundreds of leading construction and maintenance companies, manufacturers, and national associations. The NCCER Standardized Curricula was developed by NCCER in partnership with Pearson, the world's largest educational publisher.

Some features of the NCCER Standardized Curricula are as follows:

- An industry-proven record of success
- Curricula developed by the industry, for the industry
- National standardization providing portability of learned job skills and educational credits
- Compliance with the Office of Apprenticeship requirements for related classroom training (*CFR 29:29*)
- Well-illustrated, up-to-date, and practical information

NCCER also maintains the NCCER Registry, which provides transcripts, certificates, and wallet cards to individuals who have successfully completed a level of training within a craft in NCCER's Curricula. *Training programs must be delivered by an NCCER Accredited Training Sponsor in order to receive these credentials.*

For information on NCCER's credentials and the NCCER Registry, contact NCCER Customer Service at 1-888-622-3720 or visit **https://www.nccer.org**.

Special Features

In an effort to provide a comprehensive and user-friendly training resource, this curriculum showcases several informative features. Whether you are a visual or hands-on learner, these features are intended to enhance your knowledge of the construction industry as you progress in your training. Some of the features you may find in the curriculum are explained below.

Introduction

This introductory page, found at the beginning of each module, lists the module Objectives, Performance Tasks, and Trade Terms. The Objectives list the knowledge you will acquire after successfully completing the module. The Performance Tasks give you an opportunity to apply your knowledge to real-world tasks. The Trade Terms are industry-specific vocabulary that you will learn as you study the module.

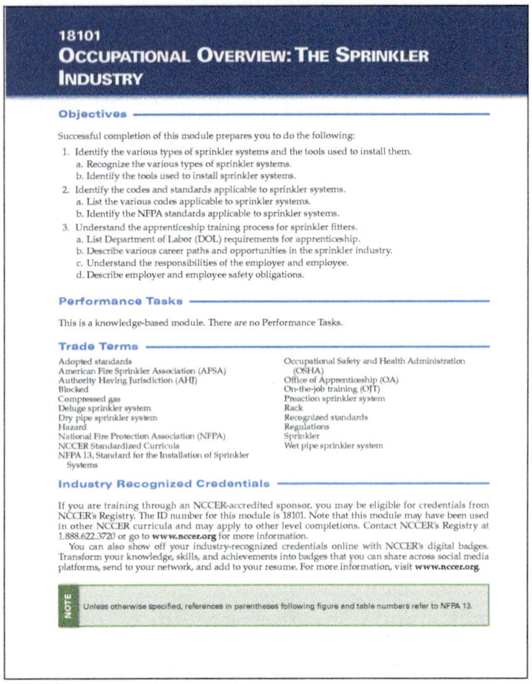

18101
OCCUPATIONAL OVERVIEW: THE SPRINKLER INDUSTRY

Objectives

Successful completion of this module prepares you to do the following:

1. Identify the various types of sprinkler systems and the tools used to install them.
   a. Recognize the various types of sprinkler systems.
   b. Identify the tools used to install sprinkler systems.
2. Identify the codes and standards applicable to sprinkler systems.
   a. List the various codes applicable to sprinkler systems.
   b. Identify the NFPA standards applicable to sprinkler systems.
3. Understand the apprenticeship training process for sprinkler fitters.
   a. List Department of Labor (DOL) requirements for apprenticeship.
   b. Describe various career paths and opportunities in the sprinkler industry.
   c. Understand the responsibilities of the employer and employee.
   d. Describe employer and employee safety obligations.

Performance Tasks

This is a knowledge-based module. There are no Performance Tasks.

Trade Terms

Adopted standards
American Fire Sprinkler Association (AFSA)
Authority Having Jurisdiction (AHJ)
Blocked
Compressed gas
Deluge sprinkler system
Dry pipe sprinkler system
Hazard
National Fire Protection Association (NFPA)
NCCER Standardized Curricula
NFPA 13, Standard for the Installation of Sprinkler Systems
Occupational Safety and Health Administration (OSHA)
Office of Apprenticeship (OA)
On-the-job training (OJT)
Preaction sprinkler system
Rack
Recognized standards
Regulations
Sprinkler
Wet pipe sprinkler system

Industry Recognized Credentials

If you are training through an NCCER-accredited sponsor, you may be eligible for credentials from NCCER's Registry. The ID number for this module is 18101. Note that this module may have been used in other NCCER curricula and may apply to other level completions. Contact NCCER's Registry at 1.888.622.3720 or go to **www.nccer.org** for more information.

You can also show off your industry-recognized credentials online with NCCER's digital badges. Transform your knowledge, skills, and achievements into badges that you can share across social media platforms, send to your network, and add to your resume. For more information, visit **www.nccer.org**.

NOTE: Unless otherwise specified, references in parentheses following figure and table numbers refer to NFPA 13.

Trade Features

Trade features present technical tips and professional practices based on real-life scenarios similar to those you might encounter on the jobsite.

Bowline Trivia

Some people use this saying to help them remember how to tie a bowline: "The rabbit comes out of his hole, around a tree, and back into the hole."

Figures and Tables

Photographs, drawings, diagrams, and tables are used throughout each module to illustrate important concepts and provide clarity for complex instructions. Text references to figures and tables are emphasized with *italic* type.

Figure 31 Marking a cutting line.

Notes, Cautions, and Warnings

Safety features are set off from the main text in highlighted boxes and categorized according to the potential danger involved. Notes simply provide additional information. Cautions flag a hazardous issue that could cause damage to materials or equipment. Warnings stress a potentially dangerous situation that could result in injury or death to workers.

NOTE: Nameplates must be posted on each material handling device. The nameplate must indicate

CAUTION: It is essential to note the revision designation on a construction drawing and to use only the latest

WARNING! Saw teeth are very sharp. Use gloves and do not handle the saw teeth with bare hands. When cutting with a saw, ensure that your fingers remain clear of the teeth at all times.

Going Green

Going Green looks at ways to preserve the environment, save energy, and make good choices regarding the health of the planet. Through the introduction of new construction practices and products, you will see how the greening of America has already taken root.

> **Going Green**
>
> **Big Horn Wind Farm**
>
> The Big Horn wind farm uses 133 GE turbines, each rising 119 meters (389 feet) from the ground to the top of the blade tips. The wind farm uses only a

Did You Know?

The *Did You Know?* features offer hints, tips, and other helpful bits of information.

> **Did You Know?**
>
> **How Blueprints Started**
>
> The process for making blueprints was developed in 1842 by an English astronomer named Sir John F. Herschel. The method involved coating a paper with a special chemical. After the coating dried, an original hand drawing was placed on top of the paper. Both papers were then covered with a piece of glass and set in the sunlight for about an hour. The coated paper was developed much like a photograph. After a cold-water wash, the coated paper turned blue, and the lines of the drawing remained white.

Step-by-Step Instructions

Step-by-step instructions are used throughout to guide you through technical procedures and tasks from start to finish. These steps show you not only how to perform a task but also how to do it safely and efficiently.

> Perform the following steps to erect this system area scaffold:
>
> *Step 1* Gather and inspect all scaffold equipment for the scaffold arrangement.
>
> *Step 2* Place appropriate mudsills in their approximate locations.
>
> *Step 3* Attach the screw jacks to the mudsills.
>
> *Step 4* Adjust the screw jacks to near their lowest position.
>
> *Step 5* Determine the location of the highest base.

Trade Terms

Each module presents a list of Trade Terms that are discussed within the text and defined in the Glossary at the end of the module. These terms are denoted in the text with **bold, blue type** upon their first occurance. To make searches for key information easier, a comprehensive Glossary of Trade Terms from all modules is located at the back of this book.

> During a rigging operation, the **load** being lifted or moved must be connected to the apparatus, such as a crane, that will provide the power for movement. The connector—the link between the load and the apparatus—is often a sling made of synthetic, chain, or **wire rope** materials. This section focuses on three types of slings:
>
> - Synthetic slings
> - Alloy steel chain slings
> - Wire rope slings

Section Review

Section Review questions can be found at the end of each section to test your knowledge of the content.

> **1.0.0 Section Review**
>
> 1. The type of sprinkler system that might be found in an aircraft hangar is a _____.
>   a. dry pipe sprinkler system
>   b. wet pipe sprinkler system
>   c. preaction sprinkler system
>   d. deluge sprinkler system
> 2. A type of ladder that provides a platform for standing is a(n) _____.
>   a. A-frame
>   b. extension trestle ladder
>   c. podium ladder
>   d. movable leg ladder

Review Questions

Review Questions are provided to reinforce the knowledge you have gained. This makes them a useful tool for measuring what you have learned.

> **Review Questions**
>
> 1. Identification tags for slings must include the _____.
>   a. type of protective pads to use
>   b. type of damage sustained during use
>   c. color of the tattle-tail
>   d. manufacturer's name or trademark
> 2. The type of wire rope core that is susceptible to heat damage at relatively low temperatures is the _____.
>   a. fiber core
>   b. strand core
>   c. independent wire rope core
>   d. metallic link supporting core
> 3. Synthetic slings must be inspected _____.
>   a. once every month
>   b. visually at the start of each work week
>   c. before every use
>   d. once wear or damage becomes apparent
> 4. An alloy steel chain sling must be removed from service if there is evidence that _____.
>   a. the sli...
> 7. Chain hoists are able to lift heavy loads by utilizing a _____.
>   a. rope and pulley system
>   b. rigger's strength
>   c. stationary counterweight
>   d. gear system
> 8. Before attempting to lift a load with a chain hoist, make sure that the _____.
>   a. hoist is secured to a come-along
>   b. load is properly balanced
>   c. tag lines are properly anchored
>   d. tackle is connected to its power source
> 9. A hitch configuration that allows slings to be connected to the same load without using a spreader beam is a _____.
>   a. double-wrap hitch
>   b. choker hitch
>   c. bridle hitch
>   d. basket hitch
> 10. To make the emergency stop signal that is used by riggers, extend both...

NCCER Standardized Curricula

NCCER's training programs comprise more than 80 construction, maintenance, pipeline, and utility areas and include skills assessments, safety training, and management education.

Construction and Maintenance

Boilermaking
Cabinetmaking
Carpentry
Concrete Construction
Construction Craft Laborer
Construction Technology
Core: Introduction to Basic Construction Skills
Drywall
Electrical
Electronic Systems Technician
Heating, Ventilating, and Air Conditioning
Heavy Equipment Operations
Heavy Highway Construction
Hydroblasting
Industrial Coating and Lining Application Specialist
Industrial Maintenance Electrical and Instrumentation Technician
Industrial Maintenance Mechanic
Instrumentation
Ironworking
Manufactured Construction Technology
Masonry
Mechanical Insulating
Millwright
Mobile Crane Operations
Painting
Painting, Industrial
Pipefitting
Pipelayer
Plumbing
Reinforcing Ironwork
Rigging
Roofing
Scaffolding
Sheet Metal
Signal Person
Site Layout
Sprinkler Fitting
Tower Crane Operator
Welding

Maritime

Maritime Industry Fundamentals
Maritime Electrical
Maritime Pipefitting
Maritime Structural Fitter
Maritime Welding
Maritime Aluminum Welding

Green/Sustainable Construction

Building Auditor
Fundamentals of Weatherization
Introduction to Weatherization
Sustainable Construction Supervisor
Weatherization Crew Chief
Weatherization Technician
Your Role in the Green Environment

Energy

Alternative Energy
Introduction to the Power Industry
Introduction to Solar Photovoltaics
Power Generation Maintenance Electrician
Power Generation I&C Maintenance Technician
Power Generation Maintenance Mechanic
Power Line Worker
Power Line Worker: Distribution
Power Line Worker: Substation
Power Line Worker: Transmission
Solar Photovoltaic Systems Installer
Wind Energy
Wind Turbine Maintenance Technician

Pipeline

Abnormal Operating Conditions, Control Center
Abnormal Operating Conditions, Field and Gas
Corrosion Control
Electrical and Instrumentation
Field and Control Center Operations
Introduction to the Pipeline Industry
Maintenance
Mechanical

Safety

Fall Protection Orientation
Field Safety
Safety Orientation
Safety Technology

Supplemental Titles

Applied Construction Math
Tools for Success

Management

Construction Workforce Development Professional
Fundamentals of Crew Leadership
Mentoring for Craft Professionals
Project Management
Project Supervision

Spanish Titles

Acabado de concreto: nivel uno (*Concrete Finishing Level One*)
Aislamiento: nivel uno (*Insulating Level One*)
Albañilería: nivel uno (*Masonry Level One*)
Andamios (*Scaffolding*)
Carpintería: Formas para carpintería, nivel tres (*Carpentry: Carpentry Forms, Level Three*)
Currículo básico: habilidades introductorias del oficio (*Core Curriculum: Introductory Craft Skills*)
Electricidad: nivel uno (*Electrical Level One*)
Herrería: nivel uno (*Ironworking Level One*)
Herrería de refuerzo: nivel uno (*Reinforcing Ironwork Level One*)
Instalación de rociadores: nivel uno (*Sprinkler Fitting Level One*)
Instalación de tuberías: nivel uno (*Pipefitting Level One*)
Instrumentación: nivel uno, nivel dos, nivel tres, nivel cuatro (*Instrumentation Levels One through Four*)
Orientación de seguridad (*Safety Orientation*)
Paneles de yeso: nivel uno (*Drywall Level One*)
Seguridad de campo (*Field Safety*)

Acknowledgments

This curriculum was revised as a result of the vision and leadership of the following sponsors:

ABC Pelican Chapter
Air Liquide Large Industries U.S. LP
Cianbro Corporation
Dow Chemical
FLSmidth USA Inc.
Great Basin College
H&H Engineering & Supply Co.
Industrial Parts Specialties
Lee College
TIC
Turner Industries

This curriculum would not exist were it not for the dedication and unselfish energy of those volunteers who served on the Authoring Team. A sincere thanks is extended to the following:

Jack Baker
Thomas Bruns
Kelly French
Maurice Gould
Paul Howard
Ronald Jones
James Maloney
Kim Panepinto
Dave Roberts
Neil Steagall
Michael Stirrat

NCCER Partners

NCCER partnering organizations are national associations and organizations that share a common interest in the goals and objectives of NCCER. To learn more about NCCER business partners, go to **https://www.nccer.org/about-us/partners**.

You can also scan this code using the camera on your phone or mobile device to view these partnering organizations.

Contents

Module Six

Field Sketching (Elective)

Teaches the basic skills needed to make a good field sketch to convey information about how parts should be made or assembled. (Module ID 15202; 10 Hours)

Glossary

This course map shows all of the modules in *Millwright Level One*. The suggested training order begins at the bottom and proceeds up. Skill levels increase as you advance on the course map. The local Training Program Sponsor may adjust the training order.

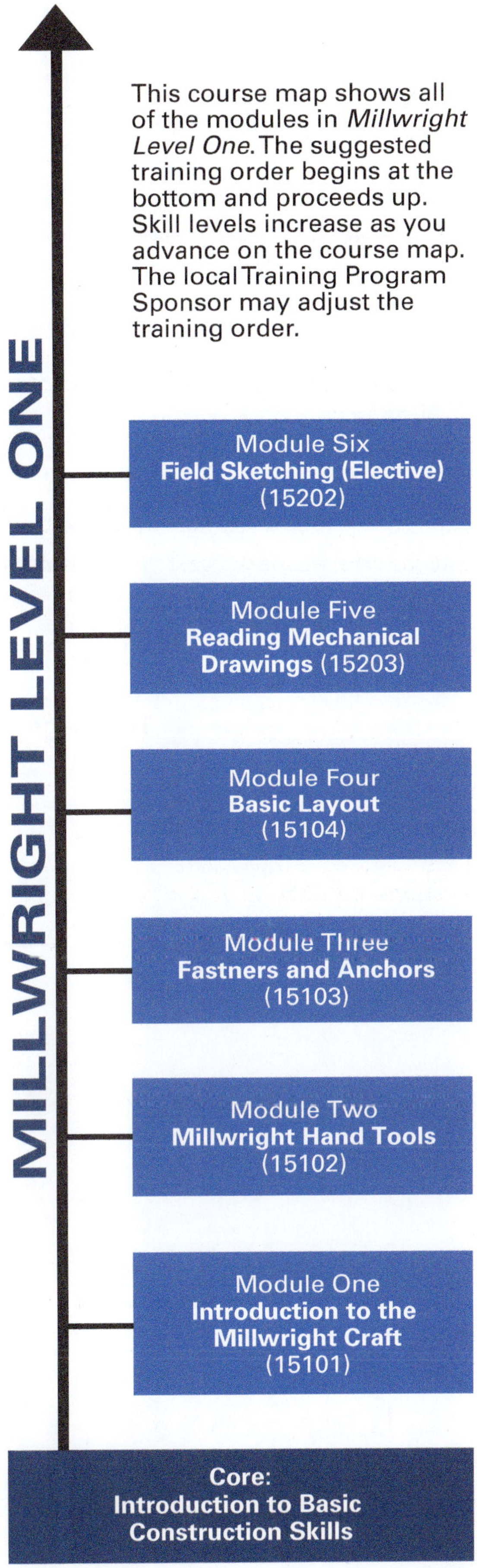

This page is intentionally left blank.

Introduction to the Millwright Craft

Overview

The millwright craft developed from the need for skilled craftworkers to build and maintain water-driven flour and grain mills. Millwrights needed to master several related crafts to construct the mills and keep them running. Since then, the size and complexity of industrial machinery has grown tremendously. The millwright craft has expanded and grown to accommodate the changing needs of today's industrial and manufacturing facilities.

Millwrights need to know how to safely move machinery and equipment, assemble and install machinery, read construction and mechanical drawings, and maintain various types of equipment. A millwright is a proud, respected problem solver who uses many different tools and skills to accomplish tasks with precision. This module provides an overview of the millwright craft, identifies the personal characteristics of successful craftworkers, and reviews the structure of craft apprenticeship programs.

Module 15101

Trainees with successful module completions may be eligible for credentialing through the NCCER Registry. To learn more, go to **www.nccer.org** or contact us at 1.888.622.3720. Our website, **www.nccer.org**, has information on the latest product releases and training.

Your feedback is welcome. You may email your comments to **curriculum@nccer.org**, send general comments and inquiries to **info@nccer.org**, or fill in the User Update form at the back of this module.

This information is general in nature and intended for training purposes only. Actual performance of activities described in this manual requires compliance with all applicable operating, service, maintenance, and safety procedures under the direction of qualified personnel. References in this manual to patented or proprietary devices do not constitute a recommendation of their use.

15101 V4.0

15101

INTRODUCTION TO THE MILLWRIGHT CRAFT

Objectives

Successful completion of this module prepares you to do the following:

1. Summarize the history of the millwright craft and the opportunities it offers.
   a. Recount the history of the millwright craft.
   b. Describe the work and responsibilities of millwrights.
   c. Identify career opportunities for millwrights.
   d. State general safety guidelines for the construction workplace.
2. Identify the traits of a successful craftworker and describe modern apprenticeship training.
   a. Identify the personal traits and characteristics of a successful craftworker.
   b. Describe modern apprenticeship training.

Performance Tasks

This is a knowledge-based module. There are no Performance Tasks.

Trade Terms

Accredited
Apprentice
Computer numerical control (CNC) machines
Dies
Lathe
Micrometers
Plasma arc cutters
Shutdowns
Taps
Turbine
Ultrasonic

Industry Recognized Credentials

If you are training through an NCCER-accredited sponsor, you may be eligible for credentials from NCCER's Registry. The ID number for this module is 15101. Note that this module may have been used in other NCCER curricula and may apply to other level completions. Contact NCCER's Registry at 1.888.622.3720 or go to **www.nccer.org** for more information.

You can also show off your industry-recognized credentials online with NCCER's digital badges. Transform your knowledge, skills, and achievements into badges that you can share across social media platforms, send to your network, and add to your resume. For more information, visit **www.nccer.org**.

Contents

Figures

Section One

1.0.0 The Millwright Craft

Objective

Summarize the history of the millwright craft and the opportunities it offers.

a. Recount the history of the millwright craft.
b. Describe the work and responsibilities of millwrights.
c. Identify career opportunities for millwrights.
d. State general safety guidelines for the construction workplace.

Trade Terms

Apprentice: A person learning a craft through practical experience with skilled workers, as well as through instruction and study, while earning a wage.

Computer numerical control (CNC) machines: Equipment used in a subtractive manufacturing process based on computerized controls and machine tools that remove material from a workpiece to produce a component. Many materials can be machined in this manner, including metals, plastics, and composites.

Dies: Threading tools typically shaped like a nut, used to cut external (male) threads on properly sized round stock.

Lathe: A machine tool designed to shape metal, wood, and similar materials using a rotating drive that turns the workpiece in opposition to handheld or stationary cutting tools.

Micrometers: Precision measuring tools that measure small distances or thicknesses between two faces, adjusted to contact the workpiece(s) by turning an adjusting barrel.

Plasma arc cutters: Equipment used like a cutting torch to cut through electrically conductive materials using a jet of superheated plasma at a temperature of roughly 40,000°F (~22,200°C) generated by a handheld or automated torch head. Plasma cutting offers improved speed and precision compared to traditional oxyfuel cutting processes.

Shutdowns: Periods of time when facility production is largely halted. Although the term can be applied to the cessation of production for any reason, industrial crafts often use the term *shutdown* to identify pre-planned production interruptions that may last for days or even weeks to allow for scheduled maintenance and significant changes to production equipment.

Taps: Threading tools used to cut internal (female) threads in properly sized round holes.

Turbine: A machine for producing power in which a wheel or rotor fitted with a series of vanes is forced to rotate by the flow of steam, gas, water, air, or similar fluids.

Ultrasonic: Related to the use of sound waves moving at a frequency above the upper auditory limit of the human ear (about 20,000 hertz).

Millwrights dismantle, move, install, repair, and replace machinery and heavy equipment used in many industries. The work of today's millwright is much broader in scope than that of millwrights in the seventeenth and eighteenth centuries. The wide range of potential workplaces and the development of new technology require millwrights to continually train and upgrade their skills. To find success as a millwright, the journey must begin with training in a variety of topics, providing a foundation on which to build a career. This program provides the knowledge and skills that today's industries require of millwrights.

1.1.0 History of the Millwright Craft

In the mid-eighteenth century, millwrights were not part of an independent craft, but instead functioned as a division of the carpenter's craft. A millwright's primary duties were to build and maintain the water mills used to grind grain into flour. A young millwright might begin as a carpenter's apprentice or learn the craft directly from a family member or mentor.

By the end of the 1780s, an inventor by the name of Oliver Evans had devised an automated mill to produce flour. This new automated process was thought by some to be a replacement for humans in the manufacturing process. Of course, as we now know, removing humans from the process entirely is not easy. Even when the process worked as advertised, maintenance and repair work for millwrights became steadier and more complex.

In England and Scotland, the original millwright was recognized as a person with knowledge of many different crafts. Many of the mills developed in this part of the world were powered by wind rather than water. As a result, energy had to be transmitted to the process from a different and less reliable source. Millwrights were called upon to develop new methods and techniques for power transmission.

Masters of the craft were thought of as engineers, people who were equally comfortable at the **lathe**, the blacksmith's anvil, and the carpenter's bench (*Figure 1*).

The early millwright was knowledgeable about math and geometry as well, further supporting their recognition as engineers. They could calculate the power of various machines; draw mechanical plans; and even construct buildings, conduits, or watercourses when the need arose. Indeed, the millwright craft is considered by many to be the forerunner of the mechanical engineering profession.

Over the years, industrial and manufacturing processes flourished and became more complex. The added complexity, among other things, has changed the nature of the craft. Today, the millwright craft has expanded beyond the days when millwrights were merely responsible for grain mills. However, millwrights remain proud of their heritage and have the respect of other craftworkers who work in their midst.

Figure 1 Artisan at work on a wood lathe.

1.2.0 Today's Millwright

A millwright's responsibilities often begin before machinery arrives at a jobsite, as some pieces must be moved into service immediately. New equipment must be unloaded, inspected, and then moved into position. To lift and move light machinery, millwrights use rigging and hoisting devices such as jacks, pulleys, and chain falls. To handle heavier equipment, they often require the assistance of forklifts or mobile cranes. Because millwrights often choose the rigging devices for a task, they must understand and respect the loadbearing properties and limitations of ropes, cables, hoists, and cranes.

Equipment must be accurately positioned relative to established points or other machinery. Engineers typically determine with precision where the equipment must be located to function properly. This information is documented on drawings, and additional information is often found in the written specifications for a job. Millwrights are then responsible for interpreting the project documents and accurately establishing the physical location.

Did You Know?

Oliver Evans

Oliver Evans, an American inventor born in Delaware and residing in the Northeast, was a true visionary and pioneer in industrialization. His accomplishments include designing and building the first high-pressure steam engine and developing the first automated industrial process.

Oliver began as a wheelright apprentice—a craftworker skilled in the repair of wooden wheels. Later in life, he went into business with his brothers and spent years developing an automated mill. He later became a strong proponent of high-pressure steam as an energy source and developed his first high-pressure steam engine. He was also the first to describe the vapor-compression refrigeration cycle that is still in use today, although 30 more years passed before a working example was successfully constructed by others.

Oliver Evans eventually published a manual covering his invention of the new milling process, with updates spanning multiple editions. The manual, *The Young Millwright & Miller's Guide*, presented what a millwright of that era needed to know and the details of what was known as the Evans process. Surprisingly, the 12th edition of his book, published in 1848, is still available, having been reprinted in 2018.

In some cases, the equipment requires a new foundation. A machinery foundation serves several purposes, and its design largely depends upon the equipment's weight and operating characteristics. The base upon which the foundation will be constructed is also a factor. Millwrights may prepare the foundation themselves or lay out and supervise its construction by others. As a result, they must know how to read construction drawings and be familiar with a variety of building materials.

To assemble machinery, millwrights must fit bearings, align gears and wheels, attach motors, and connect drive components according to a manufacturer's drawings and instructions. Precision leveling and the alignment of moving parts are crucial in the assembly process, and millwrights are the best of the best when precision is essential. They work with precision measuring instruments such as **micrometers** (*Figure 2*), routinely measuring values to the nearest 0.001" (0.0254 millimeters). Laser alignment systems and **ultrasonic** measuring devices are also common to the field. Because of the critical and precise nature of their work, millwrights are an essential part of company quality assurance systems and processes.

Millwrights also work with hand and power tools, such as thread **taps** and **dies** (*Figure 3*), special wrenches, grinders, and cutting torches or **plasma arc cutters** (*Figure 4*). In addition to installing and dismantling machinery, many millwrights work as part of a team with industrial maintenance mechanics and instrumentation specialists to repair and maintain equipment.

Increasingly sophisticated processes and automation means more complicated equipment for millwrights to install and maintain. For example, millwrights may install and maintain **computer numerical control (CNC) machines**—machine tools that fabricate parts with little human intervention. This machinery requires special care and training to operate, so millwrights must work closely with programmers, electricians, engineers, and manufacturers' representatives to complete the task.

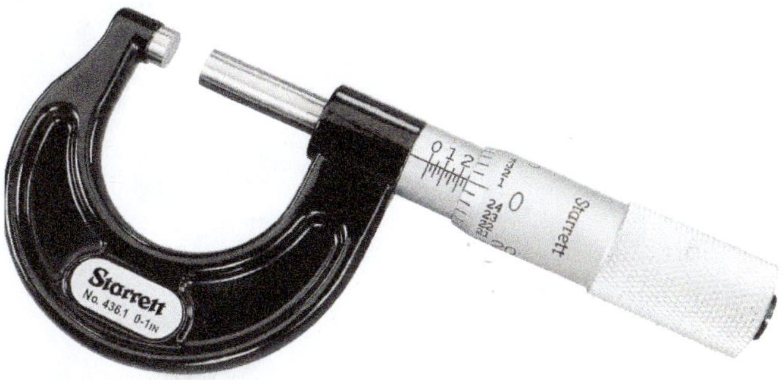

Figure 2 A micrometer.

(A) Thread Tap

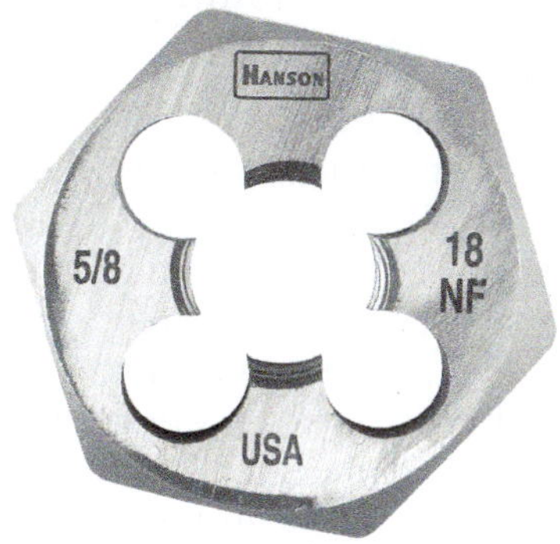

(B) Thread Die

Figure 3 Thread tap and die.

Figure 4 Working with a plasma arc cutter.

1.2.1 Training and Advancement

Millwrights often receive training for four to five years through apprenticeship programs that combine on-the-job learning (OJL) with classroom instruction. Others learn through community college coursework coupled with informal OJL. Apprenticeship programs include classroom training and field participation in dismantling, moving, erecting, and repairing machinery. Classroom training typically includes topics such as the following:

- Mathematics
- Mechanical and architectural drawings
- Hydraulics
- Pneumatics
- Bearings and seals
- Pump design and repair (*Figure 5*)
- Rotating shaft alignment
- Optical alignment for site layout
- **Turbine** design and construction (*Figure 6*)

Trainees routinely receive instruction in related skills such as pipefitting, welding, and vibration analysis. Since the role of a millwright can vary significantly from one industry to another, skills that are critical in one place may not be very important at another location. All millwrights, especially trainees, should seize every opportunity to learn a new skill. A broad skill set enhances your potential for advancement and allows you to take advantage of the best employment opportunities.

Employers prefer applicants with a high school diploma or the equivalent, as well as some vocational training or experience. High school or technical school courses in science, mathematics, mechanical drawing, and machine shop practices are always helpful. Even experienced millwrights are expected to maintain their basic skills and attend additional training on new equipment and procedures. A millwright's educational process should never come to an end.

Figure 5 Centrifugal water pump.

Figure 6 Industrial gas turbine.

Because millwrights assemble, disassemble, and repair complex equipment, mechanical aptitude is very important. Reasonable strength and agility are also necessary for common lifting and climbing tasks. Millwrights need good interpersonal and communication skills to work as part of a team and to effectively mentor team members with less experience.

1.2.2 Working Conditions

Working conditions vary by industry. Millwrights employed in the manufacturing environment work in a typical shop setting and on the manufacturing floor. Those engaged in new construction may work outdoors in more challenging weather conditions. Like all construction and industrial craftworkers, millwrights must use personal protective equipment (PPE) to avoid common hazards. Safety harnesses, safety glasses or goggles, safety shoes, and hard hats are just a few of the many examples of PPE. All craftworkers must make safety their highest priority, regardless of the environment.

Millwrights may work independently or as part of a team. They must perform tasks quickly and precisely, as **shutdowns** and disabled machinery cost companies a great deal of time and money. Many millwrights work overtime to complete critical tasks, especially during a shutdown. They may also work in shifts around the clock when the schedule demands it. Although paying overtime is expensive for an employer, the financial losses from an idle production line are generally far greater. Many craftworkers look forward to the extra income and a chance to prove they are masters of their craft.

1.3.0 Millwright Career Paths

Once you complete your initial training and gain some experience, many career opportunities become available. If you have leadership skills, you can become the foreman of a millwright crew or the superintendent of a larger group. Some experienced millwrights become instructors for contractors, industrial companies, or career and technical education (CTE) schools. Depending on your level of skill, experience, and desire to be an entrepreneur, you may even wish to own and operate your own contracting business. *Figure 7* provides some examples of millwright career paths. However, there are many more possible paths to success, and success in the workplace can be defined in a variety of ways.

Setting Up an Equipment Baseplate

Baseplates for equipment such as motors and pumps must be established with precision. Such tasks are all in a day's work for a millwright.

1.3.1 Employment

More than 47,000 millwrights were on the job in 2019, according to the US Bureau of Labor Statistics (BLS). The mean annual wage for millwrights at that time was $57,000. Millwrights employed in the electric power generation, transmission, and distribution industries earned the highest wage, with a mean annual wage exceeding $86,000. Other industries with higher annual wages for millwrights included the following:

- *Scientific research and development services* — $76,000
- *Utility system construction* — $73,000
- *Other heavy and civil engineering construction* — $72,000
- *Aerospace product and parts manufacturing* — $71,000

Texas employed the most millwrights in 2019, along with Michigan, Ohio, California, and Indiana. The mean annual wage for millwrights in New Mexico was the highest among the states, although the cost of living must always be factored into local and regional wages. Alaska, New York, California, and Washington round out the top five in annual wages for millwrights.

Most millwrights work in manufacturing facilities, refineries, petrochemical plants, and durable goods industries. The durable goods industry includes motor vehicle assembly and parts plants, as well as iron and steel mills. Although millwrights work in every state, employment is typically concentrated in industrialized areas.

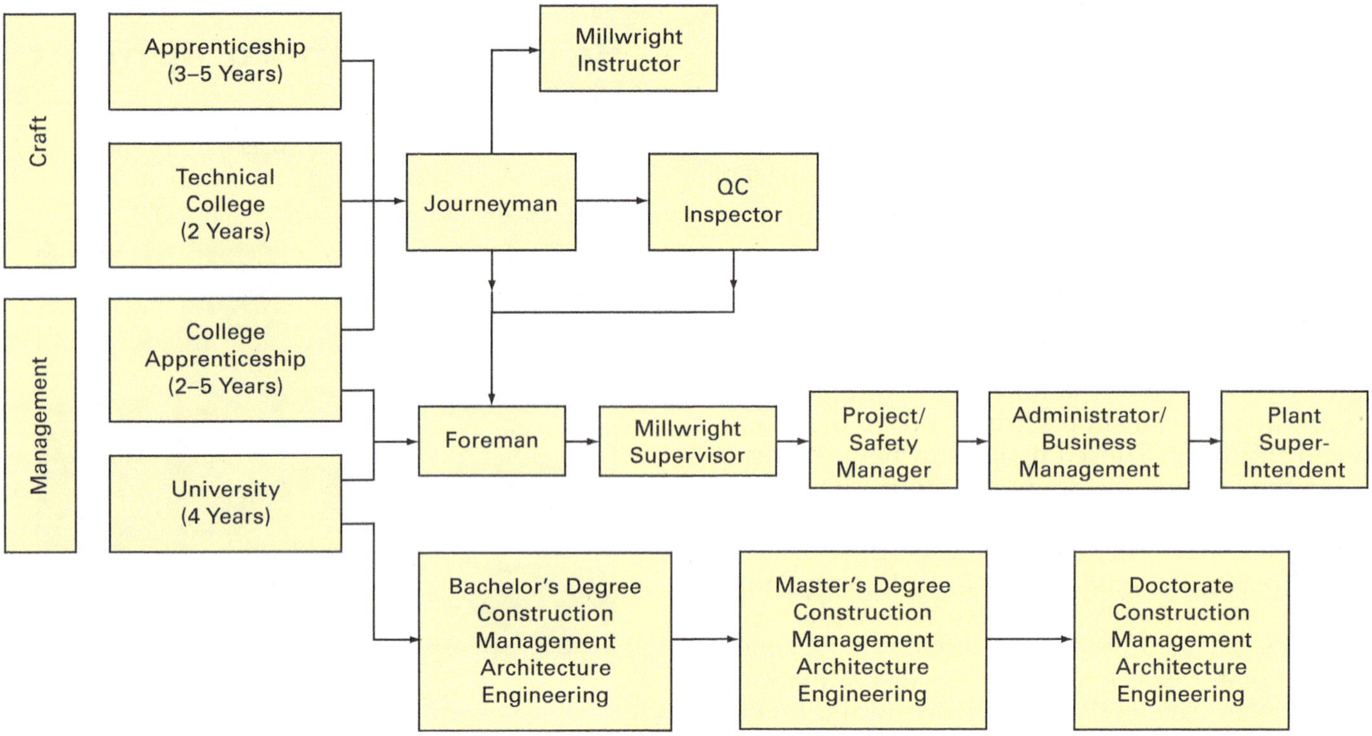

Figure 7 Example career paths for millwrights.

The outlook for the millwright craft is bright, as they are currently in demand, and demand is expected to increase in the future.

Because millwrights maintain and repair existing machinery, dismantle old machinery, and install new equipment, properly trained and skilled applicants have excellent job opportunities and are in constant demand. The employment of millwrights has historically been cyclical, rising and falling in line with investments in automation by the nation's industrial and manufacturing sectors. While industrial growth and expansion are at the heart of many employment opportunities, additional opportunities come from the need to replace retiring millwrights.

1.3.2 Related Occupations

Millwrights require broad knowledge and a variety of skills to accomplish their daily tasks. They must also be able to work with precise measurements and the related tools (*Figure 8*). Other workers with similar jobs and responsibilities include the following:

- Industrial maintenance mechanics
- Industrial maintenance electrical and instrumentation technicians
- Aircraft and avionics equipment mechanics and service technicians
- Structural and reinforcing iron and metal workers, such as boilermakers
- Heavy-vehicle and mobile-equipment service technicians and mechanics

An African proverb, now familiar to many across the world, proclaims that "it takes a village to raise a child." It also takes a village to build a home, an office building, or an industrial facility. No single craft can manage these tasks alone. As millwrights ply their craft, they must often work closely with other crafts, including the following:

- Mobile crane operators
- Riggers
- Pipefitters
- Boilermakers
- HVACR technicians
- Welders
- Electricians

1.4.0 General Construction Safety Guidelines

Construction work can be hazardous. The millwright craft works with a variety of metals, tools, and powered equipment. If these items are handled improperly, injury or even death can result. As your training progresses, you will be taught many safety rules and guidelines that are specific to the millwright craft.

Figure 8 Laser alignment instruments.

Did You Know?

Occupational Statistics and Information

There are numerous resources that collect and provide valuable information regarding occupations in the United States. Three important resources are described here:

- The O*NET Resource Center (**www.onetcenter.org**) is an important source of occupational information. Valid and pertinent information is essential to understanding the rapidly changing nature of work and how it impacts the workforce and US economy. Based on this information, plans are developed to facilitate the development and maintenance of a skilled workforce. Central to the effort is the O*NET database, containing hundreds of standardized and occupation-specific descriptors on almost 1,000 occupations covering the entire US economy. Input from workers in each occupation continually updates the database, which is available to the public at no cost.
- The BLS measures labor market activity, working conditions, price changes, and productivity in the US economy to support public and private decision making. As an agency of the US Department of Labor (DOL), BLS provides statistical guidance to the department and works in partnership with other agencies to support their data needs. The organization maintains partnerships with state and territorial governments, coordinates with other federal statistical agencies, and contracts with private vendors for data collection, information technology, and related services. For more information, visit the BLS website at **www.bls.gov**.
- NCCER's Build Your Future initiative (**www.byf.org**) aims to be the catalyst for recruiting the next generation of craft professionals. The initiative works to equip industry and education with the resources needed to effectively reach students, educate them about the industry, and promote careers in construction. The site's offerings include a craft labor demand map, interactive construction career path, and a blog targeted to industry and students.

However, many safety guidelines apply to every craft. Here are some general construction safety guidelines that everyone must learn and follow:

- Wear the appropriate PPE for the task (*Figure 9*). PPE protects workers from a wide variety of job hazards, such as cuts, falls, eye injuries, hearing loss, fume inhalation, and electric shocks. In addition to the most common PPE encountered, such as safety glasses and hard hats, there are many other items that are required for specific environments and situations. Failure to wear the proper PPE on today's jobsites may be grounds for immediate dismissal.
- Wear snug-fitting clothing and tie back long hair. Never wear wristwatches, rings, bracelets, necklaces, earrings, long scarves, or neckties when working around rotating machinery and power tools. It is best to keep these and similar items off the jobsite altogether.
- Wear gloves whenever practical. Many craftworkers avoided gloves in the past, since older glove styles were clumsy at best. Today's work gloves are designed for the craftworker in mind, with many different types to suit specific needs. Once you become used to wearing gloves on the job, you will feel unprepared without them.
- If a task involves the use of a potentially hazardous substance, always read and understand the relevant Safety Data Sheet (SDS). This information is developed and provided by the product manufacturer to inform users of the potential hazards and safety guidelines associated with a product. The SDS is part of the Globally Harmonized System of Classification and Labeling of Chemicals (GHS). The GHS specifies the information that must be included on an SDS, as well as the information included on product labeling. The SDS, however, typically provides more information than a product label and in a globally consistent format.
- Learn how to lift properly using your legs instead of your back (*Figure 10*). Crouch as close to the load as possible, keeping the back muscles locked and leg muscles ready to do the work. Keep the load as close to your chest as possible while lifting. If the load is heavier on one side than the other, keep the heavier side of the load closest to the body. Always ask for assistance if a load is too heavy to lift alone. Heroes on the jobsite often have shorter careers.
- Immediately report injuries. Ignoring injuries or failing to report them could lead to serious health problems, or even death. Failing to report an injury that develops into a serious problem could, in some cases, also lead to the loss of worker's compensation benefits. Organizations must be made aware of injuries suffered on the job so incidents can be properly investigated, and steps taken to avoid them in the future. When an incident or injury occurs, report it to your immediate supervisor in a timely manner, even if medical care is not required. Near misses should also be reported.
- Do not carry tools in pants or shirt pockets. Tools should be carried in tool pouches, bags, or boxes.
- Keep tools sharp and in good repair. Dull tools are typically more dangerous than sharpened tools.
- Immediately stop using damaged or otherwise hazardous tools and machinery, such as a table saw without a guard. Attach a *Danger/Do Not Operate* tag (*Figure 11*) to the equipment to alert others, then report the issue to a supervisor.
- Never remove guards from machines and power tools.
- Use a brush or broom instead of your hands to remove metal and wood chips from around machines to avoid cuts and embedded debris. Gloves should be worn during clean up as well. Blowing the debris away with compressed air is also not recommended, as the debris can easily find its way into someone's eyes. Wearing basic PPE is essential for every jobsite task.

Figure 9 Common PPE.

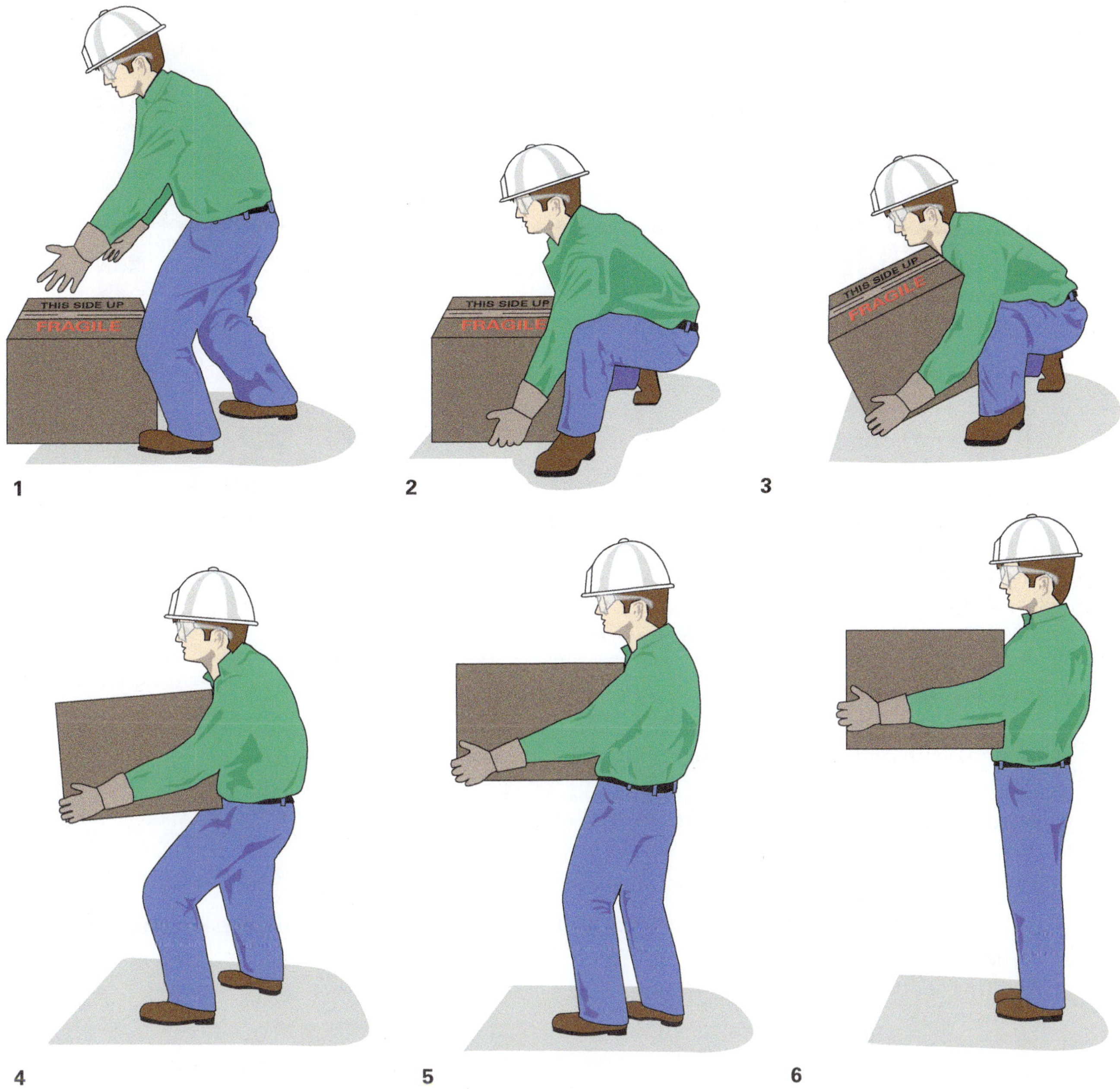

Figure 10 Proper lifting technique.

- Keep oily rags and other materials in self-closing metal containers designed for their disposal (*Figure 12*). Such materials can spontaneously combust.
- Read and follow all safety instructions on ladders and scaffolds to protect yourself from falls and to prevent injury to those working below.
- Always familiarize yourself with local safety procedures and know the location of exits, fire extinguishers, and first-aid kits. Jobsites often have different safety requirements that are unique to the work environment.
- Never engage in horseplay on the job.

Figure 11 Danger tag for inoperative or hazardous equipment.

Figure 12 Container for oily rags and waste.

1.0.0 Section Review

1. In the eighteenth century, the millwright craft developed as a division of the _____.
   a. millers craft
   b. carpentry craft
   c. blacksmith craft
   d. wheelwright craft

2. Which of the following topics is a millwright *least* likely to study during apprenticeship training?
   a. Structural ironworking
   b. Rotating shaft alignment
   c. Mathematics
   d. Hydraulics

3. Generally, the industry offering the *highest* annual wage for millwrights is the _____.
   a. steel industry
   b. public transportation industry
   c. aerospace parts manufacturing sector
   d. electric power generation, transmission, and distribution industry

4. The Safety Data Sheet (SDS) for a product is developed and provided by _____.
   a. OSHA
   b. the employer
   c. the product manufacturer
   d. the Build Your Future Foundation

SECTION TWO

2.0.0 SUCCESSFUL CRAFTWORKERS AND APPRENTICESHIP PROGRAMS

Objective

Identify the traits of a successful craftworker and describe modern apprenticeship training.

a. Identify the personal traits and characteristics of a successful craftworker.
b. Describe modern apprenticeship training.

Trade Term

Accredited: In the context of NCCER craft training programs, a status granted to an organization or site recognized as having met the established minimum standards and criteria for the administration of NCCER training and assessment programs.

A career as a millwright offers solid pay, opportunities for advancement, and training that translates to a variety of other crafts and fields. By learning the basics, applying a good work ethic, and continually researching industry changes and methods, it is possible to build an impressive resume that clearly demonstrates your value to employers.

Doing a job with precision and excellence provides a sense of satisfaction and pride in craftsmanship that is hard to find in other careers. It is important to understand, though, that your personal reputation as a millwright and a team member means a great deal in the workplace, beyond the skills listed on your resume. Your personal characteristics are as important as your mechanical skills. Exceptional mechanical skills are of little value if no one wants to work with you.

2.1.0 Characteristics of Successful Craft Professionals

Construction operations are inherently hazardous. Every site, every contractor, and every craftworker must make safety the highest priority. Grading the success of a project begins with an accounting of injuries; there can be no true success without every worker going home intact. Beyond that, the success of a project or team relies largely on the individual craftworkers experiencing success in their daily tasks.

A successful project requires coordinating each worker's unique skills and knowledge. This is where leadership comes in. Foremen, supervisors, superintendents, and site leaders are strategically positioned to manage the efforts of the workforce. All craftworkers have responsibilities on a project and are expected to produce high-quality work in a timely manner. This is a component of *craftsmanship*, and it is the cornerstone of a successful construction team. As you continue your progression through the modules in this program, review the *Cornerstone of Craftsmanship* found in each *Millwright Level One* module. Millwright craft professionals that supported this program as subject matter experts (SMEs) offer their thoughts in these special features on what the word *craftsmanship* means to them.

Craftworkers must take pride in their work and be motivated to display craftsmanship in every task. This does not mean that mistakes never happen, as they certainly will (*Figure 13*). One difference between an experienced craftworker and a novice is that an experienced craftworker can determine what to do when an error happens. However, true craft professionals do not take chances when it comes to their personal safety and the safety of others. This is a critical lesson to learn.

The word *professionalism* is a broad term that describes the behavior and attitude expected in the workplace. Professionalism is too often absent from the construction site. Many would argue that professionalism must start at the top for craftworkers to take it seriously. It is true that management should display professionalism to set an example for the workforce. But it is equally important that individuals recognize their own responsibility to act professionally, regardless of the example set by others. As an apprentice craftworker, you must find your own inspiration to embrace craftsmanship and professionalism. A good place to start is by identifying leaders and co-workers who provide good examples and follow their lead.

Professionalism encompasses an array of personal characteristics, many of which are discussed in the sections that follow. As your career progresses, you will undoubtedly observe unprofessional behavior at some point. Do not embrace the unprofessional behavior of those around you too casually. Instead, observe and learn from it, demonstrate professional behavior personally, and serve as an example to effect change.

The sections that follow provide guidance related to the personal characteristics that will help you find success in the workplace.

Figure 13 Mistakes will be made, but training and experience will help you identify solutions.

2.1.1 Honesty

Honesty and personal integrity are important traits. Professionals pride themselves in performing a job well, and in being punctual and dependable. Complete each job in a professional way, never cutting corners or wasting materials. A professional protects property such as the tools and materials belonging to employers, clients, and craftworkers from damage or theft at the shop or jobsite.

Honesty and success go hand in hand. Do not consider it a choice between good and bad only, but also as a choice between success and failure on the job. Dishonesty will always catch up with you. Whether it is stealing materials, tools, or equipment from the jobsite or simply lying about work accomplished, the employer and other craftworkers will eventually find out. Perhaps you can find another employer, but this option will eventually disappear too. A craftworker's reputation tends to have a life of its own, and you must protect yours through desirable behavior and performance. Your reputation is much harder to repair than it is to damage, and a good reputation can open many doors.

Honesty, however, means more than just refraining from theft. It means giving a fair day's work for a fair day's pay. It also means accepting responsibility for errors made. It means honoring your part of an agreement. Employers place a high value on an employee who prioritizes honesty and offers these traits.

2.1.2 Attitude

A positive attitude is essential to a successful career. Being positive means being energetic, highly motivated, attentive, and alert. Workers with a positive attitude bring the best version of themselves to the workplace.

A positive employee contributes to the productivity of others. Both negative and positive attitudes affect others on the job. A persistent negative attitude can spoil the positive attitudes of others. It is very difficult to maintain a high level of productivity while working next to a person with a negative attitude.

People favor a person who is positive. Being positive makes a person's job more interesting and exciting. The attitude you display has a great deal to do with your future success in the company.

Supervisors can determine a subordinate's attitude by their approach to the job, reactions to directives, and the way they handle problems. Those that respond well are more likely to move into supervisory positions themselves.

There is far more to a positive attitude than just presenting a smile. A smile is just one physical sign of a positive attitude. As a matter of fact, people often convey a positive attitude even though they seldom smile. They do this by the way they treat others, the way they embrace their responsibilities, and the approach they take when faced with problems. Here are some suggestions to help you maintain a positive attitude:

- Remember that your attitude follows you wherever you go. If you try to be a more positive person in your personal life, it will likely transfer to the jobsite.
- Negative comments are seldom welcomed by fellow workers on the job. Talk about positive things and be complimentary. Constant complainers rarely build healthy and fulfilling relationships. On the other hand, positivity should not be used to justify ignoring genuine problems that must be solved.
- Look for the good things in people on the job. Nobody is perfect, and almost everyone has a few worthwhile qualities. If you focus on a person's positive characteristics, it will be easier to work with them.
- Look for the good things in your work. It is doubtful that you will like everything; this is normal. No work assignment is perfect, but if you concentrate on the good things, the negative factors will seem less important.
- Look for the good things in the company. Just as there are no perfect assignments, there are no perfect companies. Nevertheless, most organizations offer some good features. You cannot expect everything to be ideal, but the company should have a variety of positive attributes. If you feel change is necessary internally, work toward change in a positive and professional manner, starting with your immediate supervisor.
- You may not be able to change the negative attitude of another employee, but you can protect your own attitude from becoming negative. Use those moments when negativity is encountered to remind yourself of how it hinders one's progress and success.

2.1.3 Loyalty

Employees expect their employers to be loyal and look out for their best interests, provide them with steady employment, and promote deserving craftworkers to better positions whenever possible. Employers feel that they too have a right to expect loyalty from their employees. Employees should make every effort to keep the company's interests in mind, to speak well of it to others, to avoid spreading negative information, and to keep internal matters confidential. Both employers and employees must remember that loyalty is not something that develops through demands; rather, it is something to be earned and returned enthusiastically.

2.1.4 Willingness to Learn

Each employer and team of craftworkers has their own way of doing things. Employers expect their workers to be willing to learn their ways and adhere to their accepted methods. Being willing to adapt and learn new methods and procedures quickly are essential to every worker's success. Changes in safety regulations or the purchase of new equipment, for example, makes it necessary for even experienced employees to learn something new (*Figure 14*).

Employees may sometimes resent change, especially when it comes as a surprise. However, employers will rightfully expect employees to make the necessary effort. Construction methods must progress for a business to compete, to stand out as a reliable and competent option, and to earn a profit. It is profit that allows an owner to expand and provide secure jobs and benefits for more employees as time marches on.

2.1.5 Willingness to Take Responsibility

Most employers expect workers to see what needs to be done, then go ahead and do it. It is very tiresome to have to ask again and again for a task to be done. It is obvious that, having been asked once, an employee should assume responsibility for it. The employer should clearly state a craftworker's primary responsibilities and provide the necessary guidance to successfully fulfill them. The craftworker must then embrace and take ownership of those responsibilities.

Figure 14 Demonstrating new equipment.

Craftworkers also need to be engaged enough in their role to know when to assume responsibility for something without direction. Employers are pleased when employees take the initiative and embrace new responsibility, as it indicates the presence of leadership skills.

All craftworkers must also take responsibility for their errors. Avoiding responsibility for an error that was clearly yours is one way to damage an otherwise solid reputation. Stepping up and taking responsibility for an error, on the other hand, is a good way to further enhance a good reputation. Be willing to own your mistakes as well as your successes.

2.1.6 Willingness to Cooperate

To cooperate means to work together, in partnership or with a team. Cooperation is the key to getting many things done on the job. Learn to work as a member of a team with your employer, supervisor, and fellow craftworkers towards the common goal of getting a job done safely, efficiently, and on time. Sharing workplace success as a team is especially fulfilling (*Figure 15*). Look for ways to contribute to the team's goals, not just your own.

People can work well together only if they understand what is to be done, when and how it will be done, and who will be responsible for individual tasks. Watching a team of craftworkers work seamlessly to accomplish a challenging task can be an inspiring experience.

2.1.7 Present and On Time

Tardiness refers to being late for work, while absenteeism means being away from the job. Consistent tardiness and frequent absences suggest poor personal habits and planning, a weak work ethic, and/or a lack of commitment to the craft or the team.

Usually, the clock regulates your working life. Every worker is required to be at work at a specified time. Failure to get to work on time results in confusion, lost time, and resentment from team members who do arrive on time. In addition, tardiness can lead to disciplinary action, including dismissal. You are obligated to be at work at a specified time. In many cases, a missing team member cripples the team and stalls the project. The company experiences a significant loss when an entire crew is unable to get started.

Figure 15 Sharing workplace success as a team is especially fulfilling.

Consider the matter from the point of view of a supervisor. Supervisors cannot keep track of workers who arrive at random times. It is not fair to others for the tardiness of one worker to be ignored. Acting in the best interests of the team and the company, a supervisor must address a worker's tardiness. Employers often resort to docking pay, demotion, and even dismissal to control tardiness and absenteeism. No employer likes to punish employees. However, to be fair, an employer is often forced to discipline those who fail to meet the standard.

Get enough rest before going to work each day (*Figure 16*). Sufficient rest is essential for everyone to be at their best on the job. A practiced, consistent morning routine helps prevent rushing and breathless, late arrivals. In fact, arriving a little early shows your respect for the team and enthusiasm for the work, which everyone appreciates. Tardiness is one of those habits that can easily stand in the way of personal success. Workers who cannot manage their own time are rarely chosen for leadership positions.

The most frequent causes of absenteeism are illness, death in the family, accidents, personal business, and dissatisfaction with the job. Some of these causes are legitimate and unavoidable, while others can be managed. One can tend to many personal affairs after working hours, and each worker must plan carefully to handle those issues that must be addressed during the workday. A respected, reliable worker will not generally have a problem getting time off to address something important.

No one is expected to work when they are legitimately sick or when there is a serious personal matter at hand. However, it is also possible to get into the habit of allowing unimportant and unnecessary matters get in the way of your job. Your absence can result in lost production and hardship for those who try to carry on the work without your help. If you are ill, use the time at home to recover as quickly as possible. Each of us would expect no less from someone hired to do an important job. Notify your supervisor as soon as possible when you are ill, so planning for your absence can begin.

If dissatisfaction with the job is a problem, plan carefully for a change in employment. However, do so in a professional manner. During your last days on a job, make every effort to perform reliably and be as productive as possible. Put forth as much effort on your last day as you did on your first. Handling these matters professionally will help you build a positive reputation in the workplace. Also remember that it is rarely a good idea to burn a bridge behind you—one never knows with certainty how important that bridge may be in the future.

2.1.8 Establishing and Pursuing Goals

Establishing and pursuing goals is very important to achieving personal success. Having both short-term and long-term goals is also critical to enhancing your value on the jobsite. An employee's wages largely depend on performance. Setting goals and working toward them are essential to improving your performance and being rewarded with improved wages and job satisfaction.

Figure 16 Get the proper rest and show some enthusiasm for your work, regardless of the assignment.

The acronym SMART applies to personal goal setting (*Figure 17*). It is a reminder that goals should have the following characteristics:

- **S**pecific
- **M**easurable
- **A**ttainable
- **R**elevant
- **T**imely

Setting major goals is a good place to begin, but significant goals are best accomplished in steps. Once you form the big picture and set a major goal, break it down into smaller goals that are measurable and easier to achieve. Write them down so they can be viewed and considered with clarity. Then plan the steps needed to achieve each goal and cross them off the list as you complete them. It is also important to set realistic target dates for the completion of each step.

If establishing realistic goals for yourself seems intimidating, ask more experienced workers and supervisors for assistance.

Figure 17 The SMART acronym.

Did You Know?

Active Apprenticeships

According to the US Department of Labor, more than 252,000 individuals nationwide entered an apprenticeship in 2019. There were over 633,000 active apprentices developing the craft skills to succeed while earning wages to build financial security. Also, in 2019, 81,000 apprentices graduated from an apprenticeship system. There are 38 percent more active apprentices today than the previous 10-year average.

George Washington (surveyor), Benjamin Franklin (printer), and Paul Revere (silversmith) all served as craft apprentices.

2.2.0 Modern Apprenticeship Training

Many craftworkers learn their skills through a registered apprenticeship program, which this training material supports. Apprenticeship combines OJL with related classwork. There are also many who enter a craft and learn on the job from a more experienced worker but do not participate in classroom training. This generally produces a craftworker who is unable to match the work of those with better training. Properly executed apprenticeship programs produce a more knowledgeable, skilled, and well-rounded craftworker.

2.2.1 The Office of Apprenticeship

The Office of Apprenticeship, a division of the DOL, sets the minimum standards for registered apprenticeship programs in the United States. These programs require both classroom instruction and OJL. Each craft program typically requires at least 144 hours of classroom (or similar) instruction and 2,000 hours of OJL per year (*Figure 18*). Most craft apprenticeship programs are four years in length. During that time, trainees spend a minimum of 576 hours in the classroom and 8,000 hours in OJL before receiving certificates issued by registered apprenticeship programs. If you are part of a registered apprenticeship program, the path forward and the requirements for successful completion are provided to you during the registration process.

The Office of Apprenticeship administers apprenticeship programs at the federal level. It works closely with State Apprenticeship Councils (SACs) that administer the programs at the state level. The apprenticeship system has grown up with the United States and, like the country, it is still growing and changing. Programs must generally follow these guidelines:

- Apprentices are typically 18 years or older. However, 16- and 17-year-old students may qualify under the Fair Labor Standards Act (FLSA) if certain conditions are met.
- There must be a schedule of work processes through which an apprentice is to receive training as well as experience on the job.
- The program must include organized instruction designed to provide apprentices with technical knowledge in topics related to their craft.

Figure 18 An apprentice on the job.

- Each apprentice must be provided with a documented, progressively increasing schedule of wages for the duration of the program.
- The OJL must be properly supervised, and adequate facilities to train both apprentices and instructors must be provided.
- The progress of apprentices, both on the job and in the classroom, must be evaluated periodically and the appropriate records must be maintained.
- There must be no discrimination in any phase of selection, employment, or training.

2.2.2 The Apprenticeship Advantage

For workers seeking employment in the construction industry, apprenticeship has important advantages. One significant advantage is the ability to earn a wage as the training progresses with a clearly documented wage scale for the duration of the program. It also offers an efficient way to learn because the training is well planned and organized.

An adequate supply of skilled workers is vital to the construction industry. Apprenticeship programs produce competent journey-level workers who generally offer the following qualities:

- They are well trained and understand not only how to do a task, but also why a task is done in a particular way.
- They are flexible and can adapt quickly to the changing demands of their jobs.
- They provide productive work and assistance from the start and are well prepared to accept greater responsibility upon completion of their training.

Around the World

From Journeyman to Master

To be considered a master in days gone by, an apprentice had to produce a workpiece equal to what a master craftsman could produce. That would be the apprentice's masterpiece and the key to the future.

Masters could start their own business, or they could choose to work as journeymen for daily wages. The term *journeyman* comes from the French word *jour*, which means day. Thus, journeyman literally means *day man*. Today, the word describes someone who has completed the required training and has gained the appropriate knowledge and expertise to be productive in a craft.

Informal OJL provided by employers is usually less thorough than that provided through a formal apprenticeship program. Without the proper classroom instruction, a worker learns only what co-workers and supervisors know. They may also lack a proper education in the craft. While self-study can certainly be helpful, the lack of a structured approach and a plan for assessing and documenting the knowledge gained limits the value of the effort.

During an apprenticeship, trainees practice basic skills under the supervision of an instructor or journeyman, who assigns the specific tasks. A good work ethic and consistent attention to safe work practices are essential to an apprentice's success throughout the process.

Classroom training supports OJL. In the classroom, the knowledge needed to successfully complete assigned tasks is provided. In many cases, the training facility offers an opportunity to practice specific tasks before the trainee attempts the skill in the field (*Figure 19*).

2.2.3 The NCCER Millwright Program

National Center for Construction Education and Research (NCCER) programs are built on the knowledge that SMEs bring to the development process. SMEs take part in the programs' design and review their content periodically. Like every craft training program offered by NCCER, this program was developed by and for the construction industry. A continuous review process ensures that the correct information is provided consistently, and the materials remain current and relevant to the craft.

To support training in the construction and industrial crafts, NCCER has developed comprehensive craft training materials and training programs. NCCER uses the standards developed by the Office of Apprenticeship as a foundation to build the curricula for each craft. The programs provide trainees with focused craft training using a competency-based teaching approach in a modular format. This means that trainees must show a qualified instructor or performance evaluator that they possess the knowledge and skills needed to safely perform the hands-on tasks presented in each training module.

This module provides a structural example of all the NCCER modules developed for millwright training. *Trainee Guide*, or TG, is the proper term for a training module in print. TGs represent the bulk of the learning resources. While each module can be purchased separately in printed form, they are normally bundled together in *levels*.

Figure 19 Practicing laser alignment in an educational setting.

A level is a collection of modules that represent the total number of instructional hours for one year of the program. In other words, each level contains a minimum of 144 hours of instruction, as outlined by the DOL Office of Apprenticeship.

At the beginning of this module, you will find its learning objectives, further broken down into sub-objectives. The objectives are followed by a listing of the *Performance Tasks*. Each trainee is required to complete the Performance Tasks to the satisfaction of an instructor or performance evaluator. Since there are no specific skills presented in this module, there are no Performance Tasks listed. Going forward, you will find that most modules do require you to complete Performance Tasks to physically demonstrate learned skills.

Trade Terms are a collection of relevant terms and their definitions, many of which are unique to a craft or to construction in general. Each Trade Term is highlighted in the TG text, indicating that it is defined in a list at the beginning of the relevant section, and again in a comprehensive listing near the end of the module.

At the end of each major section associated with a learning objective, *Section Review* questions are provided. More review questions are provided at the end of the module to further support your studies. Level One modules, those that are designated for the first year of training, generally contain more Trade Terms than modules in the other levels. As a result, Level One modules have a *Trade Terms Quiz* at the end to help you commit the terms to memory.

While the Performance Tasks assess skills developed during training, a trainee's knowledge must also be assessed. To successfully complete each module, you must pass a module exam with a minimum score of 70 percent. This is in addition to the successful completion of the Performance Tasks. The number of questions on each module exam varies, although most will contain 15 to 30 questions.

You will also note that each module includes features with various titles, such as *Around the World* and *Going Green*. They are separate from the main text. The features are a means of sharing interesting information and insights relative to a craft. It is important to note that the content of features is not tested. No review or exam questions are based on the information found in features. To enhance the learning process, an instructor may discuss or even expand on a feature topic but, again, no exam questions are based on their content.

When the instructor is satisfied that a trainee has demonstrated the required knowledge and skills for a given module, that information is sent to NCCER and recorded in the NCCER Registry. The Registry provides a transcript and certificates for successful trainees learning through an **accredited** training sponsor. These documents provide confirmation of a craftworker's knowledge and skill as they move from state to state and company to company. Trainees in an accredited NCCER program are added to the Registry and receive a certificate for each level of training completed. Examples of NCCER's credential documents can be examined in the *Appendix* of this module. The valuable credentials offered by NCCER through the successful completion of its training and assessment programs are unique in the construction industry and are recognized at the national level. Millions of craftworkers have already earned credentials through NCCER training programs. With the appropriate amount of effort and commitment, your name will be added to the list of successful craftworkers with exciting careers ahead of them.

2.0.0 Section Review

1. Employers and employees alike *must* remember that loyalty _____.
   a. should not be expected from anyone
   b. is not something that develops through demands
   c. is a concept that no longer applies
   d. is only important to employers

2. A registered apprenticeship program requires *at least* _____.
   a. 4 weeks of instruction annually
   b. 6 weeks of instruction annually
   c. 112 hours of instruction annually
   d. 144 hours of instruction annually

Review Questions

1. By the end of the 1780s, Oliver Evans had devised an automated _____.
   a. turbine
   b. lathe
   c. CNC machine
   d. flour mill

2. An ultrasonic measuring device determines distance using _____.
   a. ultraviolet light
   b. sound waves
   c. a laser beam
   d. thermal data

3. All craftworkers, regardless of the environment where they work, *must* _____.
   a. complete more than one apprenticeship program
   b. make safety their highest priority
   c. take annual written exams relevant to their craft
   d. be willing to work overtime without extra pay

4. If a task involves the use of a potentially hazardous substance, a craftworker should acquire and read the relevant _____.
   a. Safety Data Sheet (SDS)
   b. GHS list
   c. NCCER module
   d. product reviews

5. Which of the following statements about safety and injuries is *correct*?
   a. Every near miss and injury must be reported to OSHA by the worker.
   b. Failing to report an injury can lead to the loss of workmen's compensation benefits.
   c. Although injuries must be reported, reporting a near miss is just a waste of time.
   d. Minor jobsite injuries do not need to be reported to your employer.

6. Only a persistent smile is needed to convey a positive attitude to your teammates.
   a. True
   b. False

7. The SMART acronym applies to _____.
   a. encouraging loyalty from employers
   b. the development of a positive attitude
   c. establishing goals
   d. the acceptance of responsibility for errors

8. The minimum standards for apprenticeships in the United States are established by _____.
   a. the Office of Apprenticeship, a division of the US Department of Labor (DOL)
   b. OSHA
   c. the O*Net Resource Center
   d. The US Bureau of Labor Statistics (BLS)

9. Which of the following statements about apprenticeship is *true*?
   a. Apprentices only need to work four weeks each year in their chosen craft.
   b. A four-year apprenticeship program typically requires 8,000 hours of OJL.
   c. Apprentices working full time in their chosen craft are not required to study or take any tests.
   d. In a registered apprenticeship program, trainees never know how much they will earn from year to year.

10. NCCER programs are built on the knowledge brought to the development process by _____.
    a. the US government
    b. college professors
    c. technical writers
    d. subject matter experts

Trade Terms Quiz

Fill in the blank with the correct term learned from your study of this module.

1. When a manufacturer plans to upgrade production machinery and accomplish significant maintenance, the company will likely schedule a series of _______________.

2. To create threads on a piece of round stock, a millwright uses _______________.

3. _______________ are tools used to create internal (female) threads in round holes.

4. A(n) _______________ organization is one that has been recognized by NCCER to have met the established minimum standards for the administration of NCCER training and assessment programs.

5. _______________ tools and instruments operate using high-frequency sound that is beyond the range of the human ear.

6. _______________ offer increased speed and precision for cutting electrically conductive materials.

7. Machines used in a subtractive manufacturing process that are based on computerized controls are referred to as _______________.

8. _______________ are precision measuring tools.

9. A trainee learning a craft through both OJL and a prescribed amount of study while also earning a wage is considered a(n) _______________.

10. A machine tool that rotates a workpiece in opposition to handheld or stationary tools to remove material is called a(n) _______________.

11. A(n) _______________ is a machine that typically has a series of vanes along a shaft that rotates from the input of water pressure, steam, or similar fluids to produce power.

Trade Terms

Accredited
Apprentice
Computer numerical control (CNC) machines
Dies
Lathe
Micrometers
Plasma arc cutters
Shutdowns
Taps
Turbine
Ultrasonic

Cornerstone of Craftsmanship

Thomas Bruns
Great Basin College
Industrial Millwright Technology Instructor

How did you choose a career in the industry?

The company I was working for offered to pay for my training. I'd attend two days of school and then spend three days on the job. Ultimately, I chose the millwright craft due to the variety of work and the opportunities. The other fields didn't offer the same possibilities.

Who inspired you to enter the industry?

My wife and daughters inspired me. I wanted a career that allowed me to be home in the evenings with my family, and this industry allowed me to educate myself and build a sustainable lifelong career.

What types of training have you been through?

I earned my Associate of Applied Science in industrial millwright technology, and I've also completed many specialized training programs, such as lubrication technology, laser alignment, and pump construction. I'm also a certified NCCER Master Trainer.

How important is education and training in construction?

It's essential not only in construction but with everything else in life as well. Without an education, you do not have a foundation for success. Think about your high school education. You hated math, but you need it every single day. Reading and writing? That's how you communicate on the jobsite. And then there is science, where the fundamentals of pneumatics and so much more are found. A basic education is just the starting point.

How important are NCCER credentials to your career?

The NCCER credentials have allowed me to expand my horizons and career. Not only do I have a college degree, but I have nationally recognized credentials for my knowledge and skills. If I apply for another job, other applicants will have a hard time matching everything the way those credentials stack up. The credentials open a lot of doors for me.

How has training/construction impacted your life?

I've seen a whole new perspective because of teaching and training. I work with people fresh out of high school, and it takes years of experience and the opportunity to make some mistakes from which to learn. Showing the new people how to do things safely and efficiently, and to stop and think before you jump in, is a monumental step to build on. Training the younger generation also makes me better at what I do as well.

What kinds of work have you done in your career?

I grew up on a farm and started working for living accommodations at an early age. That work included everything from planting and harvesting crops to dirtier work. I also worked in restaurants and as a laborer. Once I started training and entered the millwright craft, I started working as a process maintenance mechanic and worked my way up to the journeyman level. I continued to the master mechanic level, and today I'm a CTE teacher.

Tell us about your current job.

My present job is teaching industrial millwright technology at Great Basin College in Elko, Nevada.

The program allows students to graduate with an Associate of Applied Science degree and includes all millwright levels. We're giving students the opportunity to complete what is normally a much longer program in one year, so they can get into the workforce sooner and apply hands-on training as quickly as possible.

What do you enjoy most about your job?

When the light comes on for students. It's the moment when they've figured out how it all works. It's incredible when a student, two years later, thinks back on something I've taught and calls me to say it helped him solve a problem. That's the reason why I teach. I think of it a lot like coaching youth sports. You get to see them succeed and know you made a difference.

What factors have contributed most to your success?

My parents taught me very important values, including the value in a hard day's work. You get up and you go every single day because they are expecting you. That's such an important, basic foundational part of one's character, and my parents always held me to the standards they taught.

Would you suggest construction as a career to others? Why?

Obviously, yes! It is so rewarding at the end of the day to see the progress you've made. Even if you don't feel it, the outside world sees it. You can hang your hat on something at the end of the day and say, "I did this today," and no one can ever take that away from you.

We have very smart people out there that have no idea how to fix and build things. Construction is so rewarding in that respect. Even though we are moving toward more complex digital technology, there still needs to be someone to turn the wrenches on the machinery. Digital control doesn't matter if the machinery won't function mechanically.

What advice would you give to those new to the field?

You'll always be learning. I'm considered a subject matter expert, and I'm still learning. When you stop learning, you stop living, so pay attention and soak up all the info. You never know when you will need it.

Interesting career-related fact or accomplishment?

Being nominated to serve as an SME for my craft is a huge accomplishment to me. If you told me 30 years ago that I'd be teaching, I'd say you must have bumped your head. Being recognized by my peers as a valued addition to my team and the NCCER team is the ultimate compliment.

How do you define craftsmanship?

Craftsmanship is not taking shortcuts. It is making sure the project is done correctly every time and making sure everything is cleaned up and back in its place and better than you found it. I used to tell the youth sports teams I coached that it's all about being honest and doing the right thing when no one is looking. There may be no praise or attention when you get done, but leaving it better than when you started, well, that's the ultimate display of craftsmanship. And that's what we're trying to teach people in these programs.

Trade Terms Introduced in This Module

Accredited: In the context of NCCER craft training programs, a status granted to an organization or site recognized as having met the established minimum standards and criteria for the administration of NCCER training and assessment programs.

Apprentice: A person learning a craft through practical experience with skilled workers, as well as through instruction and study, while earning a wage.

Computer numerical control (CNC) machines: Equipment used in a subtractive manufacturing process based on computerized controls and machine tools that remove material from a workpiece to produce a component. Many materials can be machined in this manner, including metals, plastics, and composites.

Dies: Threading tools typically shaped like a nut, used to cut external (male) threads on properly sized round stock.

Lathe: A machine tool designed to shape metal, wood, and similar materials using a rotating drive that turns the workpiece in opposition to handheld or stationary cutting tools.

Micrometers: Precision measuring tools that measure small distances or thicknesses between two faces, adjusted to contact the workpiece(s) by turning an adjusting barrel.

Plasma arc cutters: Equipment used like a cutting torch to cut through electrically conductive materials using a jet of superheated plasma at a temperature of roughly 40,000°F (~22,200°C) generated by a handheld or automated torch head. Plasma cutting offers improved speed and precision compared to traditional oxyfuel cutting processes.

Shutdowns: Periods of time when facility production is largely halted. Although the term can be applied to the cessation of production for any reason, industrial crafts often use the term *shutdown* to identify pre-planned production interruptions that may last for days or even weeks to allow for scheduled maintenance and significant changes to production equipment.

Taps: Threading tools used to cut internal (female) threads in properly sized round holes.

Turbine: A machine for producing power in which a wheel or rotor fitted with a series of vanes is forced to rotate by the flow of steam, gas, water, air, or similar fluids.

Ultrasonic: Related to the use of sound waves moving at a frequency above the upper auditory limit of the human ear (about 20,000 hertz).

Appendix

Samples of NCCER Training Recognition Documents

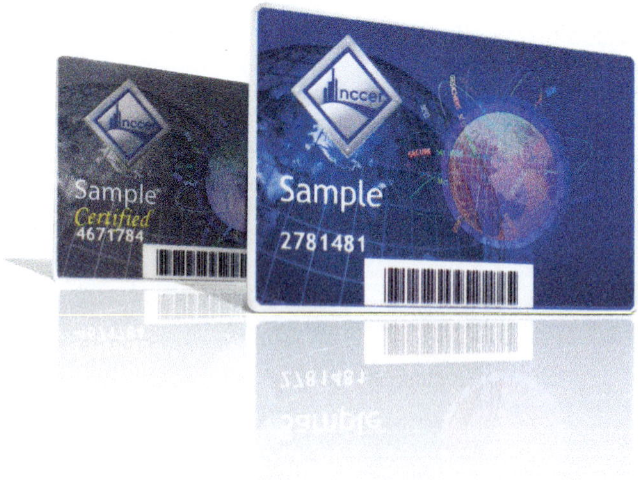

Additional Resources

This module presents thorough resources for task training. The following reference material is suggested for further study.

Apprenticeship.gov, an official website of the US Government. **https://www.apprenticeship.gov/**

NCCER's Build Your Future initiative **(www.byf.org)** aims to be the catalyst for recruiting the next generation of craft professionals. The initiative works to equip industry and education with the resources needed to effectively reach students, educate them about the industry, and promote careers in construction. The site's offerings include a craft labor demand map, interactive construction career path, and a blog targeted to industry and students.

The O*NET Resource Center, the nation's primary source of occupational information. **www.onetcenter.org/**

US Department of Labor, Office of Apprenticeship. **www.dol.gov/apprenticeship/**

US Bureau of Labor Statistics. **www.bls.gov/**

Figure Credits

iStock@kali9, Module Opener, Figure 15

"The Blockmaker", Smithsonian American Art Museum, Gift of William T. Evans. Licensed under Creative Commons Zero (CC0) license., Figure 1

The L.S. Starrett Company, Figure 2

Image property of Stanley Black & Decker. Used with permission., Figure 3

The Lincoln Electric Company, Cleveland, OH, USA, Figure 4

iStock@industryview, Figure 6

Pruftechnik Americas, a Fluke Reliability company, Figure 8

iStock@bonezboyz, Figure 9

Courtesy of Justrite Safety Group, Figure 12

iStock@Cligatron, Figure 13

iStock@izustun, Figure 16

iStock@BrianAJackson, Figure 17

iStock@industryview, Figure 18

Section Review Answer Key

Section 1.0.0

Answer	Section Reference	Objective
1. b	1.1.0	1a
2. a	1.2.1	1b
3. d	1.3.1	1c
4. c	1.4.0	1d

Section 2.0.0

Answer	Section Reference	Objective
1. b	2.1.3	2a
2. d	2.2.1	2b

NCCER CURRICULA — USER UPDATE

NCCER makes every effort to keep its textbooks up-to-date and free of technical errors. We appreciate your help in this process. If you find an error, a typographical mistake, or an inaccuracy in NCCER's curricula, please fill out this form (or a photocopy), or complete the online form at **www.nccer.org/olf**. Be sure to include the exact module ID number, page number, a detailed description, and your recommended correction. Your input will be brought to the attention of the Authoring Team. Thank you for your assistance.

Instructors – If you have an idea for improving this textbook, or have found that additional materials were necessary to teach this module effectively, please let us know so that we may present your suggestions to the Authoring Team.

NCCER Product Development and Revision

13614 Progress Blvd., Alachua, FL 32615

Email: curriculum@nccer.org

Online: www.nccer.org/olf

❑ Trainee Guide ❑ Lesson Plans ❑ Exam ❑ PowerPoints Other ____________

Craft / Level: Copyright Date:

Module ID Number / Title:

Section Number(s):

Description:

Recommended Correction:

Your Name:

Address:

Email: Phone:

This page is intentionally left blank.

Millwright Hand Tools

Overview

A millwright uses specialized tools ranging from heavy-duty pullers to delicate precision measuring and leveling instruments. Millwright work involves tolerances that are often measured in tiny fractions of normal units of measure. Even a slight error in measurement or assembly can affect the operation of the equipment being serviced.

Millwrights are required to lay out equipment locations, assemble and disassemble mechanical equipment, align rotating components, and ensure the calibration of moving parts and components. Becoming familiar with the tools presented in this module and their proper use will prepare you to accomplish these challenging tasks safely and effectively.

Module 15102

Trainees with successful module completions may be eligible for credentialing through the NCCER Registry. To learn more, go to **www.nccer.org** or contact us at 1.888.622.3720. Our website, **www.nccer.org**, has information on the latest product releases and training.

Your feedback is welcome. You may email your comments to **curriculum@nccer.org**, send general comments and inquiries to **info@nccer.org**, or fill in the User Update form at the back of this module.

This information is general in nature and intended for training purposes only. Actual performance of activities described in this manual requires compliance with all applicable operating, service, maintenance, and safety procedures under the direction of qualified personnel. References in this manual to patented or proprietary devices do not constitute a recommendation of their use.

15102 V4.0

15102
MILLWRIGHT HAND TOOLS

Objectives

Successful completion of this module prepares you to do the following:

1. Identify and describe how to use various millwright hand tools.
   a. Recognize and describe how to use various gauge standards.
   b. Identify and describe how to use pipe and tubing cutters.
   c. Identify and describe how reamers and honing tools are used.
   d. Recognize and describe how to use thread taps, dies, and extractors.
   e. Identify and describe how to use unique wrenches and pullers.
   f. Recognize types of metal snips and describe their use.
   g. Identify and describe how to use other tools important to millwrights and similar crafts.

Performance Task

Under supervision, you should be able to do the following:

1. Demonstrate the proper use and care of eight instructor-selected hand tools presented in this module.

Trade Terms

Blind hole
Bonnet
Cheater bar
Chuck
Desiccant
Dress
Drive coupling
Ferrous
Fillet
Grit
Honing
Jig
Lapping plates
Packing
Pipe scale
Pitch
Polyband belts
Sheave
Shim stock
Stuffing box
Tolerance

Industry Recognized Credentials

If you are training through an NCCER-accredited sponsor, you may be eligible for credentials from NCCER's Registry. The ID number for this module is 15102. Note that this module may have been used in other NCCER curricula and may apply to other level completions. Contact NCCER's Registry at 1.888.622.3720 or go to **www.nccer.org** for more information.

You can also show off your industry-recognized credentials online with NCCER's digital badges. Transform your knowledge, skills, and achievements into badges that you can share across social media platforms, send to your network, and add to your resume. For more information, visit **www.nccer.org**.

NOTE

This module uses US standard and metric units in up to three different ways. This note explains how to interpret them.

Exact Conversions

Exact metric equivalents of US standard units appear in parentheses after the US standard unit. For example: "Measure 18" (45.7 cm) from the end and make a mark."

Approximate Conversions

In some cases, exact metric conversions would be inappropriate or even absurd. In these situations, an approximate metric value appears in parentheses with the ~ symbol in front of the number. For example: "Grip the tool about 3" (~8 cm) from the end."

Parallel but not Equal Values

Certain scenarios include US standard and metric values that are parallel but not equal. In these situations, a slash (/) surrounded by spaces separates the US standard and metric values. For example: "Place the point on the steel rule's 1" / 1 cm mark."

Contents

Figures and Tables

SECTION ONE

1.0.0 MILLWRIGHT HAND TOOLS

Objective

Identify and describe how to use various millwright hand tools.

a. Recognize and describe how to use various gauge standards.
b. Identify and describe how to use pipe and tubing cutters.
c. Identify and describe how reamers and honing tools are used.
d. Recognize and describe how to use thread taps, dies, and extractors.
e. Identify and describe how to use unique wrenches and pullers.
f. Recognize types of metal snips and describe their use.
g. Identify and describe how to use other tools important to millwrights and similar crafts.

Performance Task

1. Demonstrate the proper use and care of eight instructor-selected hand tools presented in this module.

Trade Terms

Blind hole: A hole that does not go completely through the material and therefore has a bottom.

Bonnet: The removeable cover of a valve body, through which the valve stem usually passes. Valve repair typically requires removal of the bonnet to gain access to the valve components beneath it.

Cheater bar: A length of pipe or similar material slipped over a wrench or other tool to substantially increase leverage. A cheater bar creates a hazard, as tools and hardware are not constructed to withstand the added stress.

Chuck: A special type of clamp equipped with jaws in a radial pattern that is typically used to hold rotating tools, such as drill bits and thread taps. Some chucks require a chuck key, which is a small, geared tool used to loosen and tighten the jaws of the chuck.

Desiccant: A material such as silicon dioxide (silica gel) that attracts and captures moisture from the surrounding air or nearby moist materials.

Dress: The process of forming, shaping, and changing the finish of stone to a desired shape or condition.

Drive coupling: A mechanical component used to connect the shaft of a driver, such as an electric motor, to the shaft of a driven load, such as a pump or compressor. The two shafts must be precisely aligned so that they share a common centerline. Drive couplings are typically classified as either rigid or flexible.

Ferrous: Containing or consisting of iron.

Fillet: A concave, rounded surface that serves as a transition between two intersecting surfaces. In welding, a fillet is formed when two pieces of metal are joined at an inside corner.

Grit: A number indicating the density of particles for grinding stones, honing stones, sandpaper, and similar materials. High numbers indicate smaller particles packed more closely together (fine), while low numbers indicate larger particles spaced farther apart (coarse).

Honing: A process that uses abrasive stones to create a smooth, precise surface on a metal workpiece in a controlled, consistent manner.

Jig: A device constructed or assembled to maintain a positional relationship between a workpiece and a tool or between components of an assembly.

Lapping plates: Coarse stones used to rub other stones and restore their flat surface.

Packing: Rope-like material with a consistent shape used to prevent fluid leakage by compressing it around a shaft or valve stem.

Pipe scale: Deposits remaining on new ferrous pipe products left over from the heating and forming process. In operational piping systems, scale refers to the buildup of corrosion and/or mineral deposits on the inside walls of the piping.

Pitch: Relative to threaded hardware and components, refers to the number of threads found over a unit of distance, e.g., a $\frac{1}{4}$" bolt with coarse threads has a pitch of 20 threads per inch.

Polyband belts: Drive belts that combine two or more belts on a common band, side by side.

Sheave: A grooved wheel, also called a *pulley*, mounted to a shaft and used to transmit power from a driver to a driven load through drive belts that ride in its grooves.

Shim stock: Thin material, typically metal, placed between two objects to fill space and allow for the precise positioning of one or both objects. Shims are often used in *shaft alignment* to adjust the position of the motor relative to the driven shaft. Shim stock typically refers to the raw material from which individual shims are cut.

Stuffing box: A casing or chamber in which packing material is placed and compressed around a moving shaft or axle to reduce or eliminate gas or liquid leakage.

Tolerance: A term that quantifies the allowable difference or deviation from specified limits.

Millwrights use a variety of specialized hand tools. There are hand tools that have multiple uses and others that serve a unique purpose. You must be able to recognize, select, inspect, maintain, and safely use hand tools to perform any job safely and efficiently. This module presents and describes the use of hand tools common to the millwright craft. Although you will have an opportunity to practice and work with these tools while completing this module, some tools require a significant amount of practice to master.

> **NOTE**
> It is assumed that you have completed NCCER Module 00103, *Introduction to Hand Tools*, before this point in your training, or that you are already familiar with hand tools common to many crafts.

Hand tools are a necessity for millwrights to perform their work. Good hand tools are expensive, although the investment can certainly pay off. Tools are a vital part of every task, regardless of whether you or your employer pays for them. For the investment to pay off, tools must be respected, used, and maintained properly. Hand tools must also be used safely.

1.0.1 General Hand Tool Safety

Although hand tools do not present the same hazards as power tools, the improper use of hand tools can cause serious personal injury. Properly using hand tools requires following some basic safety rules. Some hand tools are more hazardous than others and require additional personal protection and care when using them.

When working with hand tools, follow these basic safety rules:

- Choose the proper tool for the job. Every tool becomes more hazardous when it is not used for the intended purpose. For example, do not use a screwdriver as a prybar.
- Use adjustable wrenches only when necessary. They should not be used when a box or open-end wrench that fits the fastener is available. Adjustable wrenches are best used when the torque requirement is low. Remember to pull against and apply the greatest force to the fixed jaw, not the adjustable jaw.
- Keep your hands and fingers away from the sharp edges of cutting tools.
- Dull tools require greater force to do the job and provide poor results. A properly sharpened tool is a safer tool.
- Work away from your body when using cutting tools.
- Inspect hand tools before they are used and ensure each tool is in good condition. Damaged tools are a common cause of incidents and accidents.
- Keep tools clean, dry, and free of corrosion. Millwrights often place silica gel bags or disks (*Figure 1*) in their toolbox to absorb moisture and keep their tools dry. Silica gel is a **desiccant**, meaning that it aggressively attracts and absorbs moisture.
- Always wear eye protection, such as safety glasses or goggles, when using hand tools. In most cases, you should wear eye protection throughout the course of your day on the job, not just when using hand tools.
- Wear gloves appropriate for the task. There are far more reasons to wear gloves than to avoid them. Today's work gloves offer excellent fit, with many designed for specific tasks.
- Do not modify hand tools unless directed by a supervisor. Many employers and jobsites prohibit tool modifications for safety reasons.
- Give some thought to how you use every tool. Pull wrenches toward you instead of pushing them away from your body. This provides better control over the tool and maximizes the force that you can apply.

Figure 1 Silica gel.

Most injuries related to hand tools can be avoided. It is part of your job as a millwright to make safety the highest priority. As you progress through your training, you will be expected to adhere to these guidelines regardless of whether you are in the classroom or on the jobsite. Working with the tools presented in this module represents a good opportunity to demonstrate your commitment to safety.

WARNING!

Flammable atmospheres exist in some plants and work areas. For that reason, non-sparking tools may be required. Keep in mind, however, that non-sparking tools are made of metal and can still produce an occasional spark. Your employer or jobsite will establish specific safety procedures for these and similar hazardous working environments. For your own safety and the safety of others, follow these procedures carefully.

1.1.0 Gauge Standards

Gauge standards, also referred to simply as *gauges*, allow millwrights to quickly check small dimensions, identify threads, size drill bits, and similar tasks.

One thing all gauge standards have in common is that they are precision tools. They should never be used for any task beyond that for which they were designed. Although some may look indestructible, they can be damaged by poor handling and a lack of care.

After using any gauge, ensure that it is stored clean and dry. A very light coat of oil can help prevent corrosion on tempered steel gauges, but too much oil attracts dirt. Some craftworkers will wrap and store gauge standards in an oily cloth, rather than apply oil directly to the tools.

It is always best to store similar tools together. For example, gauge standards should not be stored with heavy wrenches and similar tools. Precision tools can be damaged easily by the heavier tools, especially when you are digging through them to find a particular tool. Whenever possible, retain the original packaging and continue to use it for storage and protection.

1.1.1 Thickness Gauges

The required clearance between components is often specified by the equipment manufacturer. A thickness gauge (*Figure 2*), also known as a *feeler gauge*, is used to measure small clearances.

A thickness gauge set contains an array of slender leaves, each of which is a precise thickness. Most sets fold like a pocketknife. The thickness is marked on each leaf. Thickness gauges are marked either in US standard or metric values. The leaves are inserted, one at a time, into the space between two parts to determine the exact clearance.

A US-standard thickness gauge normally has leaves ranging from about 0.0015" to 0.025", although this varies. Metric leaves range from about 0.05 millimeters (mm) to 1 mm. Models with much longer and/or tapered leaves are also common.

It is important to keep the gauge clean and free of dirt. One reason is to ensure the markings remain readable on each leaf. If the markings do become unreadable, a 0" to 1" micrometer can be used to measure the leaf and determine its exact thickness.

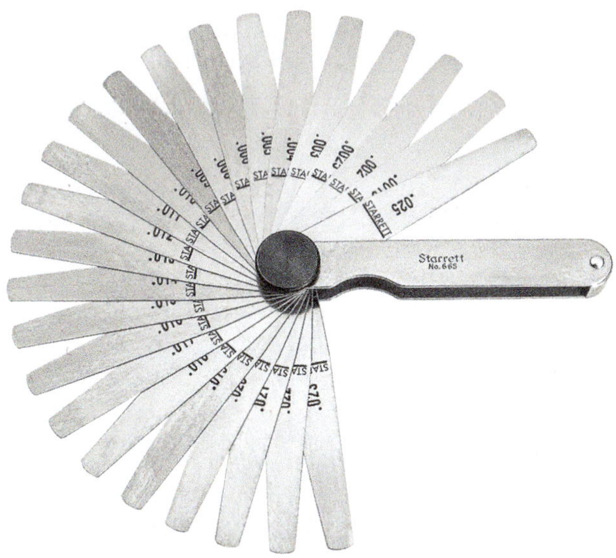

Figure 2 Thickness gauge.

1.1.2 Radius and Fillet Gauges

Radius and fillet gauges (*Figure 3*) have steel blades with accurately formed radii on their ends. They are commonly found in pocketknife-style sets, like thickness gauges. However, they are also available as individual gauges. US-standard sets typically have a range of $\frac{1}{64}$" to $\frac{1}{2}$". Metric sets commonly have a radii range of 1 mm to 15 mm.

Radius gauges are used to measure the radius of grooves and external rounded (convex) corners. Fillet gauges measure the radius of inside (concave) corners. They are used in machine work, inspection, welding, and layout tasks.

Even when acquired as a pocketknife-style set, the gauges may be removed from the set for use. To use a radius or fillet gauge, fan out the blades and compare the radii of each blade with the object being measured. Hold the appropriate gauge against the surface you are checking, preferably with a flashlight held on the opposite side of the gauge. Any points of light at the intersection between the gauge and the surface show the mismatch between the two. Once you have a match, read the number or fraction on the gauge that matches the radius or fillet.

(A) Individual Radius and Fillet Gauge (US Standard)

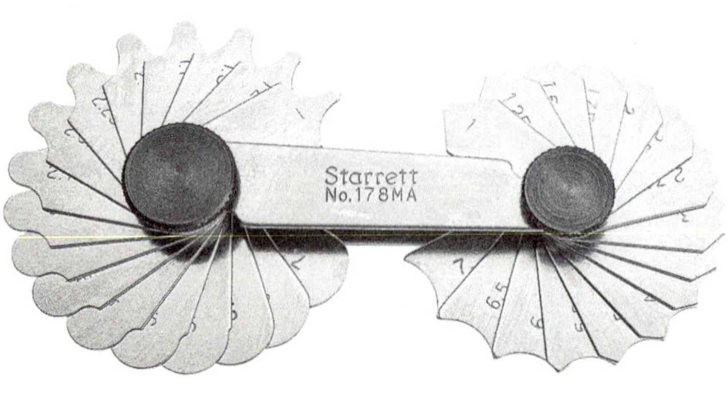

(B) Radius and Fillet Gauge Set (Metric)

Figure 3 Radius and fillet gauges.

1.1.3 Drill Gauges

Drill gauges are used to determine the exact size of a drill bit if the size is not readable. Small sizes can be very difficult to read, and the stamped information on larger drill bits can be marred in use.

Drill bits are sized using several different systems, each of which has a matching drill gauge. The most common sizing systems are as follows:

- Fractional
- Metric
- Lettered size
- Numbered size

The fractional system is the most common in the United States, as it is familiar and broad enough to cover most common needs. Metric sizing follows the metric system, as expected. Lettered and numbered sizing systems assign a letter or number to each bit, with a corresponding chart that indicates the actual decimal and/or fractional size. *Figure 4* shows two types of drill gauges.

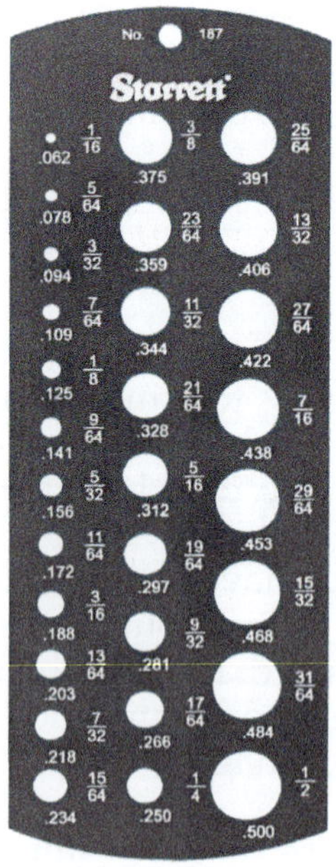

(A) Fractional Drill Gauge

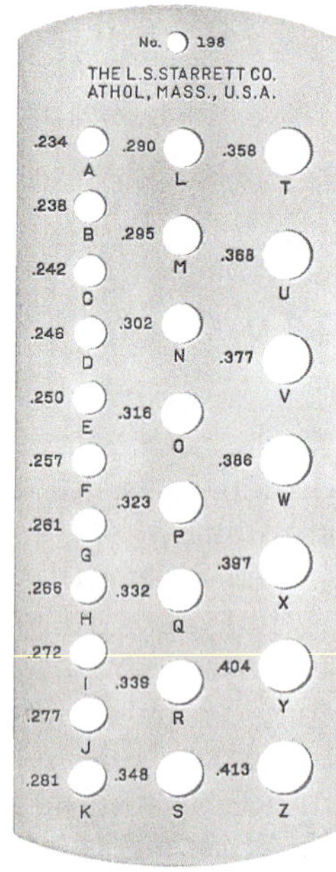

(B) Letter-Size Drill Gauge

Figure 4 Drill gauges.

To determine the size of a drill bit, first look at the smooth portion of the shaft. Even very small sizes usually have the size information engraved or stamped into the bit. If the size is unreadable, clean and remove any burrs from the shaft. Then slide the bit through the holes in the gauge from the back until it just slips through, without any play. If it is the correct hole, the bit will not fit into the next smallest hole. Read the drill size next to the appropriate hole.

1.1.4 Thread Gauges

Thread gauges, also known as *pitch gauges*, are used to measure the **pitch** of threaded fasteners (*Figure 5*). Thread gauges can be used to measure both external and internal threads.

Thread gauges have thin, flat metal leaves. Each leaf has a series of precisely spaced teeth that correspond to the spacing of common threads. *Figure 5* shows a common thread gauge. Each leaf is stamped with the number of teeth per inch or metric pitch it represents.

> **NOTE**
>
> Pitch for metric fasteners is technically defined as the distance between each thread crest, in millimeters. Most metric thread gauges are therefore labeled by metric pitch instead of threads per inch. However, some metric thread gauges are labeled in threads per inch.

To use a thread gauge, hold one of the leaves against the threads of the fastener or threaded part. Try various options, moving to smaller or larger pitches as necessary, until the teeth fit the threads perfectly. After use, ensure the gauge is clean and dry before putting it away with similar tools.

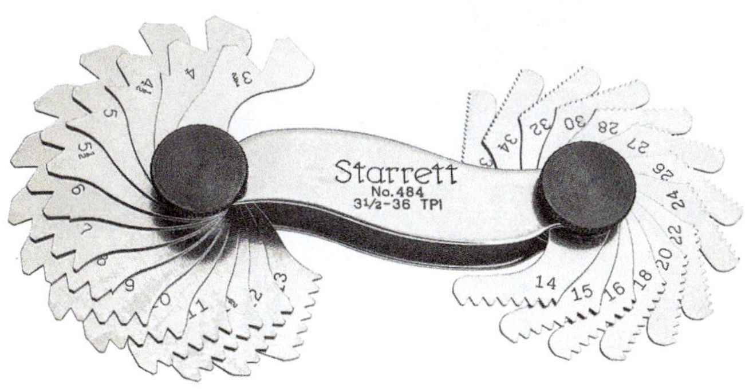

Figure 5 Thread gauge set.

1.1.5 Taper Gauges

A taper gauge set (*Figure 6*) consists of leaves that are tapered from one end to the other. They can be used to measure the diameter of a hole or the precise width of a slot (*Figure 7*). They are also used to measure misalignment in a **drive coupling**. Taper gauge sets are available in fractional inches, decimal inches, and metric measures.

To use a taper gauge for a hole, the gauge is inserted until it stops. The diameter can then be read at the graduations on the gauge. Welders also use a simple taper gauge to measure the gap between weldments before beginning a weld.

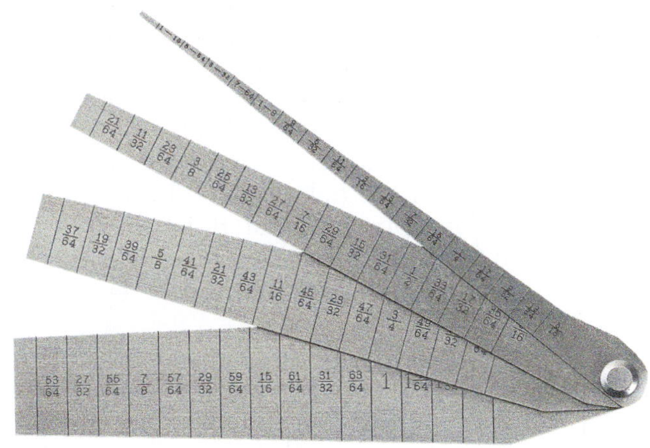

Figure 6 Taper gauge set.

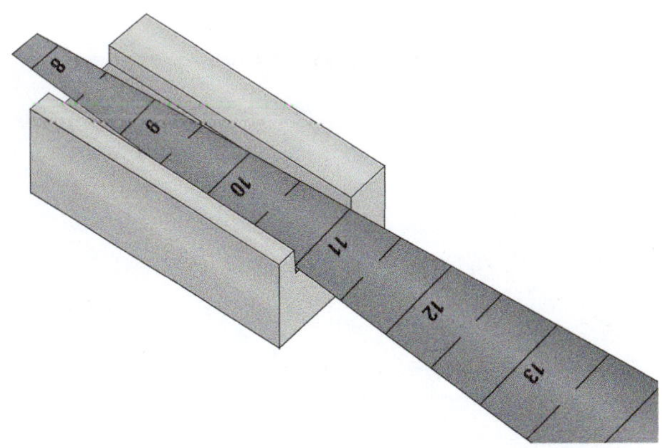

Figure 7 Using a taper gauge to measure the precise width of a slot.

1.1.6 Sheave Gauges and Belt Tension Gauges

Sheave gauges and belt tension gauges are tools associated with belt drives. Belt drives are found on air handling units, compressors, and similar equipment, most of which are driven by electric motors (*Figure 8*). The rotational speed of belt-driven equipment is determined by the speed of the drive motor and the diameter of the two sheaves.

As the belt travels around the sheaves, or *pulleys*, both the belt and grooves of the sheaves experience wear. Belts are replaced as a periodic maintenance activity; they are expected to wear more quickly. The sheaves will last much longer, but they do experience wear. Sheaves wear more quickly when the belts are slipping. Belts slip when they are too loose and when the connected load is excessive.

Belts should not ride along the bottom of the sheave groove. Only the two sides of a belt should be making contact; a small amount of clearance remains between the belt and the bottom of the groove. If it is apparent that the belt is riding on the bottom of the groove (if the paint at the bottom is worn away, for example), replace the sheave and belt set.

Sheave gauges (*Figure 9*) are used to measure wear inside the belt groove. Excess wear compromises belt grip. When the wear is excessive, the sheave is replaced.

Before using a sheave gauge, ensure that it is correct for the type of belt and sheave in use. There are many different belt styles, each of which has a different profile and dimensions. In addition to having gauges that fit a wide variety of sheaves, many sheave gauge sets also include one or more gauges that help identify the belt type. Most belts clearly indicate their type and size, but there may be cases where the information is unreadable, and confirmation is needed.

Refer to *Figure 10*. Follow these steps to use a sheave gauge:

Step 1 Acquire the technical data for the sheave. The information needed to determine which gauge to use includes the groove angle, groove dimensions, and diameter of the sheave. The dimensions will have some amount of **tolerance**, which is the amount of deviation allowed.

Step 2 Select a gauge that is correct for the sheave groove and type of drive belt in use. For example, if the belt is a B-section belt, use a sheave gauge with markings for B belts.

Step 3 Inspect the gauge to ensure that it is not bent, worn, or otherwise damaged.

Step 4 Insert the gauge into the groove of the sheave and observe the fit. Note the flashlight behind the gauge in *Figure 10*. Backlighting helps to see the fit in detail. Also note the visible gap between the gauge and the bottom of the sheave groove.

Step 5 Clean the gauge, ensure that it is dry, and store it properly with similar precision tools.

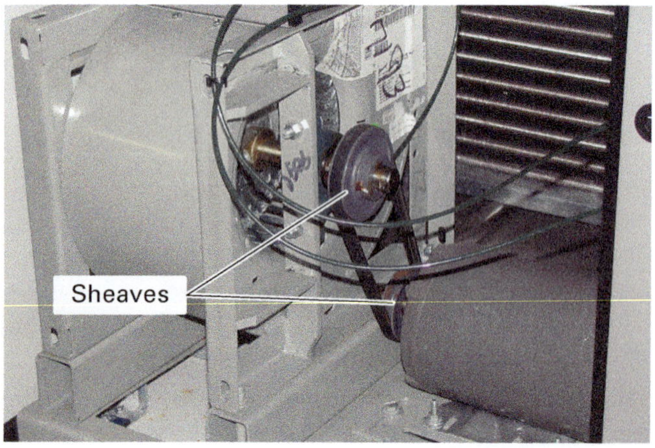

Figure 8 Small belt-driven blower.

Figure 9 Sheave gauges.

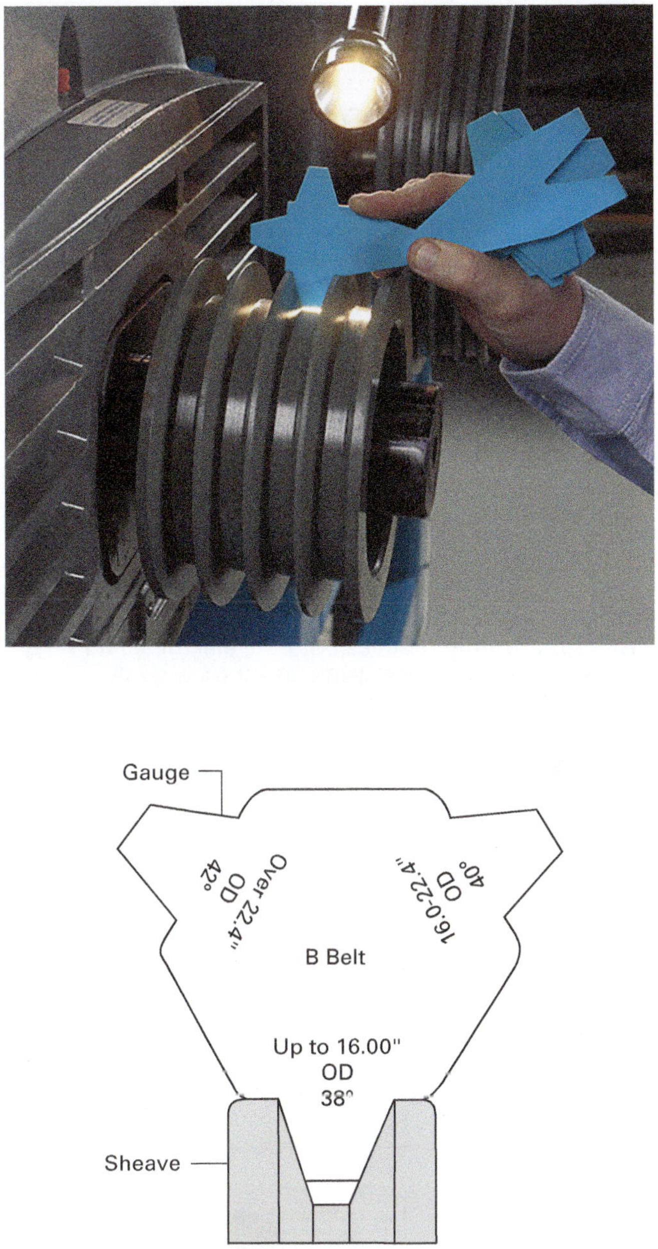

Figure 10 Using a sheave gauge.

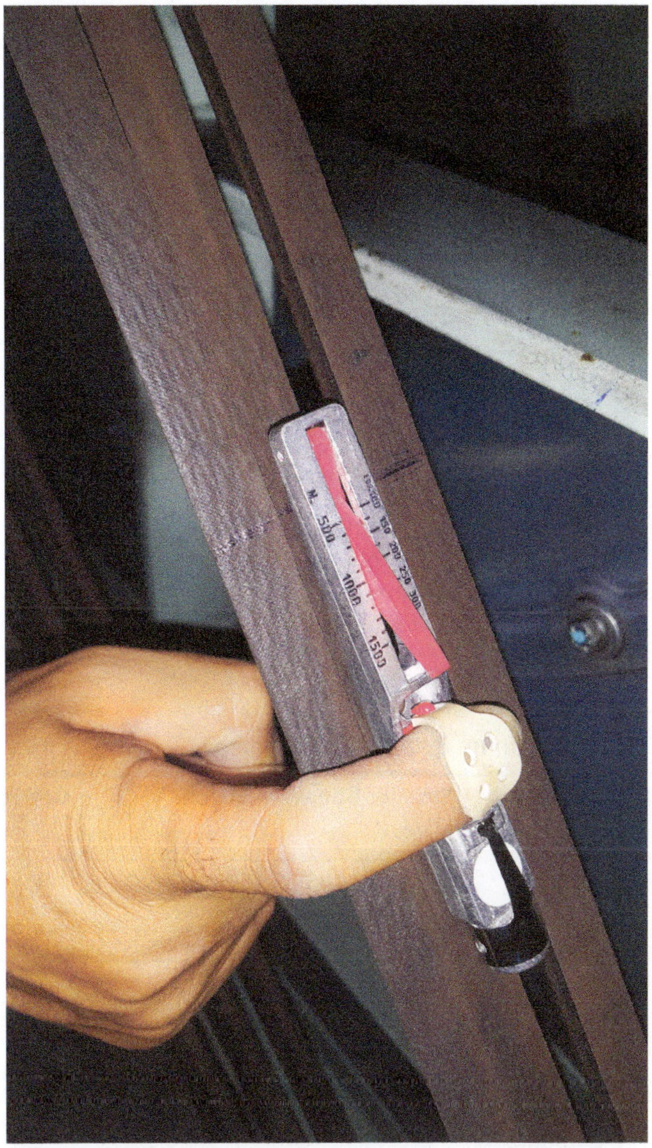

Figure 11 Mechanical belt tension gauge.

Belt Tension Gauges

Belt drive reliability and efficiency rely heavily on the proper belt tension. Loose belts slip, wasting energy and causing excessive wear on the belt and sheave. Belts that are too tight place stress on the motor bearings, as well as the bearings in the driven equipment.

A standard belt tension gauge is a mechanical device used to measure the force required, in pounds, to deflect a drive belt a given distance. Tighter belts require more force to deflect. There are several different styles, including the example shown in *Figure 11*.

For an accurate measurement, the tool must be used properly. Mechanical testers vary in their design, so it is important to review and follow the instructions carefully.

Table 1 is an abbreviated version of a belt tension table. It shows the minimum deflection, in pounds of force (lbf), required for various drive arrangements. The minimum deflection force is based on several factors:

- Belt type
- Speed range of the drive
- The diameter of the smaller sheave (typically the drive sheave on the motor)
- The ratio between the drive and driven rotational speeds

CAUTION

Table 1 provides an example only and should not be used for field work. Use the table(s) provided by the tool manufacturer.

You can also acquire belt tension information from the drive and/or belt manufacturer before beginning. It is important to request information for the right belt. **Polyband belts**, for example, will require a different tension than a single-band belt. The ideal tension is considered to be that which prevents the belt from slipping under the maximum load it will experience.

CAUTION

When servicing multiple belts on the same drive, do not mix old and new belts. Also avoid mixing belts from multiple manufacturers, even if they have the same identifying markings.

Did You Know?

Starting a Motor Slowly

Many common motors, when energized, attempt to reach their design rotational speed very quickly. When the motor is connected to a load through a belt drive, it is not only attempting to develop its own rotational speed at startup, but it is also working to bring the connected load up to speed. This creates a great deal of electrical stress on the motor. In addition, if the belts are not correctly adjusted, the lack of grip can allow the belts to slip on the motor sheave as the motor starts. This causes excessive belt and sheave wear.

Today, especially in the case of large commercial and industrial motors, motors are often "started softly." By manipulating the characteristics of the power supply, motors can be brought online slowly. This significantly reduces both the electrical and mechanical stress of starting a motor under load.

Table 1 Minimum Belt Deflection Force

Belt Section	Smaller Sheave		Deflection Force[1] (lbf) for a Drive Speed Ratio of:			
	Speed Range (RPM)	Diameter [2] (Inches)	1.0	1.5	2.0	4.0+
3VX	1200–3600	2.2	2.2	2.5	2.7	3.0
	1200–3600	2.5	2.6	2.9	3.1	3.6
	1200–3600	3.0	3.1	3.5	3.7	4.2
	1200–3600	4.1	3.9	4.3	4.5	5.1
	1200–3600	5.3	4.6	4.9	5.1	5.7
	1200–3600	6.9	5.0	5.4	5.6	6.2
5VX	1200–3600	4.4	6.5	7.5	8.0	9.0
	1200–3600	5.2	8.0	9.0	9.5	10.0
	1200–3600	6.3	9.5	10.0	11.0	12.0
	1200–3600	7.1	10.0	11.0	12.0	13.0
	900–1800	9.0	12.0	13.0	14.0	15.0
	900–1800	14.0	14.0	15.0	16.0	17.0
A or AP	1800–3600	3.0	2.0	2.3	2.4	2.6
	1800–3600	4.0	2.6	2.8	3.0	3.3
	1800–3600	5.0	3.0	3.3	3.4	3.7
	1800–3600	7.0	3.5	3.7	3.8	4.3
B or BP	1200–1800	4.6	3.7	4.3	4.5	5.0
	1200–1800	5.0	4.1	4.6	4.8	5.6
	1200–1800	6.0	4.8	5.3	5.5	6.3
	1200–1800	8.0	5.7	6.2	6.4	7.2
AX	1800–3600	3.0	2.5	2.8	3.0	3.3
	1800–3600	4.0	3.3	3.6	3.8	4.2
	1800–3600	5.0	3.7	4.1	4.3	4.6
	1800–3600	7.0	4.3	4.6	4.8	5.3

Notes:
[1]Add 30% to the deflection values to tension a new set of belts.
[2]Use closest sheave diameter for sizes not shown.

1.2.0 Tubing and Pipe Cutters

Tubing and pipe cutters (*Figure 12*) are used to cut steel, brass, copper, and iron pipe or tubing. These cutters have a hardened steel cutting wheel and pressure rollers that roll around the pipe, holding it tight against the cutting wheel. They are sized based on the range of pipe or tubing sizes they can cut.

The cutting wheel is mounted so it is stationary in the cutter frame, but free to rotate. The pressure rollers are mounted on a sliding frame. The position of the rollers and, in turn, the size of the opening, are adjusted by turning the handle.

As shown in *Figure 12 (A)*, the close-quarters, or *midget*, tubing cutter is designed to cut small tubing where little room is available to rotate the tool. A ratcheting version is also available for close-quarters work. The standard screw-feed tubing cutter is very common and available in multiple sizes. The screw-feed model shown in *Figure 12 (B)* is equipped with a folding tubing reamer for convenience.

The quick-acting tubing cutter can accommodate a wider range of tubing sizes. In *Figure 12 (C)*, note that the cutting wheel and rollers have switched positions, with the cutting wheel attached to the end of the adjusting screw. Large adjustments in the opening are made by sliding the ratcheting adjusting-screw mechanism in and out. The adjusting screw is then turned to apply pressure and make the cut.

The heavy-duty pipe cutter shown in *Figure 12 (D)* is used for steel and iron pipe, rather than tubing. While tubing can often be handheld for cutting, pipe must be mounted in a vise. A pipe cutter is a far more robust tool than a tubing cutter. Tubing cutters are only used to cut material with thin walls.

Follow these steps to use pipe and tubing cutters:

Step 1 Select the proper type and size of cutter. The effective range of sizes is generally stamped or printed on the body of the cutter.

Step 2 Inspect the cutting wheel to ensure that it is not nicked, bent, or dulled. The replaceable cutting wheel may fracture if too much pressure is applied to the cutter, especially when cutting steel pipe. When cutting tubing, the material may instead be distorted when too much pressure is applied.

Digital Belt Tension Meter

Mechanical devices to measure drive belt tension are becoming increasingly rare. They are being replaced by electronic devices, many of which measure tension through the frequency of belt vibration.

The model shown here uses a pulsed beam of red light to measure the frequency of a vibrating belt. With the instrument ready and the beam pointed at the belt surface in the correct location, the user merely strums the belt to initiate vibration. Algorithms determine the tension based on the vibration frequency. The digital display can provide readings in units of newtons (N) or pounds force (lbf).

Figure Credit: Courtesy of Checkline.

CAUTION

Do not overtighten the cutter. The cutting wheel can be broken, or the tubing can be distorted. If rotating the cutter requires excessive effort, the cutter is too tight. Allow the rotating cutting wheel to do the work and continue to rotate as you tighten the cutter in small increments.

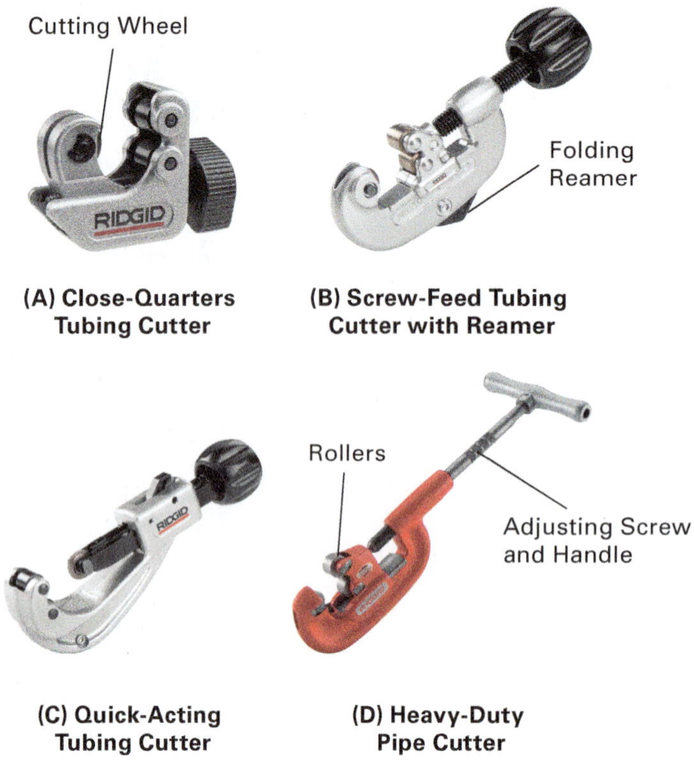

Figure 12 Tubing and pipe cutters.

Step 3 Inspect the rollers to ensure they are not damaged and there is no buildup of debris on the rollers. Pipe scale and dirt can accumulate on the rollers as they travel around the pipe.

Step 4 Inspect the cutter frame to ensure that it is not twisted, cracked, or otherwise damaged.

Step 5 Mount pipe in an appropriate vise. Larger tubing may also require a vise; smaller tubing can usually be handheld.

Step 6 Place the cutter over the pipe at the point where it is to be cut, aligning the cutting wheel directly over the mark. Then turn the handle until light pressure is felt between the rollers and the pipe.

Step 7 Grip the cutter by the handle and rotate it around the pipe one or two turns to start the cut.

Step 8 Be sure the cutting wheel traces a single groove around the material as cutting begins. In some cases, the cutter will try to cut a spiral track around the workpiece. This can be caused by a buildup of debris on the rollers, excess wear of the rollers or cutting wheel, or a twisted tool frame.

Step 9 Tighten the pressure rollers incrementally to maintain pressure while continuing to rotate the tool around the pipe.

Step 10 Continue until the cut is finished. Avoid grasping the inner or outer edges of the cut end; the edges are extremely sharp.

Step 11 Clean the cutter thoroughly to remove dirt, grease, and scale.

Step 12 Apply a light oil to rotating parts and store with similar tools in a dry location.

WARNING!

Pipe and tubing cutters create extremely sharp edges and burrs on the end of the workpiece. Wear gloves to handle pipe and tubing, and never run bare fingers around the inside or outside edge of the cut ends. All pipe and tubing should be properly reamed immediately following the cut.

All pipe and tubing cuts must be reamed. Reaming removes the sharp burr that develops on the inside edge of the cut. The raised burr is not only a safety hazard, but it also interferes with the flow of material through the pipe. Burrs increase the resistance to flow and cause turbulence.

CSST Cutters

Corrugated stainless-steel tubing (CSST) is often used in residential, commercial, and industrial buildings to provide natural gas fuel to equipment and systems. It can replace rigid steel pipe in many situations, offering a flexible solution that is preferred in areas prone to earthquakes. A special tubing cutter, shown here, is designed specifically for this purpose. Two sets of guide wheels located above and beside the cutter wheel follow the corrugations to provide a high-quality cut. The guide rollers also float and hold the two sides of the tubing together to provide a clean cut through the material before they separate.

CSST is often covered with a yellow jacket. The yellow handle on the tubing cutter provides a visual indication that it is specifically designed for CSST.

Figure Credit: Courtesy of RIDGID Tool Company.

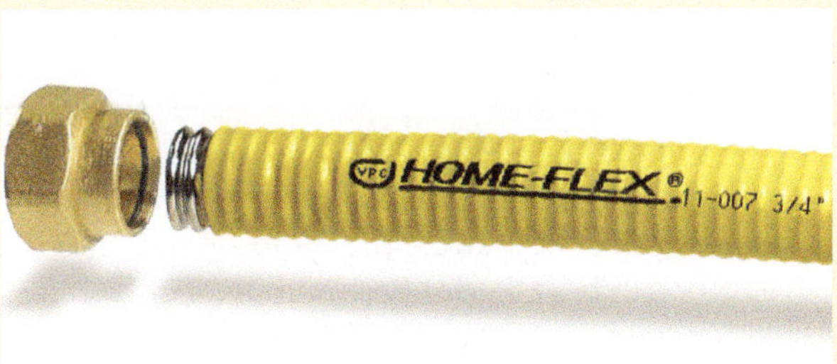

Figure Credit: Valencia Pipe Company.

Tubing cutters may have a folding reamer built in. However, they are not as effective as a separate reaming tool. A reamer that deburrs both the inside and outside edges of cut tubing is shown in *Figure 13*. The blades at the opposite end of the tool are concave instead of convex. The tool is slipped over the end of the tubing and rotated around the end to clean the inner or outer edge, then reversed and rotated to clean the remaining edge.

A pipe reamer (*Figure 14*) is a much heavier tool. The pipe remains in the vise after cutting, and the reamer is rotated around the opening with one hand while applying pressure with the other hand. The head of the reamer ratchets, so the handle does not need to rotate through a 360-degree circle. Pipe threading machines are often equipped with both a pipe cutter and a reamer.

Figure 13 Inside and outside tubing reamer.

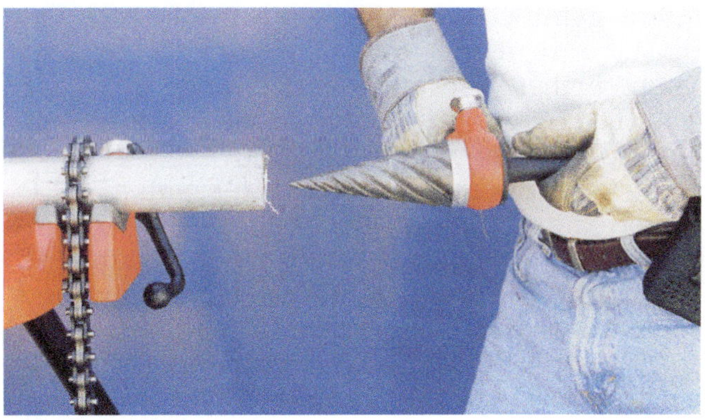

Figure 14 Pipe reamer.

1.3.0 Reaming and Honing Tools

There are a few special hand tools used for reaming and honing that millwrights should know. The use of these tools depends upon the assigned tasks. These tools will be used regularly by some millwrights, and intermittently by others.

1.3.1 Hand Reamers

Hand reamers are used to enlarge holes to a precise size or fit. A reamer consists of four parts: the body, the shank, the flutes, and the cutting edges. Hand reamers have a square end that allows the reamer to be turned with a special wrench.

There are many different styles and sizes of reamers. Reamers designed for use with machine tools and high-speed work have round shanks without the square end, or a special end shape to accommodate unique chuck designs. Hand reamers, as the name implies, are for handheld use.

Reamers are available in a wide range of sizes, including adjustable forms to incrementally increase the size of a hole. *Figure 15* shows some common reamers. Reamers are typically machined to be within 0.0001" to 0.0006" (0.0025 mm to 0.0152 mm) of their stated size. Both US standard and metric sizes are available.

Straight-flute and helical-flute reamers, like those shown in *Figure 15 (A)* and *(B)*, are commonly used for precision work by shaving off tiny amounts of metal at a time to create a precise hole size. The expansion reamer, shown in *Figure 15 (C)*, can be expanded slightly by turning the square-stemmed adjusting screw on the end. The adjustable reamer, shown in *Figure 15 (D)*, goes a step further, offering a wider range of adjustment. The range is typically around $\frac{1}{16}$" to $\frac{1}{8}$" (1.6 mm to 3.2 mm).

The taper pin reamer, shown in *Figure 15 (E)*, is tapered at a rate of $\frac{1}{4}$" per foot. This is the standard taper for a taper pin under the US standard. The standard taper for metric taper pins and reamers is 1 mm per 50 mm of length. Custom tapers are also available for special order.

It is important to use the correct tool to turn the reamer. Use a tap and reamer wrench (*Figure 16*) to avoid damaging the drive end of the reamer. Common wrenches slip off too easily and do not provide proper control over the reamer. Both the bar-style and T-handle style wrenches shown in *Figure 16* may accommodate more than one drive size.

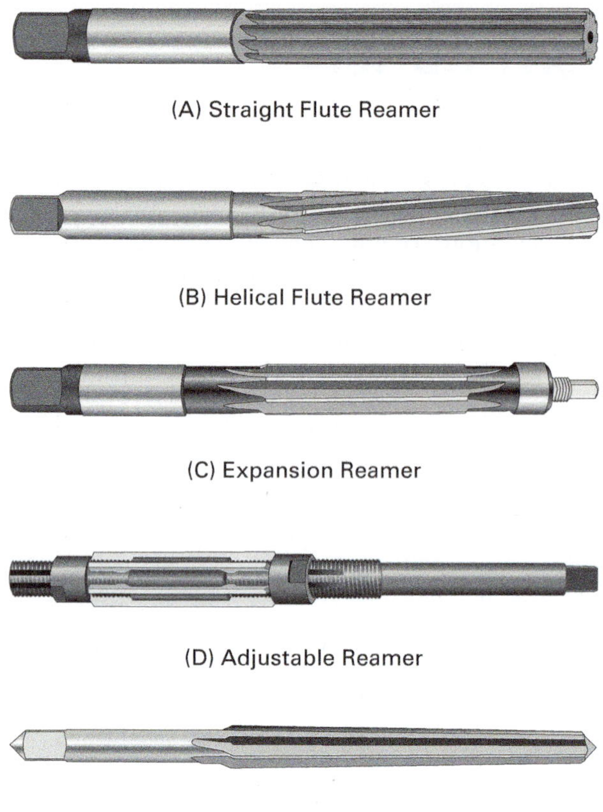

(A) Straight Flute Reamer

(B) Helical Flute Reamer

(C) Expansion Reamer

(D) Adjustable Reamer

(E) Taper Pin Reamer

Figure 15 Reamers.

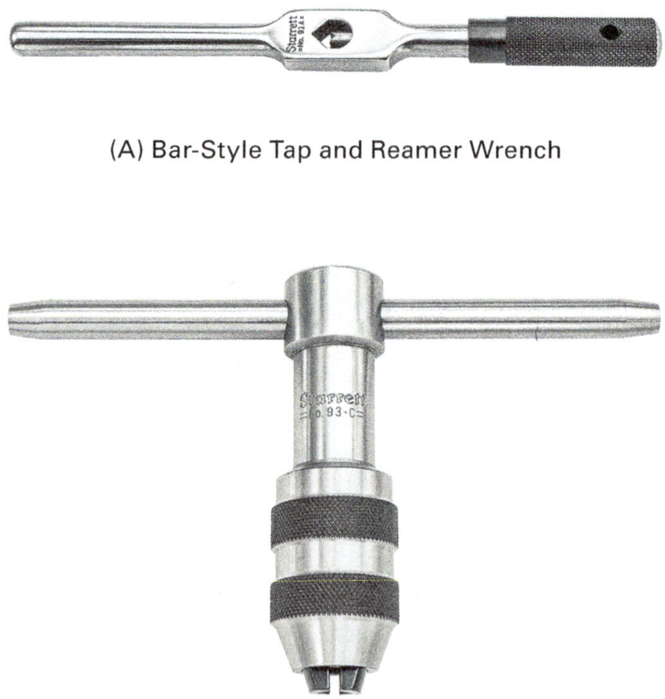

(A) Bar-Style Tap and Reamer Wrench

(B) T-Handle Tap and Reamer Wrench

Figure 16 Tap and reamer wrenches.

Cutting Plastic Tubing and Pipe

The ratcheting plastic pipe cutter shown here quickly slices through rigid or soft plastic pipe and tubing. Each time the handles are pulled together, the blade advances down slightly, eventually cutting through. The tool is indispensable when working with PVC and similar piping materials. It is easy to keep the cut square, and the tool leaves smooth edges without burrs. Unlike a cut made with a saw, no debris or burrs are produced. Similar tools are designed exclusively for soft tubing and hoses, with the cut often completed in a single stroke of the handles.

Figure Credit: Courtesy of RIDGID Tool Company.

The flutes and cutting edges of a reamer are brittle and therefore fragile. Exercise caution when using and storing the reamer to avoid chipping the blades. After use, ensure that the reamer and wrench are clean and dry. They are best wrapped in a lightly oiled cloth to protect them from other tools as well as from corrosion. Store them in shallow tool drawers. This type of tool is more easily damaged when placed in deep drawers where bulkier, heavier tools may be thrown on top of them. Whenever possible, retain the original packaging and use it for storage.

Follow these steps to select, inspect, use, and maintain a reamer:

Step 1 Select a reamer of the proper size for the hole to be reamed.

Step 2 Inspect the reamer to ensure that it is not bent and that the blades are not chipped, nicked, or dull.

Step 3 Select a reamer wrench of the proper drive size.

Step 4 Position the reamer at the opening of the hole to be reamed.

Step 5 Squirt a small amount of cutting fluid into the hole around the reamer. One example of a cutting fluid appropriate for reaming is shown in *Figure 17.*

CAUTION

Always turn the reamer in the proper cutting direction, typically clockwise. Never turn it backwards in the workpiece as this could damage the cutting edges.

Step 6 Hold the reamer square to the surface of the piece being worked on and turn clockwise to start reaming.

Step 7 Continue to turn the reamer until the hole is properly reamed. Add cutting fluid as necessary. In a **blind hole**, the reamer will strike the bottom. For a through hole, the reamer should emerge from the opposite side generously.

Step 8 Continue turning the reamer in the same direction as you withdraw it from the hole. Do not rotate the reamer in the opposite direction.

Step 9 Collect the shavings from blind holes and around the worksite. Clean the reamer thoroughly to remove shavings and cutting fluid.

Step 10 Wrap the reamer in a lightly oiled cloth and store it with similar precision tools in shallow toolbox trays.

1.3.2 Cylinder Hones

A cylinder hone (*Figure 18*) is used to **dress** the walls of cylinders when rebuilding and repairing engines, pumps, compressors, and similar equipment. Cylinders are openings that are too big for common reamers. Light-duty versions are referred to as *glaze breakers* or *cross-hatching hones*.

Figure 17 One example of cutting fluid.

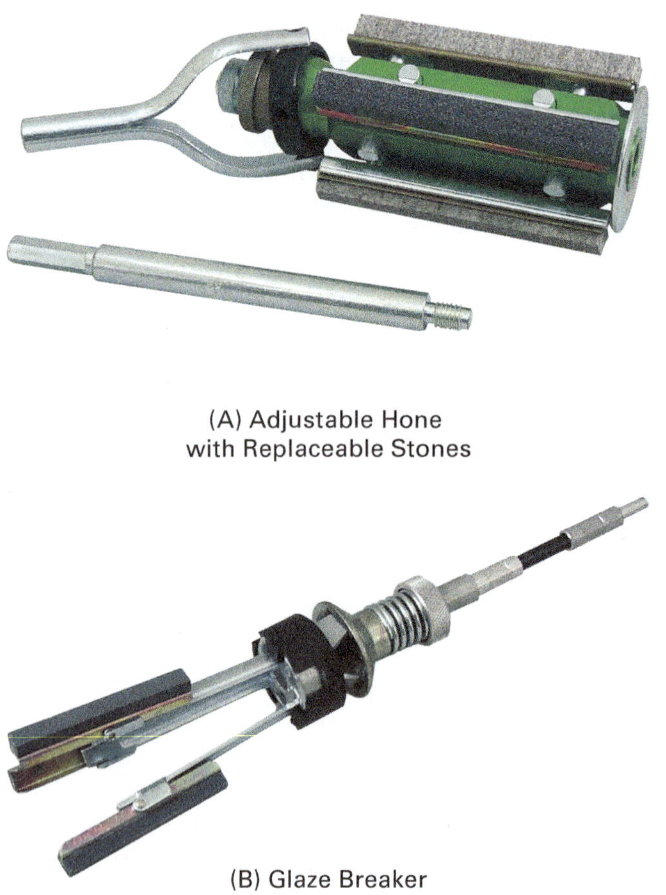

(A) Adjustable Hone with Replaceable Stones

(B) Glaze Breaker

Figure 18 Cylinder hones.

The hone is powered by an electric drill or motor. Abrasive stones cut small amounts of metal away from the cylinder wall as the hone rotates. Most stone sets are replaceable and available in different grit ratings. The grit rating of a stone refers to the density of the particles. Coarse stones have lower grit ratings, such as 80 or 100 grit. Finer stones have higher numbers, such as 400 grit. Coarse stone sets remove more material faster but leave a rougher surface. It is often necessary to begin with coarse stones and finish with fine stones.

If large amounts of metal must be removed, the cylinder must be bored instead. Boring makes use of a machine tool to remove a more significant amount of material. The amount of cylinder wear usually determines whether the cylinder should be honed or bored. Honing is a precise procedure, and the hone must be held rigid and square to the cylinder to ensure good results.

Small cylinders can be honed on a drill press, which provides excellent control over the position and movement of the hone. For larger applications, a jig may be needed to hold the hone in a precise position, directly over the cylinder centerline. If the hone is not held true to the cylinder, the cylinder can be seriously damaged.

Follow these steps to use a cylinder hone:

Step 1 Select a hone of the proper size and grit for the work.

Step 2 Inspect the hone to ensure that it is not bent or damaged and that the stones are not chipped, broken, or excessively worn.

Step 3 Install the hone into a drill press or hand-held drill chuck. For the hone to effectively reach the far end of the cylinder, the workpiece may need to be slightly elevated on the table.

CAUTION

Ensure the speed rating of the drill does not exceed the maximum speed rating of the hone. Consult the hone manufacturer's documentation for the speed rating and choose a drill accordingly.

Step 4 Secure the cylinder in a comfortable working position.

Step 5 Position the hone assembly in the center of the cylinder.

Step 6 Adjust the hone to apply light pressure to the stones against the cylinder walls.

Step 7 Lubricate the stones according to the manufacturer's recommendations, if required.

Step 8 Start the drill and move the hone up and down a few strokes in the lower half of the cylinder.

Step 9 Gradually increase the length of the stroke until the hone is travelling the full length of the cylinder bore.

Step 10 Stop the drill and readjust the pressure on the stones as the diameter of the cylinder increases and the pressure on the stones decreases.

CAUTION

Do not extend the hone through the opposite end of the cylinder more than half the length of the stones. Allowing the spinning stones to fully exit the open end can damage both the cylinder and the tool.

Step 11 Change to a finishing stone if coarse stones were used on the first part of this procedure. Hone the cylinder again to remove all rough hone marks.

Step 12 Extract the hone from the cylinder and clean both the tools and the cylinder thoroughly.

Step 13 Clean and coat the hone and stones lightly with oil and wrap in a protective cloth. Store cylinder hones with similar tools to avoid damage. Do not store them in a large box or drawer with heavy tools that could cause damage.

1.3.3 Honing Stones

Honing, or sharpening, stones are used for a variety of purposes, including surface conditioning and tool sharpening. Honing stones belong to one of two groups: natural or artificial. Artificial stones are usually made of silicon, aluminum oxide, or carbide. Artificial stones may also be enhanced with diamond particles. Some natural stones, such as the soft Arkansas oilstone shown in *Figure 19*, are treated with oil, and are referred to as *oilstones*. Oil enhances the honed edge and helps prevent the stone from becoming clogged with tiny metal particles. The term used to describe this condition is *glazed*.

Some stones require water as a lubricant instead of oil. A waterstone, also shown in *Figure 19*, is generally softer than a dry stone and may be natural or artificial. Stones are difficult to identify by color or appearance. It is therefore important to retain the packaging, which will protect the stone in storage and aid in identification.

The honing stones are made in various grades: coarse, medium, fine, and very fine are all common examples. These are broad categories, with each including a range of grit ratings. Some manufacturers have more categories or assign different names to them. Stones are more accurately identified by their grit rating, which generally ranges from 100 to 30,000 for extremely fine stones.

The stones also come in various shapes and sizes, including square, rectangular, triangular, round, and wedge shaped with rounded edges.

(A) Soft Arkansas Oilstone

(B) Combination Arkansas Waterstone

Figure 19 Examples of honing stones.

It is important to use the proper stone for any honing operation. Use a coarse stone to begin sharpening large, dull, or nicked tools. Use a medium stone to sharpen tools that do not require a finished edge or are not especially dull. Use a fine stone to sharpen a knife or tools requiring a very sharp, razor-like edge. Fine and medium stones can be used behind a coarse stone as necessary.

Honing stones are fragile and must be stored in boxes or kept in special racks to protect them from breakage. Never leave a stone on a bench or table where it may be knocked off or where heavy objects may fall on it. Since some stones are designed for use with water, make sure they are dry before storing them in a protective box (preferably their original packaging) to prevent mold and mildew from forming.

These stones do wear, and softer stones wear more quickly than others. After some use, the stone surface may no longer be flat. Other stones, called **lapping plates** or *flattening stones*, are used to dress a honing stone and restore the flat surface.

The following procedure describes sharpening a chisel. Follow these steps to use a honing stone:

Step 1 Select a honing stone with a coarse grain on one side and a fine grain on the other, or separate stones with similar characteristics. You must know whether the stones are designed for dry use or require oil or water.

Step 2 Inspect the stone to ensure that it is not cracked or broken.

Step 3 Coat the stone with a thin layer of machine oil (oilstones) or water (waterstones) to reduce glazing. Oil and water should not be used with stones designed for dry use.

Step 4 Make sure the stone is properly secured and cannot move.

Step 5 Hold the chisel in one hand with the bevel flat against the coarse side of the stone. It is important to maintain the proper angle.

Step 6 Steady the chisel and hold it down against the stone with the fingers of the other hand.

Step 7 Work the chisel back and forth parallel to the stone, using smooth, even strokes.

Step 8 Turn the chisel over and repeat for the opposite edge. Try to use as much of the stone surface as possible to balance wear across it.

CAUTION

The tips of wood chisels are ground to a single angle. When it is flipped over for sharpening, keep the flat side perfectly flat against the stone. Raising the chisel even slightly will produce a bevel on the tip. This will result in a very poor cutting edge that takes time to repair.

Step 9 Turn the chisel over again and hone the opposite side a few strokes to remove any remaining feather edge or burr.

Step 10 Clean and oil the chisel; then store it with similar tools.

Step 11 Clean the stone with a rag, small stiff brush, or steel wool using honing oil (mineral or water-based, depending on the stone) and rags. A waterstone can also be cleaned with warm soapy water.

Step 12 Cover the stone with a light coat of machine oil; clean waterstones and allow them to dry.

Step 13 Put the stone in its protective box and store it in its proper place.

Knife Sharpening System

Precision knife sharpeners like the one shown here make the task go quickly and require very little skill to get an extremely sharp edge. Using natural diamond rods, this sharpener can handle four different edge angles and is compact as well as lightweight. This three-way honing system is designed primarily for knives.

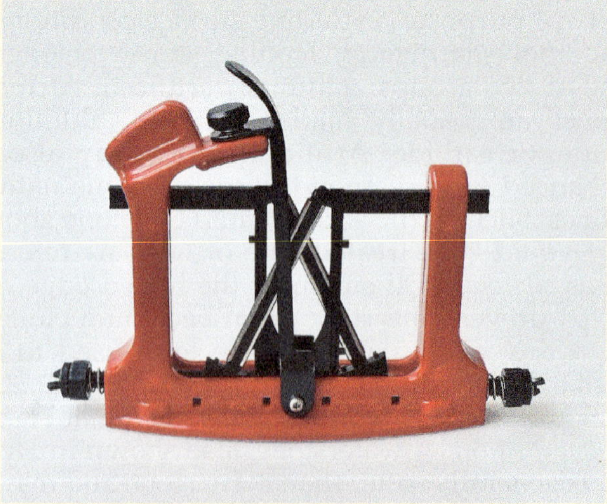

1.4.0 Thread Taps, Dies, and Extractors

Taps and dies are used to cut threads in materials such as metal, plastic, and hard rubber. Taps are used to cut internal threads in materials. Dies are used to cut external threads on smooth rod. Both are also used to clean up damaged threads. Extractors are used to remove broken taps or hardware from threaded holes.

Refer to *Figure 20*. The following is a list of the most common taps:

- *Taper taps* — Taper taps may also be referred to as *starter taps*. They are used when starting a tapping operation or when tapping through holes. They have a tapered tip that extends across 8 to 10 threads.
- *Plug taps* — Plug taps, also known as *intermediate taps*, are used after the taper tap. They have a taper length of 3 to 5 threads.
- *Bottoming taps* — Bottoming taps are used after the taper and plug taps have already been used. They are used primarily for threading to the bottom of a blind hole and have a taper length of only 1 to $1\frac{1}{2}$ threads.
- *Pipe taps* — Pipe taps are used exclusively to form internal pipe threads. Standard pipe threads are tapered, but there are also straight pipe threads used for unique purposes. The diameter of a US-standard pipe tap, from end to end of the threaded portion, is tapered at the rate of $\frac{1}{16}$" per inch of length. Metric threads are tapered at a ratio of 1:16, meaning they taper 1 mm for every 16 mm of length.

Pipe taps used by hand, like reamers, have square tips that allow them to be used with a tap wrench as shown in *Figure 16*. Do not use a common wrench to turn a tap. T-handled tap wrenches are best for small tap sizes and provide good control. Larger taps require greater leverage and power to turn. It is usually best to use a bar-style tap wrench with tap sizes of $\frac{1}{2}$" (~13 mm) and larger.

Dies are available in several styles and a wide range of both US-standard and metric sizes. Like taps, they are typically tapered slightly on one side to help get threading started. Refer to *Figure 21*. The following are common types of dies:

- *Hex dies* — Hex dies are also known as *die nuts* or *rethreading dies*. They are used for repairing damaged or rusty threads on bolts or similar hardware. They are hexagon-shaped and can be turned using any type of wrench that will fit. They are not a good choice for creating new threads on round stock; they are used primarily to clean up existing threads.
- *Round threading dies* — Round threading dies create new threads but must be used with a diestock (*Figure 22*). There are no flats for a common wrench or tap wrench to engage, preventing the use of the wrong tool. Many are adjustable, having a small slot and an adjusting screw, as shown in *Figure 21*.
- *Pipe dies* — Like pipe taps, these cut tapered or straight pipe threads.

Taps and dies are often combined in sets that include a range of sizes, as well as diestocks, tap wrenches, and the screwdrivers and wrenches necessary to loosen and tighten adjusting hardware (*Figure 23*). As tools in the kit become worn or damaged, they are replaced individually. Very large or unique taps and dies are often purchased individually as well.

The first step in threading a hole is to drill a hole of the proper size. *Table 2A, Table 2B, Table 2C, Table 2D, and Table 2E* each shows an example of a tap drill table. There are many versions available, primarily from tap and drill manufacturers. *Tables 2A–2E* provides alternate metric drill sizes as well as the US-standard drill sizes. Several types of drills are represented in the *Tap Drill Size* column—letter, number, and fractional drill sizes are needed for various tap sizes. Unified taps are based on the US standard, which are classified as either coarse (UNC) or fine (UNF) threads.

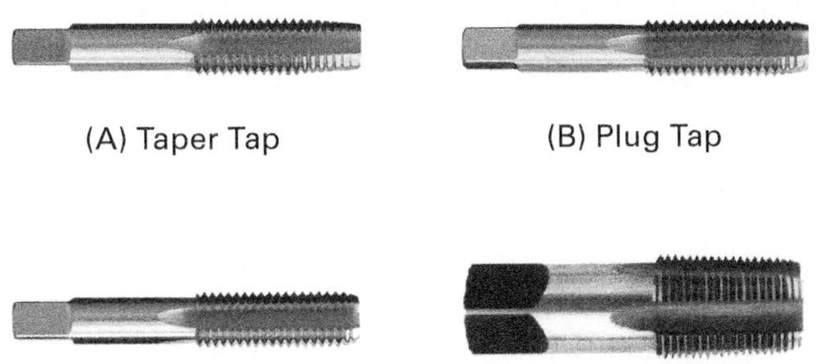

(A) Taper Tap

(B) Plug Tap

(C) Bottoming Tap

(D) Pipe Tap

Figure 20 Common tap designs.

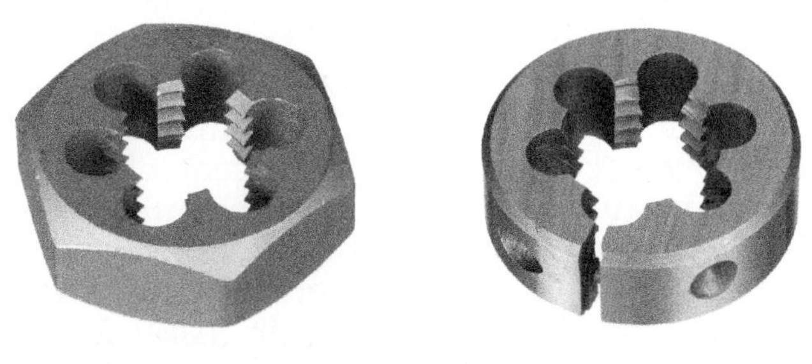

(A) Hex Die

(B) Adjustable Round Die

Figure 21 Common thread dies.

Figure 22 Thread die mounted in a diestock.

The pitch of these threads refers to the number of threads per inch (tpi). Threaded hardware begins in machine screw sizes, from #0 to #12, at the beginning of the chart. For example, a #6 machine screw has either coarse threads with a pitch of 32 tpi, or fine threads with a pitch of 40 tpi.

Continuing in *Tables 2A–2E*, fractional bolt sizes follow the machine screws, followed by tapered and straight pipe taps. Metric tap and drill sizes complete the table. Metric threads also have coarse and fine thread pitches. However, metric thread pitch is not based on the number of threads over some length. They are based on the distance between the crests of each thread. For example, an M14 (14 mm) bolt with coarse threads has a thread pitch of 2.00, or 2 mm between thread crests. An M14 bolt with fine threads has a pitch of 1.25, or 1.25 mm between thread crests.

NOTE

Additional information about threads and threaded fasteners can be found in NCCER Module 15103, *Fasteners and Anchors.*

To use *Tables 2A–2E*, simply look for the machine screw or bolt size you intend to use in the tapped hole. Then read the proper drill size in the column to the right. For example, to tap a hole for a ¼" - 20 UNC bolt, use a #7 drill from a numbered drill index. Alternatively, a metric drill size of 5.1 mm can be used.

At the top of *Tables 2A–2E*, there is a note referencing a percentage of full threads. Most threads are cut to about 75 percent of a full thread, meaning that the crest of the threads has been removed. *Figure 24* shows a set of internal threads that have about 25 percent of the thread missing. The drill sizes listed in *Tables 2A–2E* will drill holes that result in the removal of material that would otherwise form the crest of the thread. In *Figure 24*, this is shown as the tap drill diameter. This slight oversizing of the bit provides clearance and reduces the risk of damaging the threads and binding the tap as it advances. Removing the crests does not significantly reduce the strength of the threads. If you try to use a smaller drill for the hole, to create 100 percent threads, you will quickly find that it is very difficult to turn the tap without breaking it or damaging the threads.

Figure 23 Tap and die set.

The following procedure explains how to tap a blind hole. Through holes may only require the use of a taper tap, perhaps followed by a plug tap, depending on the thickness of the material. Bottoming taps are required for blind holes. Follow these steps to tap a hole:

Step 1 Determine the size of the hardware and the thread pitch needed. Threading a ¼" - 20 UNC bolt hole in ⅜" (~10 mm) plate is a good place to start.

Step 2 If the hole has not been drilled, use *Tables 2A–2E* or a similar table to determine the proper drill size, then drill the hole using a drill press. A handheld drill can be used, but the resulting hole will not be as straight and true.

Step 3 Select a taper tap of the proper size and thread pitch. Inspect it to ensure the tap is sharp and clean with no broken or nicked teeth.

Table 2A Common Tap Drill Sizes (1 of 5)

Tap Size / Pitch Form	US Tap Drill	Metric Drill	Tap Size / Pitch Form	US Tap Drill	Metric Drill
Unified Taps					
0-80	3/64"	1.25	7/16" - 14 UNC	U	9.40
1 - 64 UNF	53	1.55	7/16" - 20 UNF	W	9.90
1 - 72 UNF	53	1.55	1/2" - 13 UNC	27/64"	10.80
2 - 56 UNC	51	1.85	1/2" - 20 UNF	29/64"	11.50
2 - 64 UNF	50	1.90	9/16" - 12 UNC	31/64"	12.20
3 - 48 UNC	47	2.10	9/16" -18 UNF	33/64"	12.90
3 - 56 UNF	46	2.15	5/8" - 11 UNC	17/32"	13.50
4 - 40 UNC	43	2.35	5/8" - 18 UNF	37/64"	14.50
4 - 48 UNF	42	2.40	3/4" - 10 UNC	21/32"	16.50
5 - 40 UNC	39	2.65	3/4" - 16 UNF	11/16"	17.50
5 - 44 UNF	37	2.70	7/8" - 9 UNC	49/64"	19.50
6 - 32 UNC	36	2.85	7/8" - 14 UNF	13/16"	20.40
6 - 40 UNF	33	2.95	1" - 8 UNC	7/8"	22.25
8 - 32 UNC	29	3.50	1" - 12 UNF	59/64"	23.25
8 - 36 UNF	29	3.50	1" - 14 UNS	15/16"	23.50
10 - 24 UNC	25	3.90	1 1/8" - 7 UNC	63/64"	25.00
10 - 32 UNF	21	4.10	1 1/8" - 12 UNF	1 3/64"	26.50
12 - 24 UNC	17	4.50	1 1/4" - 7 UNC	1 7/64"	28.00
12 - 28 UNF	15	4.70	1 1/4" - 12 UNF	1 11/64"	29.50
1/4" - 20 UNC	7	5.10	1 3/8" - 6 UNC	1 7/32"	30.75
1/4" - 28 UNF	3	5.50	1 3/8" - 12 UNF	1 19/64"	32.75
5/16" - 18 UNC	F	6.60	1 1/2" - 6 UNC	1 11/32"	34.00
5/16" - 24 UNF	I	6.90	1 1/2" - 12 UNF	1 27/64"	36.00
3/8" - 16 UNC	5/16"	8.00	1 3/4" - 5 UNC	1 9/16"	39.50
3/8" - 24 UNF	Q	8.50	2" 4.5 UNC	1 25/32"	45.00

Table 2B Common Tap Drill Sizes (2 of 5)

Tap Size / Pitch Form	US Tap Drill	Metric Drill
Taper Pipe Taps (NPT)		
1/16" - 27	D	6.30
1/8" - 27	R	8.70
1/4" - 18	7/16"	11.10
3/8" - 18	37/64"	14.50
1/2" - 14	23/32"	18.00
3/4" - 14	59/64"	23.25
1" - 11.5	1 5/32"	29.00
1 1/4" - 11.5	1 1/2"	38.00
1 1/2" - 11.5	1 47/64"	44.00
2" - 11.5	2 7/32"	57.00
2 1/2" - 8	2 5/8"	67.00
3" - 8	3 1/4"	82.50

Table 2C Common Tap Drill Sizes (3 of 5)

Tap Size / Pitch Form	US Tap Drill	Metric Drill
Straight Pipe Taps (NPS)		
1/8" - 27	S	8.80
1/4" - 18	29/64"	11.50
3/8" - 18	19/32"	15.00
1/2" - 14	47/64"	18.50
3/4" - 14	15/16"	23.75
1" - 11.5	1 3/16"	30.25
1 1/4" - 11.5	1 33/64"	38.50
1 1/2" - 11.5	1 3/4"	44.50
2" - 11.5	2 7/32"	57.00

Step 4 Attach a suitable tap wrench, place the end of the tap in the hole, and turn it clockwise about one-half to one turn while holding it true and applying light downward pressure. The tapered portion of the teeth will begin to cut.

CAUTION

Do not use common wrenches or similar tools with taps, dies, or extractors. Always use the proper tap or die wrench provided by the manufacturer or an aftermarket source.

NOTE

Although right-hand threads are the most common by far, a left-hand tap must be turned counterclockwise when cutting.

Step 5 Turn the tap enough so that it can hold itself in place in the hole. After each half-turn, back the tap up slightly to break the metal chip being created. Note that the threads are very fragile at this point; handle the tap with care. Check the tap with a square at two positions to ensure it is perpendicular to the surface of the workpiece. The tap must be square to the surface to ensure the hole is tapped straight.

Step 6 Lubricate the tap with cutting fluid at this point, and continue doing so regularly during the tapping process. Proper lubrication is a must. Use a cutting fluid designed for the material being tapped, such as ferrous metal, non-ferrous metal, or plastic. Continue to turn the tap, applying light pressure until the tap feeds itself into the material. Downward pressure is then unnecessary.

Step 7 Continue the tapping process of turning in one-half turn and then backing out one-quarter turn until the tap touches the bottom of the hole or you feel it grinding against the chips in the bottom.

NOTE

For a through hole, continue tapping until the full, untapered threads of the tap emerge from the opposite side.

Table 2D Common Tap Drill Sizes (4 of 5)

Tap Size / Pitch Form	US Tap Drill	Metric Drill
	Metric ISO (Coarse)	
M1.6 - 0.35	3/64"	1.25
M1.7 - 0.35	55	1.35
M1.8 - 0.35	64	1.45
M2 - 0.40	1/16"	1.60
M2.2 - 0.45	50	1.75
M2.3 - 0.40	49	1.90
M2.5 - 0.45	46	2.05
M2.6 - 0.45	44	2.15
M3 - 0.50	40	2.50
M3.5 - 0.60	33	2.90
M4 - 0.70	30	3.30
M4.5 - 0.75	27	3.70
M5 - 0 80	19	4.20
M5.5 - 0.90	15	4.60
M6 - 1.00	9	5.00
M7 - 1.00	15/64"	6.00
M8 - 1.25	H	6.80
M9 - 1.25	5/16"	7.80
M10 - 1.50	Q	8.50
M11 - 1.50	3/8"	9.50
M12 - 1.75	Y	10.20
M14 - 2.00	15/32"	12.00
M16 - 2.00	35/64"	14.00
M18 - 2.50	39/64"	15.50
M20 - 2.50	11/16"	17.50
M22 - 2.50	49/64"	19.50
M24 - 2.50	53/64"	21.00
M27 - 3.00	61/64"	24.00
M30 - 3.50	1 3/64"	26.50
M33 - 3-50	1 5/32"	29.50
M36 - 4.00	1 1/4"	32.00
M39 - 4.00	1 3/8"	35.00

Step 8 Back the tap out of the hole and clear the chips.

Step 9 Select and inspect a plug tap for the second phase of the tapping process.

Step 10 Tap the hole, using the plug tap, following the same procedure as with the taper tap. Little cutting action will be felt until you approach the bottom of the hole.

Table 2E Common Tap Drill Sizes (5 of 5)

Tap Size / Pitch Form	US Tap Drill	Metric Drill	Tap Size / Pitch Form	US Tap Drill	Metric Drill
Metric ISO (Fine)					
M3 - 0.35	37	2.65	M13 - 1.50	29/64"	11.50
M4 - 0.35	27	3.65	M13 - 1.75	7/16"	11.25
M4 - 0.50	29	3.50	M14 - 1.25	1/2"	12.75
M4.5 - 0.45	21	4.05	M14 - 1.50	31/64"	12.50
M5 - 0.50	16	4.50	M15 - 1.50	17/32"	13.50
M5 - 0.70	18	4.30	M16 - 1.00	19/32"	15.00
M5 - 0.75	18	4.25	M16 - 1.25	37/64"	14.75
M5.5 - 0.5	9	5.00	M16 - 1.50	9/16"	14.50
M6 - 0.50	7/32"	5.50	M18 - 1.00	43/64"	17.00
M6 - 0.75	5	5.25	M18 - 1.25	21/32"	16.75
M7 - 0.75	D	6.25	M18 - 1.50	41/64"	16.50
M8 - 0.50	M	7.50	M18 - 2.00	5/8"	16.00
M8 - 1.00	J	7.00	M20 - 1.00	3/4"	19.00
M9 - 0.50	Q	8.50	M20 - 1.50	47/64"	18.50
M9 - 1.00	O	8.00	M20 - 2.00	45/64"	18.00
M10 - 0.50	3/8"	9.50	M22 - 1.00	53/64"	21.00
M10 - 0.75	U	9.25	M22 - 1.50	13/16"	20.50
M10 - 1.00	T	9.00	M22 - 2.00	25/32"	20.00
M10 - 1.25	11/32"	8.75	M24 - 1.00	29/32"	23.00
M11 - 1.00	X	10.00	M24 - 1.50	7/8"	22.50
M12 - 1.00	7/16"	11.00	M24 - 2.00	55/64"	22.00
M12 - 1.25	27/64"	10.75	M24 - 2.50	27/32"	21.50
M12 - 1.50	Z	10.50			

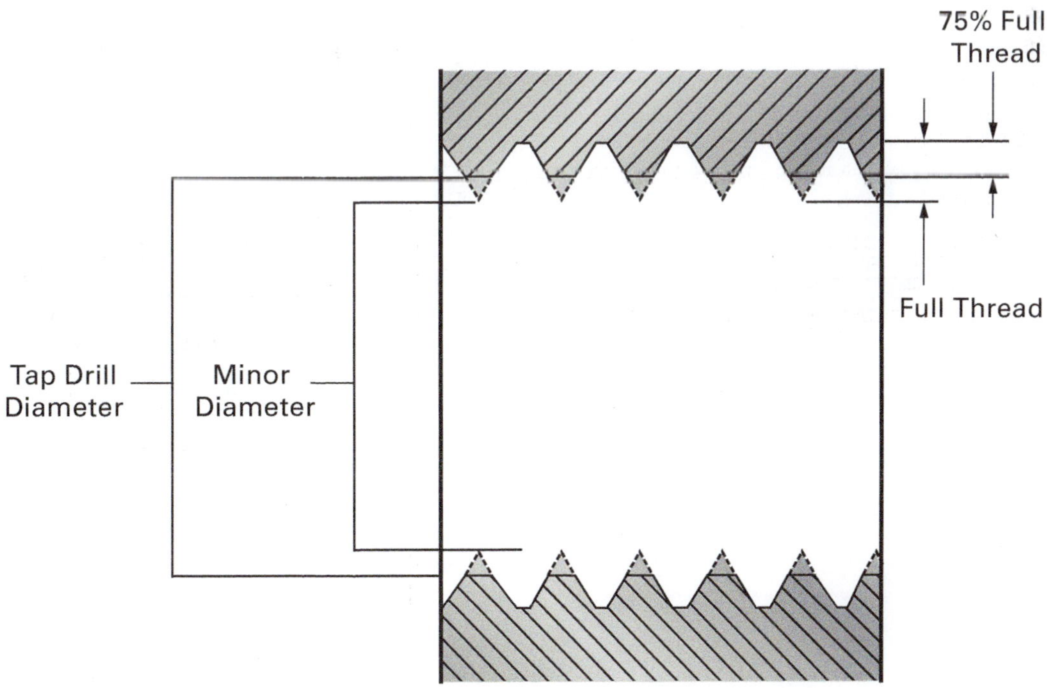

Figure 24 Thread crest removal by the tap drill.

Step 11 Back the plug tap out of the hole and clear any chips.

Step 12 Select and inspect a bottoming tap for the final phase of the tapping process.

Step 13 Tap the hole, using the bottoming tap. The tap may need to be removed several times to clear chips until it contacts the bottom of the hole.

Step 14 Back the bottoming tap out, clear the chips, and wipe up any remaining lubricant.

Step 15 Clean the taps thoroughly to remove all chips and shavings.

Step 16 Return the tools to their protective case and store them with similar tools that need protection from large, heavy tools.

Thread Files

Thread files are very handy tools, especially if you often work with and cut all-thread rod. Each side of the file is machined to fit a different thread pitch. Most files have a total of eight different faces. Thread files can quickly clean and straighten damaged threads or those clogged with debris. They are very helpful for removing burrs and aligning the threads of threaded rod at the point of the cut, making it much easier to get a nut started.

Figure Credit: "Thread restoring file.JPG" by Tenbergen is licensed under CC BY-SA 3.0.

1.4.1 Extractors

If a fastener or tap breaks off in a workpiece, the remains must be extracted. If the fastener or tap protrudes above the surface of the work, removing it may be as simple as using a pair of locking pliers to grip and turn it. If the screw or tap is broken at or below the surface of the work, however, an extractor is needed to remove it. Two common types of extractors are the spiral screw extractor and the tap extractor.

Spiral Screw Extractors

Most spiral screw extractors (*Figure 25*), also known as *easy-outs*, are reverse-fluted to remove right-hand threaded fasteners. They are made in various sizes. The manufacturer provides recommendations for the size to use to remove a given size of fastener. The recommendations also include the drill size required to engage the extractor. Follow the manufacturer's guidance with precision. An extractor that is too small may not be strong enough. If the extractor breaks off in the fastener, it will be even more difficult to remove.

It is important to consider the type of drill bit used. A standard drill bit rotates clockwise—the same direction as most hardware is turned to tighten. As a result, the drill bit is working to tighten the fastener. For fastener extraction, it is best to use a left-hand drill bit, which rotates counterclockwise. In some cases, the fastener will begin to turn counterclockwise with the drill bit, rising from the hole. In other words, using a left-hand drill bit may be all that is needed to remove the fastener. If the fastener does not move,

Figure 25 Spiral screw extractors.

continue drilling the hole to the depth needed for the extractor. At a minimum, the left-hand drill bit will not be working to tighten the fastener.

Follow these steps to use a screw extractor:

Step 1 Select an extractor of the proper size, per the manufacturer's recommendations.

Step 2 Inspect the extractor to ensure that the flutes are not damaged.

Step 3 Select a left-hand drill bit that is the proper size for the chosen extractor, again per the manufacturer's recommendations.

Step 4 Apply penetrating oil around the threads of the broken fastener and allow it to soak (overnight if practical; longer is better for heavy corrosion).

Step 5 Prick the exact center of the fastener using a center punch and ball peen hammer. Pricking the center of the fastener will help to ensure that the hole is drilled on center.

Step 6 Drill a hole in the center of the fastener.

CAUTION

For blind holes, avoid drilling all the way through the fastener and into the workpiece.

Step 7 Insert the extractor into the drilled hole and tap lightly with a ball peen hammer to ensure it is firmly seated and the edges begin to bite into the fastener. Attach a tap wrench.

Step 8 Carefully begin turning the extractor counterclockwise (for right-hand threads) until the fastener loosens and begins to rotate.

Step 9 Remove the extractor from the fastener, then clean with light oil and return it to its packaging. Use a tap to clean up any minor thread damage to the workpiece, if necessary.

Tap Extractors

Tap extractors (*Figure 26*) are designed to remove broken taps from a workpiece. Since taps and similar tools are made from steel that is very difficult to drill, another approach to removal is required. A tap extractor has fingers that fit into the flutes of the broken tap to turn it. There is a specific tap extractor for each tap size. Most taps have four flutes, but the designs vary, and they may have from two to six flutes. As a result, the tap extractor must not only be the correct size, but must also accommodate the number of flutes in the broken tap.

Follow these steps to use a tap extractor:

Step 1 Select an extractor of the proper size and number of fingers for the tap being removed.

Step 2 Inspect the extractor to ensure that it is not damaged.

Step 3 Place the tip of the extractor against the top of the broken tap, then insert the fingers of the extractor into the flutes.

Step 4 Push the sleeve of the extractor down against the top of the tap to hold the prongs securely in the flutes of the tap.

Step 5 Use a tap wrench to turn the tap back and forth slowly to break the chip.

Step 6 Turn the extractor counterclockwise to screw the tap out of the hole.

Step 7 Remove the extractor from the tap, then clean with light oil and return it to its packaging.

1.4.2 Nut Splitter

A nut splitter (*Figure 27*), also called a *nut buster* or *nutcracker*, is used when a nut cannot be removed using a wrench or socket set. Nut splitters are used to remove damaged or frozen nuts without destroying the male fastener or its threads. Hydraulic nut splitters (*Figure 28*) are used for heavy-duty applications.

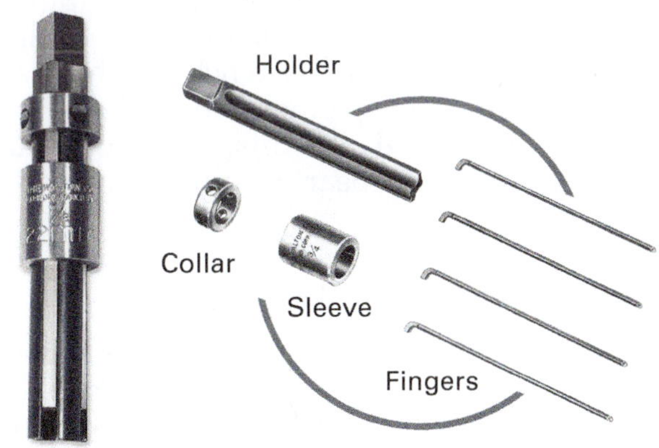

(A) Tap Extractor **(B) Tap Extractor Parts**

Figure 26 Tap extractor.

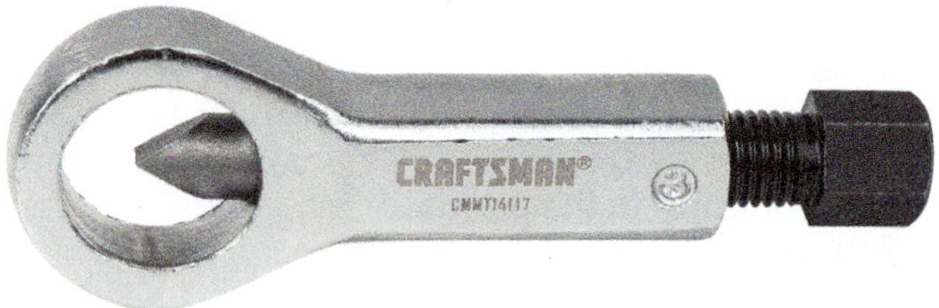

Figure 27 Nut splitter.

Follow these steps to use a nut splitter:

Step 1 Select a nut splitter of the appropriate size. Each tool will fit a range of nut sizes.

Step 2 Place the nut splitter on the nut to be removed.

Step 3 Turn the adjusting screw at the end of the handle with a wrench to drive the wedge-shaped blade into the nut.

Step 4 Tighten the adjusting screw until you hear a sharp popping sound, which is the sound of the nut splitting.

Step 5 Loosen the adjusting screw and remove the nut splitter.

Step 6 Remove the remains of the nut. Most nut splitters split the nut on one side only. If the nut cannot be spread open and removed, it must be split again. Move the nut splitter to the opposite side and repeat the procedure.

Step 7 Once the nut can be removed, remove the nut splitter. Clean the tool with light oil and store it with similar tools.

1.5.0 Unique Wrenches and Pullers

There are some unique wrenches and pulling tools that millwrights should know. These include relatives of standard pipe wrenches—chain and strap wrenches—and spanner wrenches.

Pullers introduced in this section include mechanical gear and bearing pullers as well as packing pullers.

1.5.1 Chain and Strap Wrenches

Chain and strap wrenches (*Figure 29*) are types of pipe wrenches. They are used for the same basic purposes as pipe wrenches but offer different characteristics. For example, when working on a pipe that is close to a wall, there may be enough room between the pipe and the wall to pass a chain or strap, but not the upper jaw of a pipe wrench. Both chain and strap wrenches can also be used to grip square or oddly shaped materials.

Figure 28 Hydraulic nut splitter.

Chain wrenches can accommodate large pipe sizes. They come in various sizes to handle pipe up to 18" diameter (~46 cm). The largest sizes are heavy tools that weigh roughly 90 lb, or 41 kg, and approach 4' (1.2 m) in length.

Strap wrenches can be used on pipe or other stock that may be too hard for pipe or chain wrenches to effectively grip. Most often though, they are used in place of pipe wrenches when the workpiece surface is delicate or decorative and the teeth of a pipe wrench would ruin its appearance.

Chain wrenches and strap wrenches are used in a similar manner. Follow these steps to use a chain wrench:

Step 1 Select a chain wrench of adequate size to fit the workpiece.

Step 2 Inspect the wrench to ensure that the chain is not damaged or excessively worn and that the handle is not cracked or bent.

(A) Chain Wrench (B) Strap Wrench

Figure 29 Chain and strap wrenches.

Step 3 Position the wrench on the workpiece and wrap the chain around it in the same direction the workpiece will be turned.

Step 4 Pull the chain taut and hook it on the handle at the tightest position available. The chain may not fit tightly; it will tighten on the workpiece when pressure is applied in the next step.

Step 5 Apply pressure on the wrench handle in the proper direction until the chain tightens and grips the workpiece.

Step 6 Rotate the wrench and workpiece as necessary or as far as obstructions will permit.

Step 7 Relieve pressure on the wrench, rotate it back to the starting position, and establish a new grip to rotate the workpiece again. Repeat this process until the workpiece is tight or in the desired position.

Step 8 Remove and clean the wrench thoroughly. Use light oil to prevent corrosion on the chain. Store the wrench in an appropriate place with similar tools.

CAUTION

Do not apply oil to the strap of a strap wrench. The strap must be kept clean and oil-free to provide reliable grip.

1.5.2 Spanner Wrenches

Spanner wrenches are made for working with unique hardware called *spanner nuts* (*Figure 30*). The wrenches are sometimes furnished with equipment that requires them, but millwrights encounter spanner nuts regularly. For example, many bearings have a spanner nut on the shaft to lock them in position (*Figure 31*). Several types of spanner wrenches are available, with some common types shown in *Figure 30*.

Spanner wrenches are relatively easy to use, and all are used in a similar manner. Large locknuts and comparable hardware can become difficult to loosen and may require some impact to jar loose. Common spanner wrenches are not designed to be struck with a hammer, and a **cheater bar** should not be used. Impact spanner wrenches (*Figure 32*) are designed for this purpose. A ball peen hammer is used to strike the head of the wrench and loosen stubborn hardware.

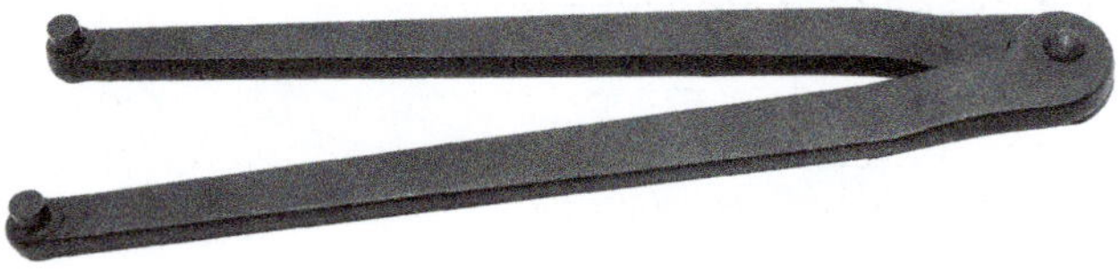

(A) Adjustable Face Spanner

(B) Adjustable Pin Spanner

(C) Adjustable Hook Spanner

(D) Example Spanner Nuts

Figure 30 Spanner wrenches and nuts.

Figure 31 Tightening a bearing locknut with a hook spanner.

Figure 32 Impact spanner wrench.

Follow these steps to use a face spanner wrench, shown in *Figure 30 (A)*. Note that, if it is not an adjustable wrench, a fixed face spanner wrench must have the correct pin size and span to fit the hardware:

Step 1 Select a spanner wrench that is suitable for the hardware. It must fit the span between the slots or holes in the hardware, and the pins or hook of the tool must properly fit the openings.

Step 2 Inspect the wrench to ensure that it is not bent or otherwise damaged. The pins or hook of the tool should not be distorted or excessively worn. Since the contact surfaces of the wrench are small and offer limited security, the fit is very important.

Step 3 Clean the holes of the hardware to remove dirt, grease, or other debris that would prevent the wrench from securely fitting into the holes.

Step 4 Insert the pins or hook of the wrench into the holes or slots of the nut. Be sure that the openings are not disfigured, and the wrench is fully engaged.

Step 5 Hold the wrench down against the nut and in the holes using pressure with one hand, and carefully turn the wrench with the other hand.

> **NOTE**
> Be sure that you are turning the wrench in the proper direction. Although most nuts are loosened by turning them in a counterclockwise direction, some nuts have left-hand threads and are loosened by turning them clockwise.

Step 6 Turn the wrench as far as clearances or reach will allow.

Step 7 Remove the wrench from the holes and return to the starting position. Reinsert the wrench and continue turning the nut until it is loose or tightened to the required position.

Step 8 Clean the wrench thoroughly and apply a light oil to prevent corrosion. Store the wrench with similar tools.

1.5.3 Gear and Bearing Pullers

Many gears and bearings are pressed into position and are difficult to remove. Attempting to remove them with a hammer or by prying them off may damage the part as well as the shaft. Mechanical pullers are made to pull gears and bearings from shafts in a way that prevents damage. A puller may also be required to remove other shaft-mounted components, such as fan blades.

Mechanical pullers (*Figure 33*) apply force to the gear or bearing in a linear and consistent manner. Pullers are usually made of steel and have two or more interchangeable jaws and a threaded shaft mounted in the middle of a yoke. Some also offer reversible jaws, allowing the hooks to face outward instead of inward.

Gear pullers come in various styles and sizes for different applications. *Figure 33* also shows a hydraulic version of the tool, with a hydraulic hand pump attached. Heavy-duty pullers with connections for an external hydraulic-power source are also available for large gears and bearings.

Follow these steps to use a mechanical puller:

Step 1 Select a gear puller with the correct jaw length and spread for the component being removed. It is also important to select a puller with a large enough shaft. The diameter of the puller screw shaft should be at least half that of the workpiece shaft. For example, to remove a bearing from a 2" (5 cm) shaft, the puller shaft should be at least 1" (2.5 cm) in diameter but less than 2".

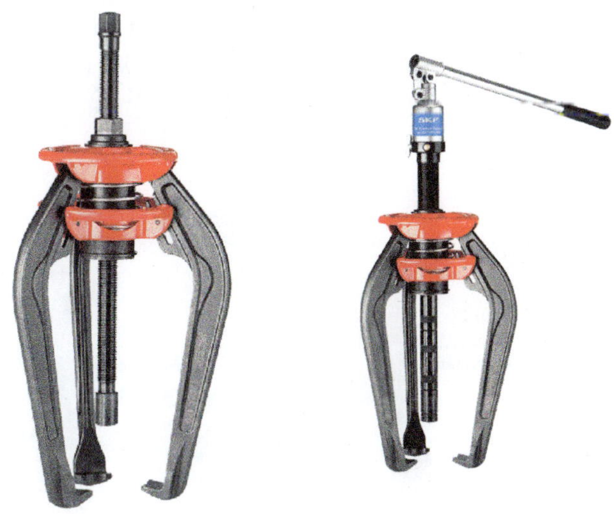

(A) Mechanical Puller (B) Hydraulic Puller

Figure 33 Mechanical and hydraulic pullers.

Step 2 Inspect the puller to ensure that the jaws are not bent or excessively worn and that the threaded puller shaft is not damaged.

Step 3 Ensure that all setscrews or locking screws have been loosened or removed from the component.

Step 4 Inspect the workpiece shaft to ensure that there are no nicks, burrs, or foreign material that would interfere with the component sliding off the shaft. Clean and smooth the shaft as necessary.

Step 5 Position the puller on the component and screw the puller shaft in (clockwise) until it presses against the precise center of the workpiece shaft (*Figure 34*). Ensure that the jaws are properly positioned on the component. The puller jaws must be positioned so that they provide a straight, even pull.

Step 6 Turn the puller shaft clockwise slowly to apply pressure to the component.

Step 7 Continue to tighten the puller until the gear or bearing is free from the shaft or can be removed by hand. It may be necessary to apply heat to the component if it will not move under reasonable pressure.

Step 8 Clean the puller thoroughly and apply light oil sparingly; then store it with similar tools.

In some cases, a bearing or gear may need to be heated for removal. Heating the component causes it to expand slightly. However, it is best to avoid heating bearings to temperatures over 250°F (120°C). In addition, the shaft needs to remain as cool as possible during the process, to prevent it from expanding along with the component.

Figure 34 Puller positioned on a bearing.

A variety of other puller configurations are available. Two other types you may work with are the clamshell puller (*Figure 35*) and the blind housing puller (*Figure 36*). The blind housing puller is used when there is no access to the back of the bearing. However, to gain access for the legs of the puller, holes must be drilled through the bearing.

1.5.4 Packing Pullers

Packing pullers are used to remove packing material from valves, pumps, and similar components. There are several types of packing pullers. Some pullers are rigid and resemble simple hooks (*Figure 37*), while others have a corkscrew tip. Flexible pullers have a spring-like shaft and are also common (*Figure 38*).

There are many different types of packing, constructed from a wide variety of materials. *Figure 39* shows just one example. Their purpose is to allow the shaft of a valve, for example, to rotate without allowing fluid to leak out. The packing is generally wrapped around a valve stem or shaft in rings, stacked as shown in *Figure 40*. The appropriate packing material is generally specified by the component or equipment manufacturer.

> **NOTE**
> Packing materials and their application is covered in detail in NCCER Module 15107, *Gaskets and Packing*.

Figure 35 Clamshell puller kit.

Figure 36 Blind housing puller.

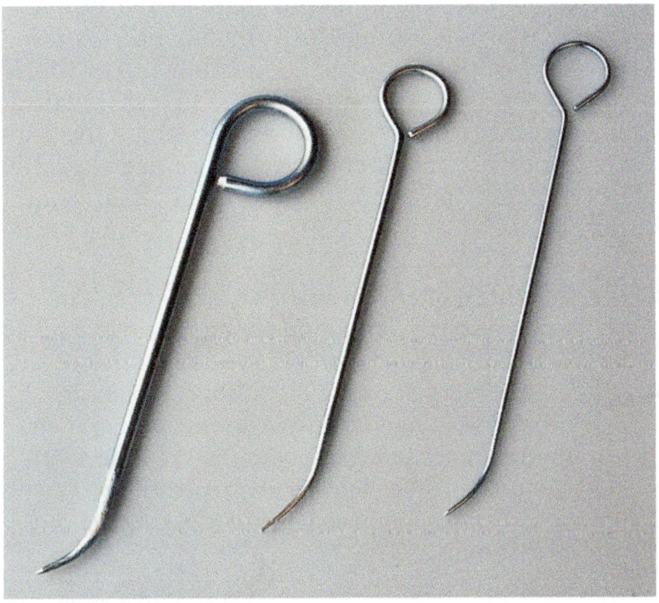

Figure 37 Rigid packing pullers.

Care must be taken when using a packing puller to prevent scarring the valve stem or shaft. Refer to *Figure 41*. Follow these steps to remove common packing materials:

Step 1 Select a corkscrew-type puller, rigid or flexible, of the proper size and length for the job.

Step 2 Inspect the puller to ensure that it is sharp and not damaged.

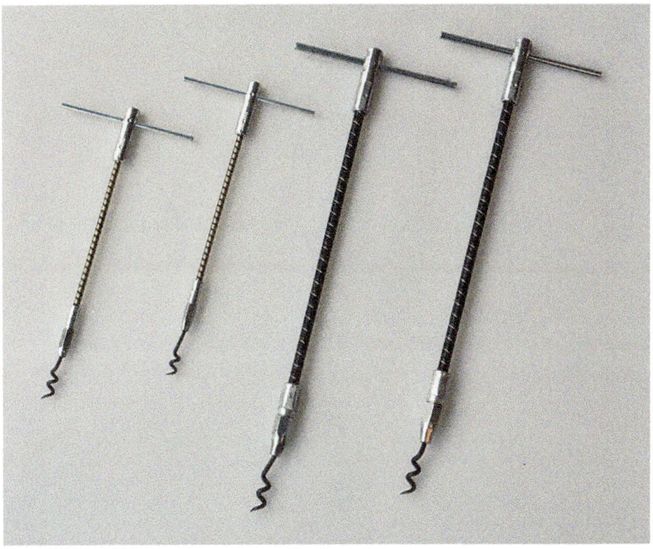

Figure 38 Flexible packing pullers.

Figure 39 An example of packing material.

Step 3 Insert the tip of the puller into the **bonnet** or **stuffing box**, turning it one or two turns clockwise to screw it into the first ring of packing.

Step 4 Pull the puller carefully to remove the packing. The packing will often break into pieces as it is removed.

Step 5 Repeat *Step 4* until all the packing rings have been removed. Some pieces may be left that the corkscrew puller cannot pick up. Choose a puller with a hook to remove stubborn pieces. Every small piece of the packing must be removed.

Step 6 Clean and dry the puller thoroughly, wipe it with an oily rag, and store it with similar small tools.

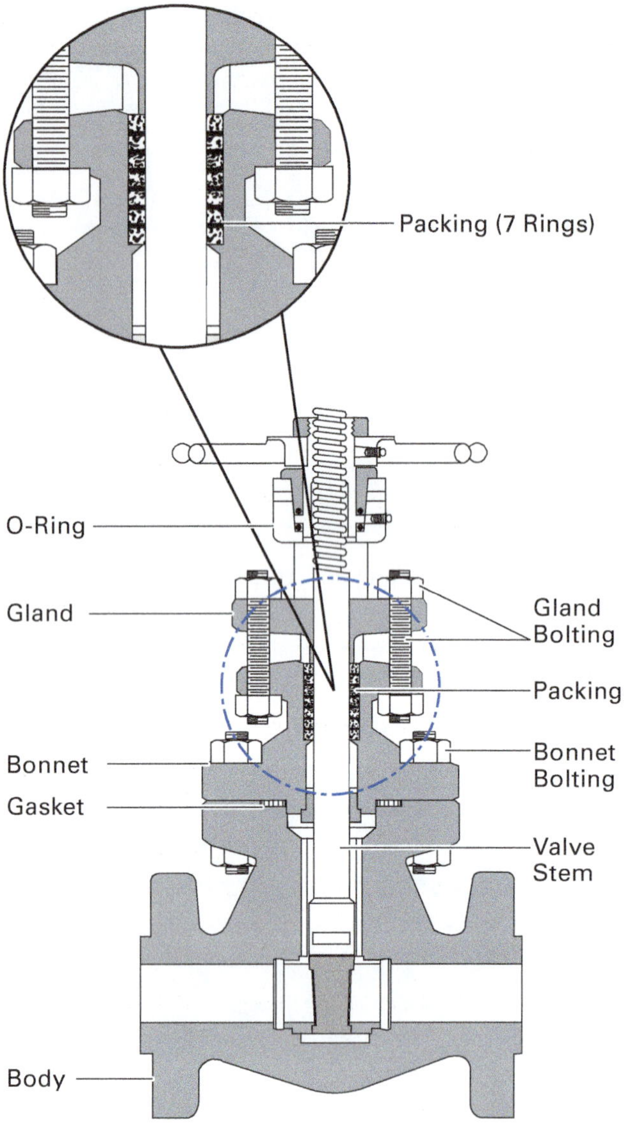

Figure 40 Large valve construction.

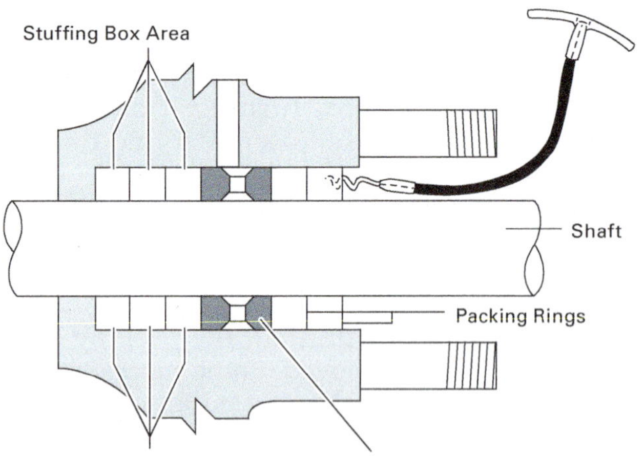

Figure 41 Capturing a ring of packing material.

1.6.0 Sheet Metal Snips

Sheet metal snips are used like scissors to cut thin metal. General-purpose snips come in two basic types: straight blade or combination blade (*Figure 42*). They are also referred to as *tin snips*.

Straight-blade snips are very strong and have relatively long and wide blades. The face of each blade runs straight up from the cutting edges. They are best for cutting a straight line.

Unlike straight-blade snips, the blade face of combination-blade snips curves back from the cutting edge. This allows the metal to slide over the top blade when cutting curves and provides greater control over the cut. Due to their shape, they also referred to as *duckbill snips*.

General-purpose snips have long handles, providing significant leverage. It is important to avoid applying excessive force to the handles. If a cut is forced by applying body weight to the handles or pounding on the blades, the tool will be permanently damaged. Forcing the blades to cut something they shouldn't drives the blades apart. Once the tips no longer meet properly, even making a simple notch cut will not be possible. Maintain snips in good condition, use them properly, and do not force them to cut beyond their capacity. If the metal is too thick or too hard to cut by hand, choose a more appropriate method, such as a power tool or a plasma arc cutter.

WARNING!

Thin metal often has sharp edges that can inflict serious wounds. Wear the appropriate gloves for handling sheet metal to increase grip and protect your hands. Never let sheet metal slip through your hands. Always keep a firm grip on the workpiece to avoid being sliced by the sharp edges. Snip off or file any sharp points and jagged edges, often called *fishhooks*, to avoid serious injury.

CAUTION

The blades and action of sheet metal snips can be damaged if used on metal that is too thick. If the metal cannot be cut by hand with reasonable effort, use another method. Snips should never be used to cut wire, rod, and similar materials regardless of their ability to do so. They are for sheet metal only.

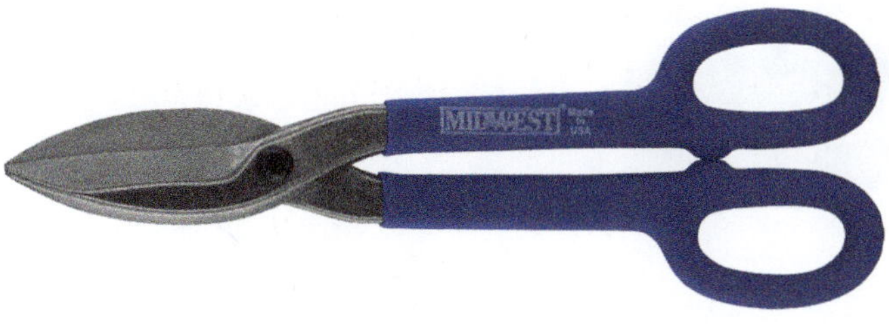

(A) Straight-Blade Snips

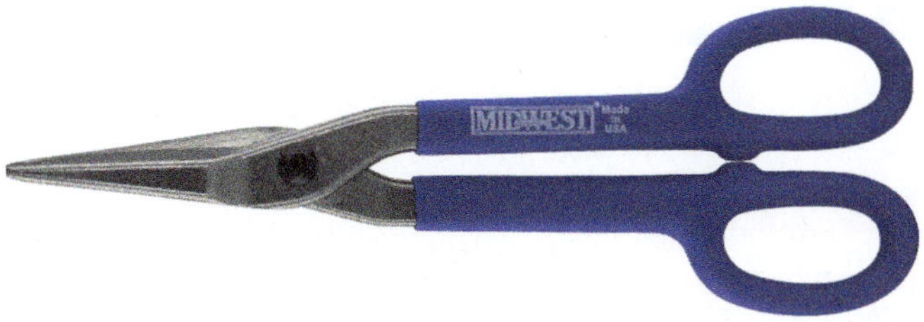

(B) Combination-Blade Snips

Figure 42 General-purpose snips.

1.6.1 Aviation Snips

Aviation snips (*Figure 43*) have a compound-lever design that enables them to cut thicker metal than general-purpose snips can typically manage. They can accurately cut straight lines, intricate curves, and inside corners. Millwrights often use aviation snips to cut **shim stock**.

Aviation snips are available in straight, left-cut, and right-cut styles. Right-cut snips cut tight curves to the right (clockwise), while left-cut snips cut tight curves to the left (counterclockwise). Straight snips are used for straight lines and gentle curves. When cutting a complex pattern, all three styles may be used.

Manufacturers also color-code the handles to make it easier to identify the type, as shown in *Figure 43*. Straight-cut snips have yellow handles; left-cut snips have red handles; and right-cut snips have green handles.

1.6.2 Offset Snips

Offset snips have the distinct advantage of allowing the user to keep the operating hand above the metal as the cut progresses. This reduces the risk of being cut by the sharp metal edges.

Like aviation snips, offset snips (*Figure 44*) have color-coded handles. Red handles identify left-cutting snips and green handles identify right-cutting snips. They can make sharp 90-degree cuts in the direction indicated by the handle color, as well as cut a straight line.

Follow these steps to cut sheet metal with aviation or offset snips:

> **WARNING!** Wear gloves, in addition to safety glasses, when handling and cutting sheet metal. Cut metal edges are sharp and may have burrs that can cause serious wounds.

Step 1 Select snips designed for the direction of your first cut.

Step 2 Inspect the snips to ensure that the cutting jaws and blades are not bent, dull, or otherwise damaged, and that the handles are in good condition.

Step 3 For practice, acquire a piece of 26-to-30-gauge steel sheet metal, roughly 12" × 12" (~30 cm × 30 cm). Trace the outline of an object that offers some straight lines as well as curved lines in both directions—even the outline of your hand offers good practice. Ensure that the cut line on the sheet metal is clearly visible.

Step 4 Begin a cut at the edge of the metal toward the cut line. Guide the snips to intersect with the cut line smoothly, rather than at a sharp angle. When cutting, do not fully close the snip blades. Much like using common scissors, this results in a jagged cut.

Step 5 Continue making left, right, and straight cuts as necessary to follow the cut line until the shape is complete.

Step 6 Properly dispose of the scrap metal.

Step 7 Use a file to remove any burrs and sharp edges.

Step 8 Clean the snips with a lightly oiled rag, then close and lock the handles together. Store snips with similar hand tools.

Figure 43 Aviation snips.

Did You Know?

Aviation Snips

How did aviation snips get their name? During the early days of aircraft manufacturing, these snips were developed to cut thin sheets of steel for the wings and fuselage.

1.7.0 Other Millwright Hand Tools

There are additional tools millwrights use that do not fit into the categories of the tools presented thus far. Tools presented in this section include the following:

- Mallets
- Cable cutters
- Retaining ring pliers
- Drift pins
- Putty knives and gasket scrapers
- Inspection mirrors

1.7.1 Mallets

Dead-blow mallets and rawhide mallets (*Figure 45*) are generally used when a steel hammer would deface or damage the workpiece. Mallets can be made of lead, brass, wood, leather, plastic, or rubber. The handles may be wooden, steel, or fiberglass.

Color Codes

Some trainees find it difficult to remember which snips cut right and which snips cut left. Oxyacetylene torch setups are also color-coded, with green related to oxygen and red related to acetylene. Oxygen fittings have right-hand threads while acetylene fittings have left-hand threads, to prevent them from being connected incorrectly. Green snips cut curves to the right, in the same direction as oxygen fitting threads. Red snips cut to the left, in the same direction as acetylene fitting threads. Perhaps this comparison will help you remember which is which, as reported by some other trainees.

Brass hammers offer a softer striking surface than steel, but still have plenty of mass. They are often used with stamping tools (*Figure 46*), since they do not mushroom the head of the stamp as easily as a steel hammer. Brass hammers still deliver a firm strike though, and do not rebound on impact like a steel hammer. Other types of mallets, such as rawhide and rubber, should not be used with stamping tools. However, they can be used sometimes to strike other tools, such as wood chisels designed for impact.

(A) Left Cut

(B) Right Cut

Figure 44 Offset snips.

(A) Dead-Blow Mallet

(B) Rawhide Mallet

Figure 45 Common mallets.

Never use a brass hammer or a mallet to drive nails or for similar tasks. They are not designed for this purpose. Mallets used to drive nails are quickly damaged and do a poor job of driving a nail. Mallets are designed to help force tight-fitting parts together and to shape or bend sheet metal. They are also used in woodworking and to set tile and pavers.

1.7.2 Bolt and Cable Cutters

Bolt and cable cutters (*Figure 47*) are used to cut wire rope and cable. Although the term *cable* is often used to describe steel cable, as well as aluminum and copper, steel cable is best referred to as wire rope. The difference is important when it comes to these tools.

(A) Letter Stamp Set (B) Letter Stamp

Figure 46 Steel stamps.

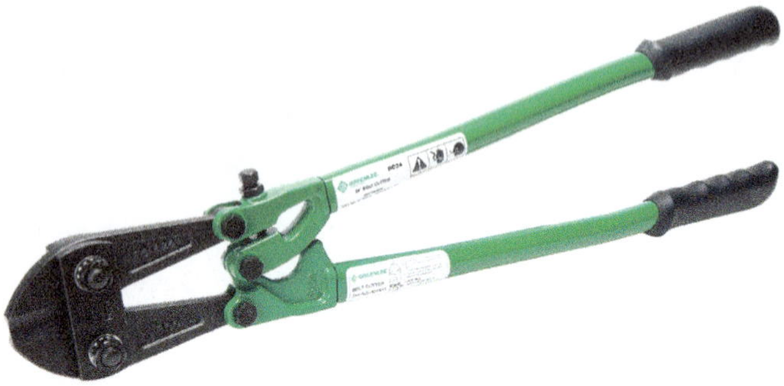

(A) Bolt Cutters

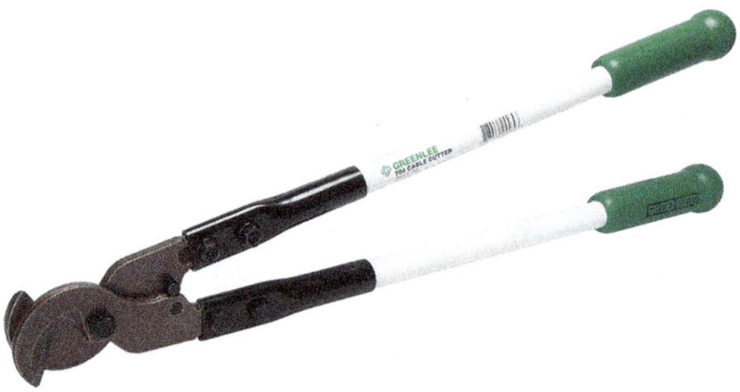

(B) Cable Cutters

Figure 47 Bolt and cable cutters.

Bolt cutters can cut wire rope, bolts, rods, chain, and even lock shackles. Cable cutters are primarily used to cut copper and aluminum products that are softer than steel. Copper and aluminum cable are most often associated with electrical work.

In *Figure 47*, note the difference in the design of the jaws. Cable cutters leave clean, smooth cuts. Bolt cutters leave a cut with a pinched look that reflects the angled shape of the jaws.

CAUTION

Do not cut wire rope and other steel materials with a cable cutter. Even if the cut is successful, the jaws of the tool will likely be damaged.

Hand tools work fine when only one or two cuts must be made. When many cuts are required, power tools reduce the effort and increase productivity. Battery-powered cable cutters (*Figure 48*) offer portability and convenience. Hydraulic cable cutters, also shown in *Figure 48*, are used to cut larger copper or aluminum cable. The model shown in the figure can develop 11 tons (97 kilonewtons, or kN) of pressure.

1.7.3 Retaining Ring Pliers

Retaining ring pliers are used to remove and install retaining rings (*Figure 49*). Retaining rings are also called *snap rings*, thus the alternate name for the tool, *snap ring pliers*. Retaining rings are also referred to as *circlips*. Both the rings and pliers come in various sizes and shapes for different applications. While most are relatively small, snap rings for large shafts can exceed 10" (~250 mm) in diameter.

Retaining ring pliers are shown in *Figure 50*. This type can be locked into position, allowing it to hold a retaining ring open on its own. Some retaining ring pliers are convertible (*Figure 51*). The action can be quickly reversed by disassembling the two halves and reassembling them using the alternate hole. This reverses the direction of the jaws when the handles are squeezed. In the configuration shown in *Figure 51*, the jaw tips move away from each other when the handles are squeezed.

Retaining ring pliers are made to use on retaining rings only and should not be used for any other purpose. The tips are fragile, and the tool should be stored with similar tools in a safe

Did You Know?

Dead-Blow Mallet

The primary mass of the dead-blow mallet is encased in the plastic head. Hidden inside is a small piece of steel or a collection of steel shot. This mass is what provides the striking power. Unlike common hammers and some other mallets, the dead-blow mallet does not rebound after striking an object, leading to its name.

place to prevent damage. To use retaining ring pliers, insert the tips of the pliers into the holes of the ring. Squeeze the handles together to open the ring.

> **WARNING!**
>
> Compressing a large snap ring stores a significant amount of energy. If a ring is released accidentally from the tool, it can travel a significant distance at high speed and cause an injury. Always wear the appropriate PPE and recognize the potential of the stored-energy hazard.

(A) Battery-Powered Cable Cutter

(B) Hydraulic Cable Cutter

Figure 48 Battery-powered and hydraulic cable cutters.

(A) Internal Retaining Ring

(B) External Retaining Ring

Figure 49 Common retaining rings.

(A) Retaining Ring Pliers

(B) Alternate Tips

Figure 50 Retaining ring pliers with replaceable tips.

Figure 51 Convertible retaining ring pliers.

Note that some rings have tapered holes. When present, insert the pliers from the tapered side. Opening the ring from the opposite side increases the chance that it will slip off under stress.

1.7.4 Drift Pins

A drift pin is a round, tapered steel dowel. Two types of drift pins are shown in *Figure 52*. A drift pin that is tapered on both ends is referred to as a *barrel pin*. Heavy-duty drift pins with a head designed for hammering are often called *bull pins*. Lighter, machined versions are called *tapered pin punches* (*Figure 53*).

Drift pins are used to align holes when connecting pipe or equipment flanges, structural steel members, and matching holes in other mating surfaces. Holes of all sizes can be aligned with these tools. The tapered pin punch, however, is used for lighter, more delicate work.

Select a pin or tapered punch having a tip small enough to fit into the holes being aligned. The proper size drift pin will fit all the way through the holes being aligned but will be large enough and strong enough to pull the mating pieces into alignment. Little force is required to align small holes and lightweight assemblies.

(A) Drift, or Bull, Pins

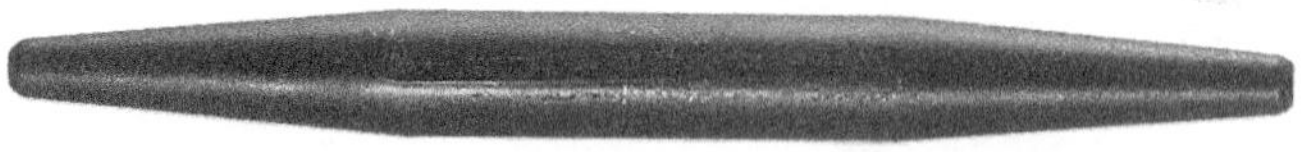

(B) Barrel Pin

Figure 52 Drift pins.

However, aligning structural steel and equipment components can place a significant amount of stress on the tool.

Once the holes are aligned, C-clamps or similar tools can be used to clamp the parts together, or a bolt can be installed in other holes to maintain the alignment. The pin is then removed, and hardware is installed in its place.

1.7.5 Putty Knives and Gasket Scrapers

Putty knives are used to scrape off caulk, spackle, grout, glue, sealants, and similar substances. Putty knives have a flexible metal blade and are available in many different lengths and widths. Plastic putty knifes can be used when a surface must be protected, although they are not durable.

Both a common putty knife and a gasket scraper are shown in *Figure 54*. Gasket scrapers have a narrower blade but are much stronger and stiffer. Both tools may be used to remove a gasket, but the gasket scraper is often needed when the gasket has hardened or is bonded to the surface.

To use a putty knife or gasket scraper on a gasket or other material, keep the blade nearly parallel to the surface. It is important to protect the mating surface of the workpiece. Deep scratches and nicks can cause leaks. If the tool is held at a steep angle, it can dig into the metal surface.

1.7.6 Inspection Mirrors

Inspection mirrors are used for the close inspection of areas that cannot be seen directly. Inspection mirrors come in a variety of shapes and sizes. The mirror is attached to the end of a rod, which often telescopes. The mount typically has a ball joint to adjust the position of the mirror. With a flashlight and the mirror held in the proper position, you can inspect most areas. Some newer mirrors have built-in lights as a convenience, thanks to the compact nature of light-emitting diodes (LEDs) and their low power requirements. Two types of mirrors are shown in *Figure 55*.

Although many inspection mirrors are logically made of glass, some are constructed of polished stainless steel. They are used to examine the flame of a gas- or oil-fired burner without damage, as well as in situations where glass could be easily broken.

Figure 53 Tapered pin punch.

(A) Putty Knife

(B) Gasket Scraper

Figure 54 Putty knife and gasket scraper.

(A) Telescoping Oval Mirror

(B) Lighted Round Mirror

Figure 55 Inspection mirrors.

Technology Enhances Inspection Devices

The simple inspection mirror has come a long way. Built-in LED lights are becoming increasingly common. Now the tool has taken another step forward, with a built-in camera that links to a smartphone or tablet. The model shown here offers 720p resolution and LED lights to illuminate the view. It also extends from roughly 15" (38 cm) to over 4' (1.2 m) in length. Now others can see exactly what you see.

Figure Credit: Courtesy of Saunders-Midwest/Ullman

Inspection mirrors must be kept clean and be stored in a place where they will not be damaged or broken. They should be kept in their original package or wrapped in a soft cloth and stored with other delicate tools.

CAUTION

Glass inspection mirrors can be broken easily through abuse or careless handling. Handle with care and ensure the mirror is well protected in storage.

1.0.0 Section Review

1. The rotational speed of belt-driven equipment is determined by the speed of the drive motor and the _____.
   a. length of the drive belt(s)
   b. diameter of the two sheaves
   c. size of the drive shaft
   d. number of drive belts

2. A generous amount of lubricant is required on the workpiece when using a tubing cutter.
   a. True
   b. False

3. The standard taper for a taper pin and reamer under the US standard is _____.
   a. ⅛" per foot
   b. ¼" per foot
   c. ½" per foot
   d. 1" per foot

4. Which of the following statements is *correct*?
   a. Round threading dies are exclusively designed for use with power tools.
   b. A bar-style tap wrench is the best choice for the smallest tap sizes.
   c. A plug tap has no taper on the first few threads.
   d. Hex dies are primarily used to clean up existing threads instead of cutting new threads.

5. When selecting a gear puller for a task, the size of the puller shaft should be smaller than the shaft of the workpiece, but *at least* _____.
   a. 10% of the workpiece shaft diameter
   b. 20% of the workpiece shaft diameter
   c. one-third of the workpiece shaft diameter
   d. one-half of the workpiece shaft diameter

6. Right-cut aviation snips have _____.
   a. blue handles
   b. red handles
   c. green handles
   d. yellow handles

7. *Cable cutter* is simply another name for a bolt cutter.
   a. True
   b. False

Review Questions

1. Moisture in toolboxes can be collected and absorbed _____.
   a. by bags of silica gel
   b. by sawdust
   c. with a vacuum
   d. with compressed air

2. Pull wrenches toward you rather than pushing them away from your body.
   a. True
   b. False

3. Which of the following is helpful when using a fillet gauge?
   a. Screwdriver
   b. Flashlight
   c. Radius gauge
   d. Strap wrench

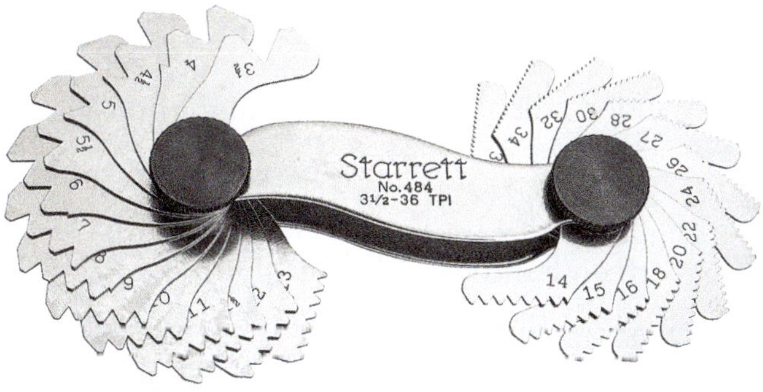

Figure Credit: The L.S. Starrett Company

Figure RQ01

4. What tool is shown in *Figure RQ01*?
   a. Radius gauge
   b. Fillet gauge
   c. Thread gauge
   d. Taper gauge

5. A tool used to determine the diameter of a hole or the width of a slot is the _____.
   a. drill gauge
   b. thickness gauge
   c. taper gauge
   d. micrometer

6. Which of the following is *not* a factor in determining the minimum deflection force for an installed drive belt?
   a. Belt type
   b. Speed range of the drive
   c. Belt width
   d. The ratio between the drive and driven rotational speeds

Figure Credit: Courtesy of RIDGID Tool Company

Figure RQ02

7. The tool shown in *Figure RQ02* is a _____.
   a. heavy-duty pipe cutter
   b. midget tubing cutter
   c. quick-acting tubing cutter
   d. plastic pipe cutter

8. When a large amount of metal must be removed from the walls of a cylinder, _____.
   a. use a portable grinder
   b. use a hand reamer
   c. boring is the best approach
   d. wrap coarse sandpaper around the stones

Figure Credit: Courtesy of Drill America, DrillAmerica.net
Figure RQ03

9. The tool shown in *Figure RQ03*, with a thread die mounted in the center of it, is called a _____.
   a. tap wrench
   b. diestock
   c. chuck
   d. thread wrench

10. A metric bolt is identified with the numbers M12-1.75. Which of the following statements is *true*?
    a. The crests of the threads are 1.75 mm apart.
    b. The size of the bolt is 1.75 mm.
    c. The thread pitch is 12 threads per inch.
    d. The thread pitch is 12 threads per centimeter.

11. The type of wrench designed for use on one unique type of fastener is the _____.
    a. strap wrench
    b. chain wrench
    c. spanner wrench
    d. tap wrench

12. A stubborn bearing can be heated to assist removal, but it should *not* be heated beyond _____.
    a. 250°F (120°C)
    b. 300°F (149°C)
    c. 375°F (191°C)
    d. 425°F (218°C)

13. Combination-blade snips are also referred to as _____.
    a. linesman's pliers
    b. duckbill snips
    c. aviation snips
    d. parrot snips

14. Aviation snips used to cut straight lines have _____.
    a. blue handles
    b. red handles
    c. green handles
    d. yellow handles

15. To avoid confusion with electrical cables made from aluminum or copper, cable made from steel should be referred to as _____.
    a. wire rope
    b. cable wire
    c. rigid cable
    d. rigid rope

Trade Terms Quiz

Fill in the blank with the correct term learned from your study of this module.

1. On a belt drive, a gauge is used to measure wear of the _______________, on which the belts ride.

2. On a threaded fastener, the number of threads per inch or the distance between thread crests is called the _______________.

3. The density of the particles that make up a honing stone is identified by the stated _______________ rating.

4. _______________ combine two or more drive belts on a single band.

5. A hole that does not go all the way through an object is called a _______________.

6. During shaft alignment, aviation snips might be used to cut pieces from _______________.

7. One use for a _______________ is to set and hold the position of a cylinder hone precisely above the centerline of the cylinder.

8. A drive and driven shaft that share a common centerline are connected using a _______________.

9. A hygroscopic material such as silica gel is an example of a _______________.

10. When a specification indicates something must be cut to a size of 1" (2.54 cm) and be within $\frac{1}{64}$" (0.4 mm) of that dimension, the $\frac{1}{64}$" represents the _______________.

11. When stones used for honing are no longer flat due to normal wear, _______________ are used to restore the flat surface.

12. A _______________ should never be used on a wrench or similar tool to gain leverage due to the hazard it represents.

13. _______________ metals contain or consist of iron.

14. A rope-like material used to reduce or eliminate fluid leakage around a shaft is referred to as _______________.

15. To reduce or eliminate leakage around a rotating shaft or axle, packing may be placed inside a _______________.

16. The removeable top of a valve assembly, through which the valve stem usually passes, is called the _______________.

17. A special clamping device used to hold drill bits, taps, and similar tools is called a _______________.

18. Deposits that remain on pipe following the manufacturing process, or accumulate inside a functional piping system, may be referred to as _______________.

19. _______________ refers to the use of an abrasive stone to create a smooth, precise surface, such as the edge of a knife.

20. A rounded concave surface that serves as a transition between two intersecting surfaces is called a _______________.

21. You can _______________ a stone to flatten its surface and compensate for wear.

Trade Terms

Blind hole
Bonnet
Cheater bar
Chuck
Desiccant
Dress
Drive coupling
Ferrous
Fillet
Grit
Honing
Jig
Lapping plates
Packing
Pipe scale
Pitch
Polyband belts
Sheave
Shim stock
Stuffing box
Tolerance

Cornerstone of Craftsmanship

Jack Baker
Millwright and Machinist Instructor
Lee College

How did you choose a career in the industry?

I started very early in school and worked evenings and weekends for F. L. Werner. I'd do anything that people in a very small town needed, from masonry and carpentry to mechanical work on cars and farm vehicles.

Who inspired you to enter the industry?

There were many colleagues offering to give a hand here and there. If it was anyone, I'd say a gentleman named Ed Charvanau. He was one of the first people I worked with, and he kept me out of trouble in those early years.

What types of training have you been through?

I went through masonry and carpentry training during my four years of high school, and then I joined the Marine Corps. The Marines put me through intensive training related to servicing track and wheeled vehicles.

How important is education and training in construction?

Both education and on-the-job training have had a great effect on the way I work on things and ascertain that the task and repairs have been completed efficiently and meet the specifications. "Close enough" has never been good enough for me!

How important are NCCER credentials to your career?

When I received my degrees in applied mechanical sciences, I also earned my NCCER certification. This allowed me to advance to supervision within Brown and Root and Wolf, Inc., as well as see increases in both pay and prestige.

How has training/construction impacted your life?

It has allowed me to make a good living. I've never been without work, and training also made what I wanted to do attainable.

What kinds of work have you done in your career?

I've had the opportunity to work for several companies. They include Todd Shipyard and Reynolds Aluminum, where I worked as a maintenance mechanic. I also worked with Missouri Pacific Railroad in maintenance of way, which means taking care of the railroad's tracks and bridges.

Tell us about your current job.

I currently work at Lee College as a millwright and machinist instructor. I'm either inside the classroom or in the lab performing the tasks.

What do you enjoy most about your job?

I enjoy teaching my students and passing along all that I have learned about this field. I also enjoy the comradery of those who choose to enter the field and improve their skills.

What factors have contributed most to your success?

Without a doubt, my willingness to do the best job I possibly can and being open to learning new things as they come my way.

Would you suggest construction as a career to others? Why?

I would tell everyone entering or thinking about this field that education and training are key traits to success. If you work to get a good education and training, it can lead you into other fields, plus afford you the opportunity for continual employment and substantive pay throughout your life.

What advice would you give to those new to the field?

Be motivated to do the very best job you can and be open to any opportunity to learn new skills.

Interesting career-related fact or accomplishment?

I'd say my most notable experience was when I traveled to the little country of Zaire in Africa, where I spent time overhauling pumps for pipeline technology, then brought them online.

How do you define craftsmanship?

People who take ownership of a task, complete it to its maximum performance and specifications, and ensure a good job is done—those people are craft professionals and embody craftsmanship.

Trade Terms Introduced in This Module

Blind hole: A hole that does not go completely through the material and therefore has a bottom.

Bonnet: The removeable cover of a valve body, through which the valve stem usually passes. Valve repair typically requires removal of the bonnet to gain access to the valve components beneath it.

Cheater bar: A length of pipe or similar material slipped over a wrench or other tool to substantially increase leverage. A cheater bar creates a hazard, as tools and hardware are not constructed to withstand the added stress.

Chuck: A special type of clamp equipped with jaws in a radial pattern that is typically used to hold rotating tools, such as drill bits and thread taps. Some chucks require a chuck key, which is a small, geared tool used to loosen and tighten the jaws of the chuck.

Desiccant: A material such as silicon dioxide (silica gel) that attracts and captures moisture from the surrounding air or nearby moist materials.

Dress: The process of forming, shaping, and changing the finish of stone to a desired shape or condition.

Drive coupling: A mechanical component used to connect the shaft of a driver, such as an electric motor, to the shaft of a driven load, such as a pump or compressor. The two shafts must be precisely aligned so that they share a common centerline. Drive couplings are typically classified as either rigid or flexible.

Ferrous: Containing or consisting of iron.

Fillet: A concave, rounded surface that serves as a transition between two intersecting surfaces. In welding, a fillet is formed when two pieces of metal are joined at an inside corner.

Grit: A number indicating the density of particles for grinding stones, honing stones, sandpaper, and similar materials. High numbers indicate smaller particles packed more closely together (fine), while low numbers indicate larger particles spaced farther apart (coarse).

Honing: A process that uses abrasive stones to create a smooth, precise surface on a metal workpiece in a controlled, consistent manner.

Jig: A device constructed or assembled to maintain a positional relationship between a workpiece and a tool or between components of an assembly.

Lapping plates: Coarse stones used to rub other stones and restore their flat surface.

Packing: Rope-like material with a consistent shape used to prevent fluid leakage by compressing it around a shaft or valve stem.

Pipe scale: Deposits remaining on new ferrous pipe products left over from the heating and forming process. In operational piping systems, scale refers to the buildup of corrosion and/or mineral deposits on the inside walls of the piping.

Pitch: Relative to threaded hardware and components, refers to the number of threads found over a unit of distance, e.g., a ¼" bolt with coarse threads has a pitch of 20 threads per inch.

Polyband belts: Drive belts that combine two or more belts on a common band, side by side.

Sheave: A grooved wheel, also called a *pulley*, mounted to a shaft and used to transmit power from a driver to a driven load through drive belts that ride in its grooves.

Shim stock: Thin material, typically metal, placed between two objects to fill space and allow for the precise positioning of one or both objects. Shims are often used in *shaft alignment* to adjust the position of the motor relative to the driven shaft. Shim stock typically refers to the raw material from which individual shims are cut.

Stuffing box: A casing or chamber in which packing material is placed and compressed around a moving shaft or axle to reduce or eliminate gas or liquid leakage.

Tolerance: A term that quantifies the allowable difference or deviation from specified limits.

Additional Resources

This module presents thorough resources for task training. The following reference material is suggested for further study.

Audel Millwrights and Mechanics Guide. Latest Edition. Thomas B. Davis, Carl A. Nelson. Hoboken, NJ: John Wiley & Sons.

NCCER Module 00103, *Introduction to Hand Tools*. Latest Edition. Alachua, FL: National Center for Construction Education and Research.

NCCER Module 15103, *Fasteners and Anchors*. Latest Edition. Alachua, FL: National Center for Construction Education and Research.

NCCER Module 15107, *Gaskets and Packing*. Latest Edition. Alachua, FL: National Center for Construction Education and Research.

Figure Credits

iStock@industryview, Module Opener

iStock@Chadawan Heemsuhree, Figure 1

The L.S. Starrett Company, Figures 2–6, 16

Courtesy of USA Roller Chain & Sprockets, Figure 9

Courtesy of Timken Belts, Figure 10 (picture)

Shutterstock.com/rungrote sommart, Figure 11

Courtesy of RIDGID Tool Company, Figures 12–14, 29

Courtesy ICS Cutting Tools, Figure 15

Courtesy of Tap Magic, Steco Corporation, Figure 17

Courtesy of Lisle Corporation, Figure 18

Courtesy of Drill America, DrillAmerica.net, Figures 20–23

Courtesy of Walton Tools, Figure 26

Image property of Stanley Black & Decker. Used with permission., Figures 27, 43, 48 (A), 50–51, 53–54

Courtesy of ENERPAC, Figure 28

Courtesy of GEARWRENCH/ Apex Tool Group, Figure 30 (A–C)

Courtesy B/A Products Co., Figure 30 (D)

Courtesy of SKF, Lincoln Industrial Products, Alemite LLC, Figures 31–34, 36

Courtesy of Richmond Gear, Figure 35

Courtesy Midwest Tool & Cutlery Co., Figure 42

Courtesy of Malco Tools, Figure 44

Courtesy of Garland Manufacturing Company, Figure 45

Courtesy Young Bros Stamp Works, Inc, www.youngbrosstampworks.com, Figure 46

Courtesy of Greenlee Tools, Inc, Figures 47, 48 (B)

Courtesy of Klein Tools/ Greenlee Tools, Inc, Figure 52

Courtesy of Saunders-Midwest/Ullman, Figure 55

Section Review Answer Key

Section 1.0.0

Answer	Section Reference	Objective
1. b	1.1.6	1a
2. b	1.2.0	1b
3. b	1.3.1	1c
4. d	1.4.0	1d
5. d	1.5.3	1e
6. c	1.6.1	1f
7. b	1.7.2	1g

NCCER CURRICULA — USER UPDATE

NCCER makes every effort to keep its textbooks up-to-date and free of technical errors. We appreciate your help in this process. If you find an error, a typographical mistake, or an inaccuracy in NCCER's curricula, please fill out this form (or a photocopy), or complete the online form at **www.nccer.org/olf**. Be sure to include the exact module ID number, page number, a detailed description, and your recommended correction. Your input will be brought to the attention of the Authoring Team. Thank you for your assistance.

Instructors – If you have an idea for improving this textbook, or have found that additional materials were necessary to teach this module effectively, please let us know so that we may present your suggestions to the Authoring Team.

NCCER Product Development and Revision

13614 Progress Blvd., Alachua, FL 32615

Email: curriculum@nccer.org
Online: www.nccer.org/olf

❏ Trainee Guide ❏ Lesson Plans ❏ Exam ❏ PowerPoints Other ____________

Craft / Level: Copyright Date:

Module ID Number / Title:

Section Number(s):

Description:

Recommended Correction:

Your Name:

Address:

Email: Phone:

This page is intentionally left blank.

Fasteners and Anchors

Overview

Fasteners come in an almost bewildering variety of types, styles, and sizes. Some are extremely versatile, while others are highly specialized. Craftworkers handle fasteners every day, so they must understand how they work. They must choose the right ones and install them correctly. Properly handling fasteners gives reliability and safety. Improperly handling them leads to failure and unsafe conditions.

Module 15103

Trainees with successful module completions may be eligible for credentialing through the NCCER Registry. To learn more, go to **www.nccer.org** or contact us at 1.888.622.3720. Our website, **www.nccer.org**, has information on the latest product releases and training.

Your feedback is welcome. You may email your comments to **curriculum@nccer.org**, send general comments and inquiries to **info@nccer.org**, or fill in the User Update form at the back of this module.

This information is general in nature and intended for training purposes only. Actual performance of activities described in this manual requires compliance with all applicable operating, service, maintenance, and safety procedures under the direction of qualified personnel. References in this manual to patented or proprietary devices do not constitute a recommendation of their use.

15103 V4.0

15103
Fasteners and Anchors

Objectives

Successful completion of this module prepares you to do the following:

1. Describe common fastener operation and torque's role behind it.
   a. Explain how many fasteners work by producing friction through torque.
   b. List common torque-based tools and explain their proper use.
2. Identify threaded fasteners and summarize their applications.
   a. List common threaded fastener types and standards.
   b. Explain how to install threaded fasteners.
   c. List common concrete anchors and describe their installation.
3. Identify non-threaded fasteners and summarize their applications.
   a. List common retainers, pins, and keys, describing their applications.
   b. Explain how blind rivets work and how to install them.

Performance Tasks

Under supervision, you should be able to do the following:

1. Install instructor-selected threaded fasteners.
2. Tighten a group of bolts to the proper torque and in the proper sequence.
3. Install instructor-selected fasteners in hardened concrete.
4. Install instructor-selected non-threaded fasteners.
5. Install a snap ring.

Trade Terms

Blind rivet
Break-away torque
Clearance hole
Countersunk
Elastic limit
Foot-pound (ft-lb)
Friction
Grade
Handedness
Inch-pound (in-lb)
Newton-meter (Nm)
Nominal size
Run-down resistance
Set/seizure
Tensile strength
Tension
Thread class
Thread pitch
Thread series
Thread standard
Thread-locking fluid
Threads-per-inch (tpi)
Torque

Industry Recognized Credentials

If you are training through an NCCER-accredited sponsor, you may be eligible for credentials from NCCER's Registry. The ID number for this module is 15103. Note that this module may have been used in other NCCER curricula and may apply to other level completions. Contact NCCER's Registry at 1.888.622.3720 or go to **www.nccer.org** for more information.

You can also show off your industry-recognized credentials online with NCCER's digital badges. Transform your knowledge, skills, and achievements into badges that you can share across social media platforms, send to your network, and add to your resume. For more information, visit **www.nccer.org**.

Contents

Figures and Tables

This page is intentionally left blank.

Section One

1.0.0 Fastener Operation

Objective

Describe common fastener operation and torque's role behind it.

a. Explain how many fasteners work by producing friction through torque.
b. List common torque-based tools and explain their proper use.

Trade Terms

Break-away torque: The torque required to loosen a fastener.

Elastic limit: The point after which stretching a material permanently deforms it.

Foot-pound (ft-lb): A US standard torque unit equal to the force of one pound applied to a lever one foot long. One foot-pound equals 12 inch-pounds. Also called the *pound-foot (lb-ft)*.

Friction: A motion-resisting force produced when two surfaces rub together.

Inch-pound (in-lb): A small US standard torque unit equal to the force of one pound applied to a lever one inch long. One inch-pound equals $\frac{1}{12}$ foot-pound. Also called the *pound-inch (lb-in)*.

Newton-meter (Nm): A metric torque unit equal to the force of one newton applied to a lever one meter long.

Run-down resistance: Additional torque required to overcome the resistance created by extra hardware, such as washers and locknuts.

Set/seizure: Fastener lockup that occurs when approaching the target torque.

Tensile strength: The point at which stretching a material breaks it.

Tension: A pulling force, such as that produced by some springs.

Torque: A twisting force that produces rotary motion.

Fasteners hold things together in a variety of ways. Some, such as pins and keys, physically prevent or transfer motion. Others, such as screws and bolts, fasten through a combination of forces. This section examines those forces and explains how they work. Understanding them will help you use fasteners correctly. You'll also avoid mistakes that lead to fastener failure.

1.1.0 How Fasteners Hold

When you press two objects together, they develop a force called **friction** that resists motion. Sliding a heavy box across carpet illustrates this effect. The amount of friction produced depends upon two things.

Material plays a major role. Sliding sandpaper across wood generates more friction than regular paper. Every pair of materials has a friction value. The higher the value, the more friction they generate. Pressure is the second factor. Tightly forcing surfaces together generates a lot of friction. Lower pressure produces less.

Many fasteners use these two factors to hold objects in place. The fastener develops friction by forcing the surfaces together. The friction itself, not the fastener, keeps the objects from moving (*Figure 1*).

Nuts, bolts, and screws also rely on friction to stay tightened. Friction between parts keeps the fastener from working loose.

1.1.1 Tension and Fasteners

Threaded fasteners like bolts use **tension** to draw objects together. Tension is a pulling force, like that produced by some springs. In fact, a tightened bolt behaves like a stretched spring. It pulls and holds the objects together, generating friction between them (*Figure 2*).

Tightening the bolt stretches it slightly. The tighter the bolt, the greater the tension. Properly tightening threaded fasteners is crucial. A loose fastener cannot generate enough friction to keep the objects from moving. The fastener relies on friction to hold itself in place too. A loose fastener can come out if the equipment vibrates.

1.1.2 Torque and Fasteners

Tightening a bolt or screw requires **torque**, a twisting force that produces rotation. When you use a wrench, you apply torque to the bolt (*Figure 3*). The torque turns the bolt and tensions it at the same time. Applying enough torque to generate the required tension is called *torquing the bolt*.

An under-torqued bolt will not do its job. But an over-torqued one creates problems too. All metals have an **elastic limit**. If you stretch them beyond this point, they deform and no longer develop tension (*Figure 4*). Tightening a bolt past its elastic limit damages it permanently. It remains stretched, so it will never hold properly. Always replace damaged fasteners immediately.

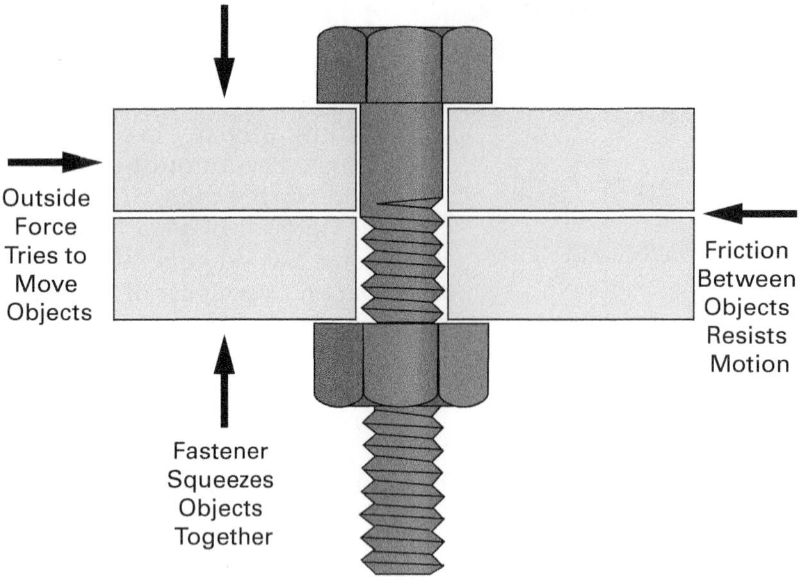

Figure 1 Fastener producing friction.

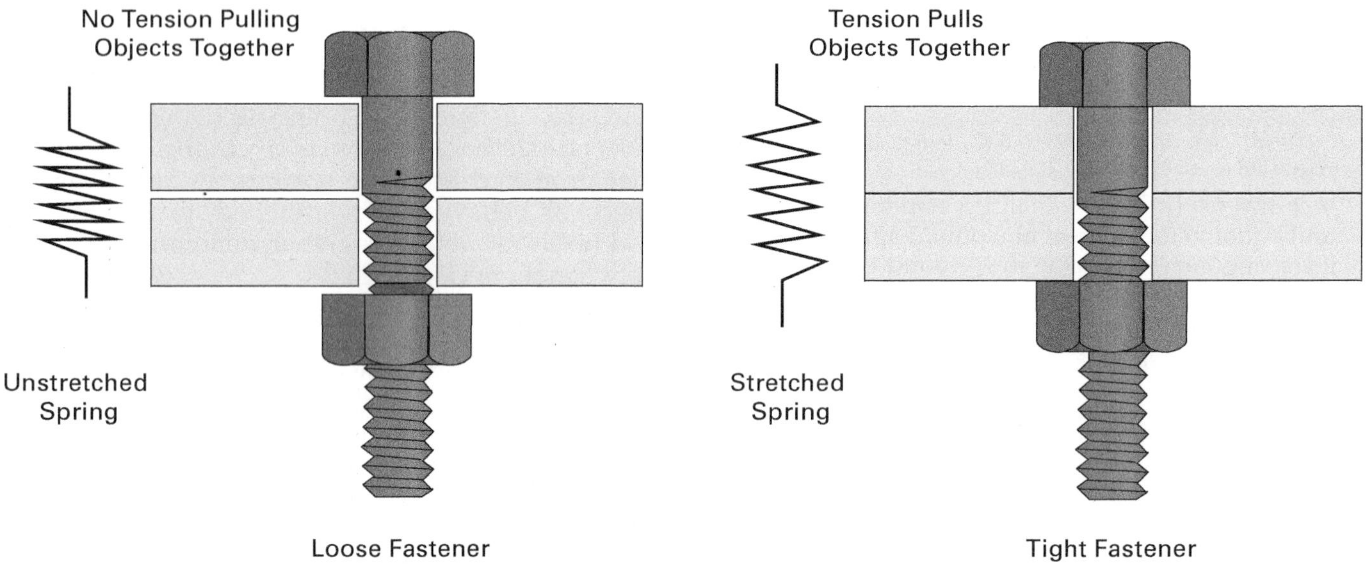

Figure 2 A tightened bolt works like a spring.

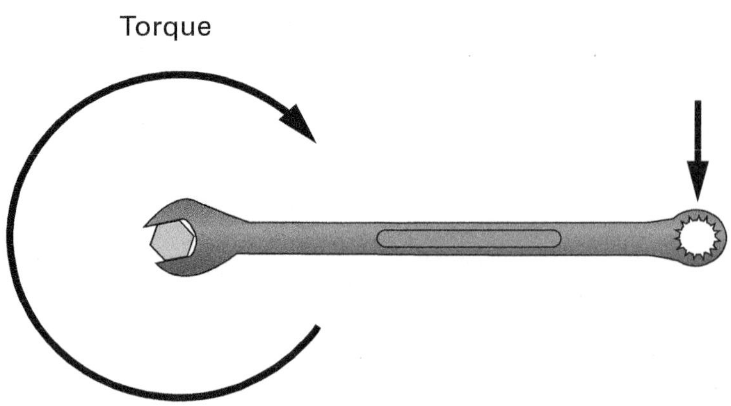

Figure 3 Torque is a twisting force.

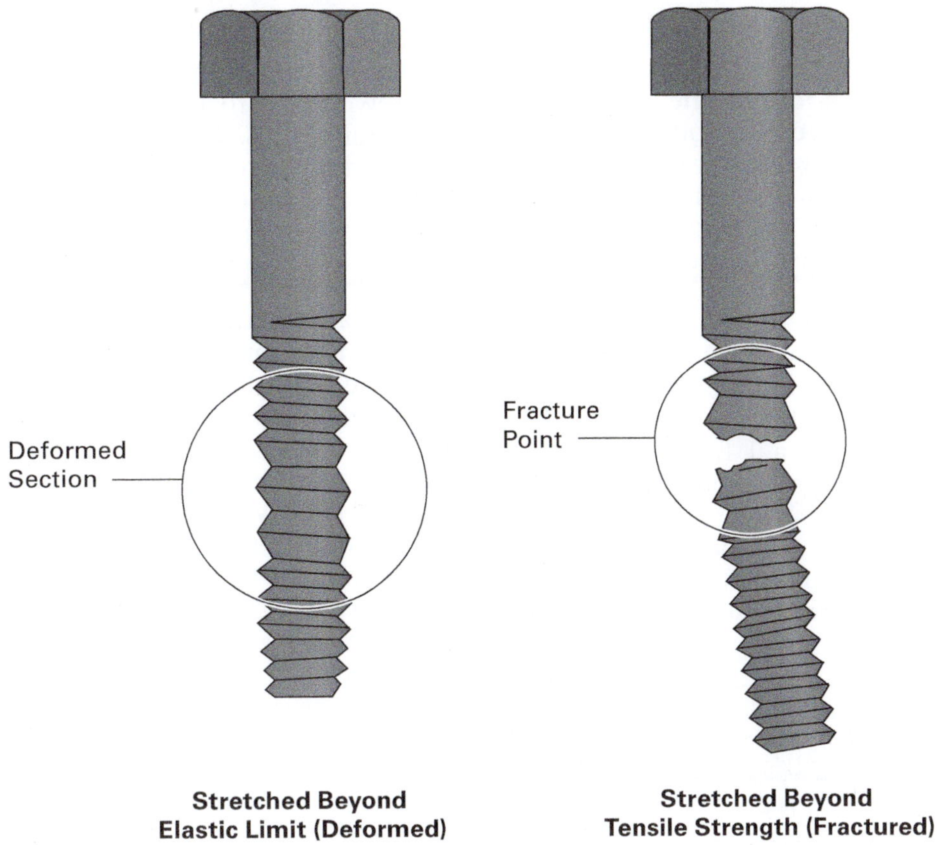

Figure 4 The consequences of over-torquing.

Metals also have a **tensile strength**. If you stretch them beyond this point, they break (*Figure 4*). Over-torquing may push a fastener to this point, snapping it in two, stripping its threads, or breaking its head. A broken fastener may be hard to remove or even damage equipment.

For these reasons, craftworkers must apply just the right amount of torque to fasteners. How much depends on the fastener and the objects being held. Manufacturers often provide recommended torque values.

WARNING!

Always follow the manufacturer's torque guidelines. Just torquing a fastener until it "feels right" is a recipe for failure. If a fastener comes apart at the wrong moment, it could cause injury or equipment damage.

CAUTION

Proper torquing is especially vital for industrial machinery that vibrates. Improperly torqued fasteners can come apart, leading to equipment failure and unsafe conditions.

1.1.3 Torque Terminology

Craftworkers should recognize the following torque terms. Equipment, tool, and fastener manufacturers use them in a variety of settings:

- **Break-away torque** — This is the torque required to loosen a fastener. Usually, it is higher than the recommended tightening torque value.
- **Set/seizure** — Sometimes during tightening, a fastener will lock up as it approaches the recommended torque. You might hear a popping sound or feel vibration when this happens. To free the fastener, loosen it slightly and then retorque to the proper value. Torque tools will not work accurately on a seized fastener.
- **Run-down resistance** — If the fastener includes extra hardware, such as washers or locknuts, these make tightening the fastener harder. You may have to consider this additional resistance when torquing the fastener and increase the target torque value. Consult the manufacturer for guidance rather than guessing. Improper torquing can have serious consequences.

1.1.4 Torque Units

When you use a wrench to torque a bolt, it acts like a lever. You apply a force at the end, and the lever multiplies that force. The longer the lever, the greater the force. This is the reason that long-handled wrenches often work better.

Torque units combine the lever length with the amount of force applied. In the United States, the **foot-pound (ft-lb)** is a common torque unit. Tightening a bolt to 1 ft-lb of torque is like hanging a 1 lb weight on the end of a 1' wrench (*Figure 5*). For smaller amounts of torque, the **inch-pound (in-lb)** unit is better. Tightening a bolt to 1 in-lb of torque is like hanging a 1 lb weight on a 1" wrench (*Figure 5*). For this reason, 1 in-lb equals $^1/_{12}$ ft-lb.

Countries that follow the metric system use the **newton-meter (Nm)** as their torque unit. Tightening a bolt to 1 Nm is like hanging a 0.1 kg weight on the end of a 1 m wrench (*Figure 5*).

> **NOTE**
> The newton (N) is the metric unit of force. It is named after Sir Isaac Newton (1643–1727), an early physicist. A 1 kg weight produces 9.8 N of force. Coincidentally, 1 N of force approximately equals the weight of an apple sitting on the table.

Fastener and equipment manufacturers specify tightening torque in one of these three units. But how do you know when you have reached the correct torque? Special tools solve this problem.

1.2.0 Torque Tools

Craftworkers use torque wrenches and torque screwdrivers to tighten bolts and screws to the proper values. These tools indicate when the applied torque reaches the target value. They come in several styles and many sizes.

1.2.1 Torque Tool Types

Manual torque wrenches simply indicate the amount of torque being applied. The beam-type torque wrench shown in *Figure 6* works this way. The beam stays in position as the wrench handle rotates. The scale on the handle moves under the beam, showing the current torque value. The operator simply tightens the fastener until the scale shows the desired torque.

Click-type torque tools (*Figure 7*) have an adjustment dial on the handle. After setting the desired torque value, the operator tightens the fastener until the tool clicks. This indicates proper torque.

> **CAUTION**
> Click-type torque tools can continue to tighten after the click, leading to over-torquing. Always stop as soon as the tool clicks.

Electronic torque tools offer the most features (*Figure 8*). These include a display and controls for setting the target torque. During tightening, the tool beeps to indicate that it is approaching the correct torque. When it reaches the target, the tool vibrates, beeps, or flashes a signal. Some tools

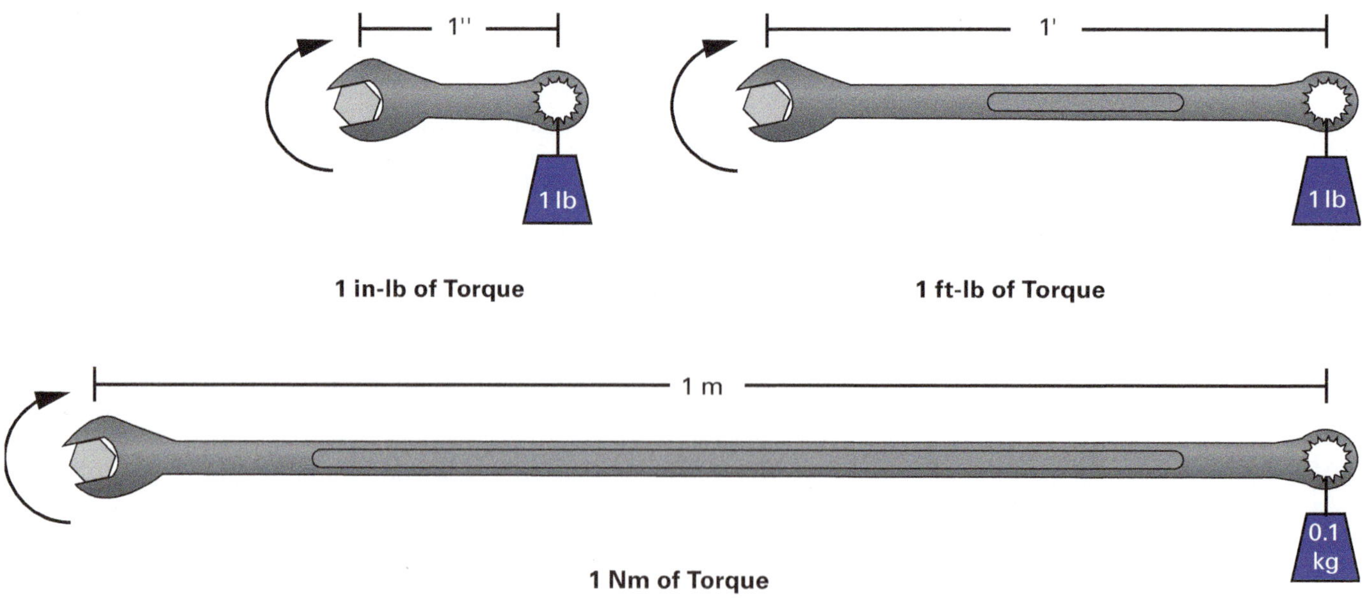

Figure 5 Common torque units (not to scale).

can store torque values for later transfer to a computer or smart device. Factories use these torque records to confirm proper equipment assembly.

Powered torque tools work like hand tools but save the operator a lot of physical effort. They're particularly useful for large fasteners with high target torques. *Figure 9* shows a hydraulically powered torque wrench.

CAUTION

Ordinary impact wrenches can save a lot of time when tightening fasteners. But they do not guarantee proper torque values and can easily over-torque a fastener or damage it.

NOTE

Torque tools require periodic calibration to ensure their accuracy. Many companies keep records and follow a rigorous schedule to guarantee regular calibrations (annually or more frequently).

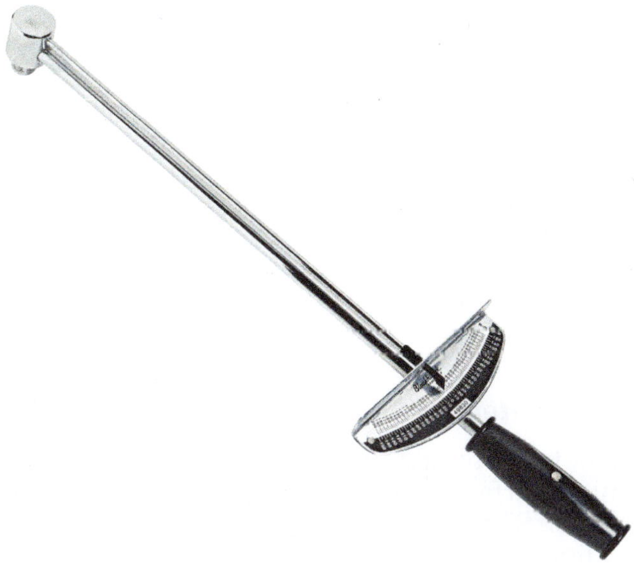

Figure 6 Beam-type torque wrench.

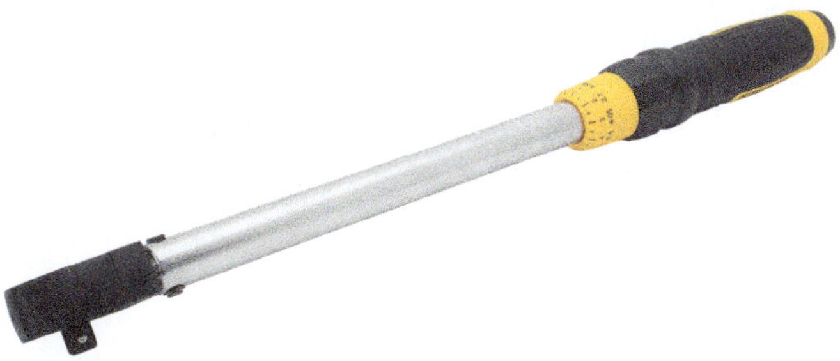

Figure 7 Click-type torque wrench.

1.2.2 Torque Tool Sizes

When torquing a fastener to the specified value, always choose the correct tool size. Torque tools have an operating range. They tend to be less accurate near the upper and lower limits. For this reason, choose a tool that places the target torque well inside the range (between 25–75 percent). When using a powered torque tool, check the size carefully as some are very powerful and can damage fasteners.

Non-electronic torque tools come rated in inch-pounds, foot-pounds, or newton-meters. If the fastener's torque specification comes in a different unit than the tool, convert it to the tool's unit first. *Table 1* shows conversion factors for all three torque units.

Figure 8 Electronic torque wrench.

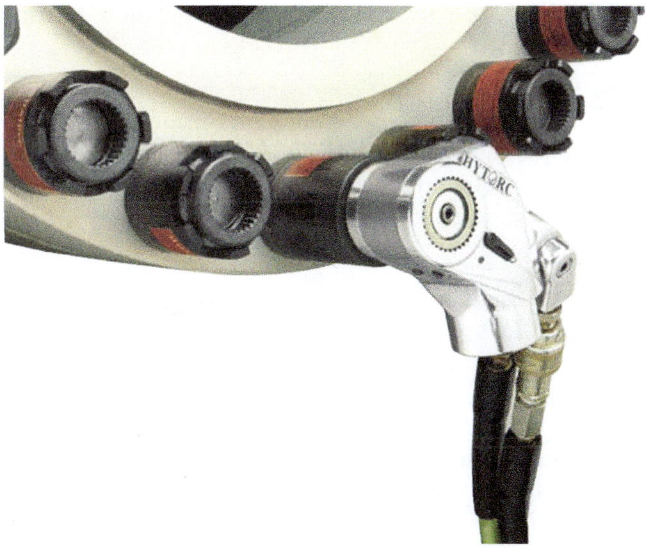

Figure 9 Hydraulic torque wrench.

Table 1 Torque Conversions

Convert from	Convert to	Multiply by
in-lb	ft-lb	0.08333
ft-lb	in-lb	12.00
in-lb	Nm	0.1130
Nm	in-lb	8.851
ft-lb	Nm	1.356
Nm	ft-lb	0.7376

For example, to use a metric torque wrench to torque a bolt to 37 ft-lb, convert the value to newton-meters as follows:

$$\text{torque (Nm)} = \text{torque (ft-lb)} \times 1.356$$

$$\text{torque} = 37 \times 1.356$$

$$\text{torque} = 50.2 \text{ Nm (rounded to the nearest 0.1 Nm)}$$

Electronic torque tools can switch torque units at the press of a button. This eliminates manual conversions.

1.2.3 Dry and Wet Torquing

If the equipment manufacturer does not provide fastener torque values, consult the fastener manufacturer. A fastener's recommended torque value depends upon the application. Another factor is whether the fastener is installed "dry" or "wet."

Dry torquing involves assembling the fastener without any kind of lubrication. Unlubricated threads rub against each other, making the fastener harder to tighten. This extra effort consumes some of the tightening torque applied by the torque tool. To compensate for the lost torque, manufacturers increase the recommended value for dry torquing.

> **NOTE**
> Torque tools measure the total torque applied to the fastener. They have no idea how much of that torque tightens (stretches) the fastener and how much is lost.

Wet torquing involves coating the fastener threads with a lubricant like oil, grease, graphite, or molybdenum. Lubricated fasteners assemble more easily and waste less of the tightening torque. Set/seizure is also less likely with lubricated fasteners. As a result, manufacturers usually specify a target wet torque value about 25 percent to 30 percent less than the dry torque value.

For example, a typical ¼-20 bolt might have a dry torque value of 101 ft-lb and a wet torque value of 76 ft-lb. A metric M22 × 2.50 fastener could have a dry torque value of 613 Nm and a wet torque value of 460 Nm.

Lubricating fasteners also makes them easier to remove in the future. Anti-seize products like Never-Seez® reduce the break-away torque. Some prevent *galling*, the tendency of certain metals to stick to each other after tightening. Some contain both lubricants and powdered metals, like nickel or copper.

A fastener may have a plating or coating to reduce friction or improve its corrosion resistance. These can act like lubricants and affect the recommended target torque value too.

Always check the fastener manufacturer's guidelines very carefully. Confirm whether they refer to dry or wet torquing since torque values differ for each. For wet torquing, be sure to use the recommended lubricant. Following the wrong specification or using a different lubricant could result in improperly torquing the fastener.

1.2.4 Tightening Sequences

When tightening multiple bolts on flanges or similar surfaces, the tightening order usually matters. Simply going around the flange can misalign or warp parts. It may also be difficult to achieve the proper torque value on certain bolts. Finally, improper tightening can damage seals and gaskets or prevent them from working correctly.

Whenever possible, follow the equipment manufacturer's guidance. Some equipment has unusual tightening sequences that go against common guidelines. If no information is available, however, sequences like the ones shown in *Figure 10* are useful.

Begin by installing all bolts. Hand-tighten them until they are just snug. Switch to a tool and begin tightening them in the correct sequence. Never tighten a bolt all at once. Tighten a small amount at a time as you work through the sequence. Keep repeating the pattern as you work towards the final torque value.

As you tighten, remember an important fact. Tightened bolts tend to relax and require several torquing cycles to ensure that they stay at the proper torque. Bring them to the final value over three cycles. First, torque to about 33 percent. Go around again and torque to about 66 percent. Finally, repeat the cycle and bring the bolts to 100 percent. Some fasteners require rest periods between cycles so they can relax.

CAUTION

Failing to follow the proper tightening sequence can lead to equipment problems immediately or in the future. Premature maintenance cycles often result from improperly tightened fasteners.

1.2.5 Using a Torque Tool

After installing fasteners, you must bring them up to their final torque using an appropriate torque tool. The following steps outline the general procedure:

Step 1 Determine the fastener's target torque value. If necessary, convert it to a different torque unit.

Step 2 Select a torque tool appropriate to the fastener type. Confirm that the fastener's target torque is well inside the tool's operating range.

Step 3 Calculate 33% and 66% of the target torque value. For example, if the target torque is 25 ft-lb (34 Nm), the following calculations give the required values:

$$\text{torque (33\%)} = 25 \times 0.33$$

$$\text{torque (33\%)} = 8.25 \text{ ft-lb (11.2 Nm)}$$

$$\text{torque (66\%)} = 25 \times 0.66$$

$$\text{torque (66\%)} = 16.5 \text{ ft-lb (22.4 Nm)}$$

NOTE

The 33% and 66% torque values do not have to be exact. Rounding to the nearest convenient unit is acceptable. Only the final torque value (100%) must be as accurate as possible.

Step 4 For adjustable torque tools, set the torque to the 33% value using the adjustment dial or controls.

Step 5 Ready the fasteners for final torquing. Follow the equipment or fastener manufacturer's recommendations.

Step 6 Place the tool on the fastener, holding it in place in a safe way. If required, use another tool to hold other fasteners, like nuts, in place.

Step 7 Tighten the fastener (*Figure 11*). For click-type or electronic tools, listen for the target torque signal. For manual tools, watch the indicator for the correct torque. As soon as you reach the target torque, stop applying force immediately.

Step 8 Following the appropriate sequence, tighten the remaining fasteners to the 33% torque value.

Step 9 For adjustable torque tools, set the torque to the 66% value using the adjustment dial or controls.

Step 10 Following the appropriate sequence, repeat the tightening process for all fasteners.

Step 11 For adjustable torque tools, set the torque to the final value (100%) using the adjustment dial or controls.

Step 12 Following the appropriate sequence, repeat the tightening process for all fasteners.

Step 13 When finished, set adjustable torque tools to zero. Clean and store the tool properly.

WARNING!

Whenever possible, pull a wrench rather than pushing it. This technique reduces the likelihood of finger and knuckle injury if the wrench slips. Always grip the wrench by the handle. Never use extensions or cheater bars, particularly on torque wrenches.

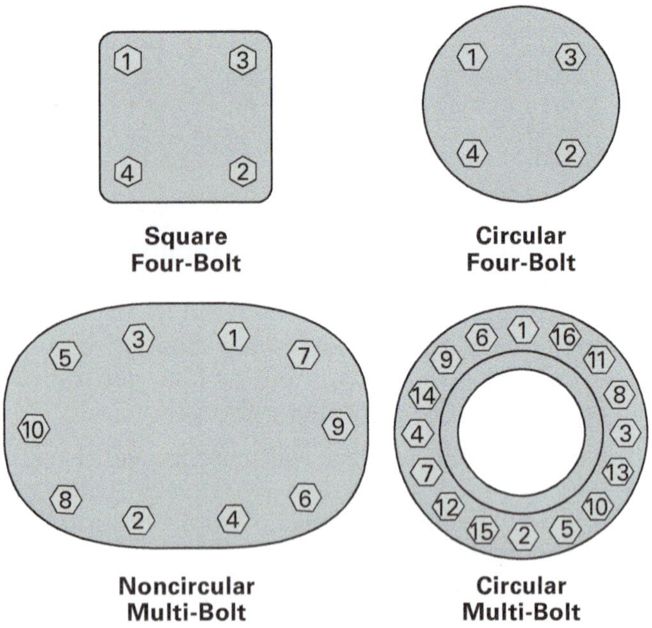

Figure 10 Common tightening sequences.

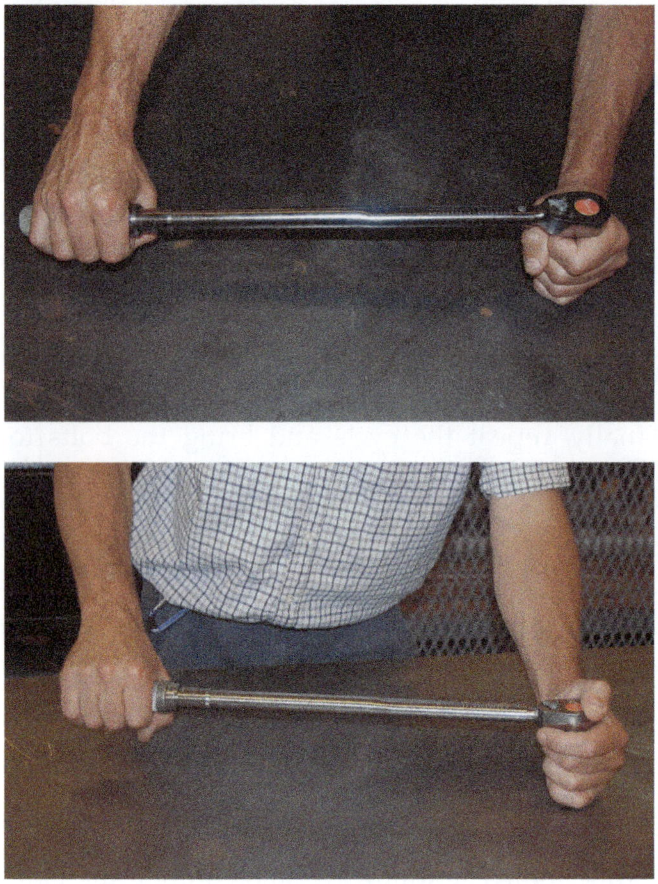
Figure 11 Using a torque wrench.

Torque Multipliers

Sometimes you need a little extra torque to get a fastener tightened properly. You might be tempted to use a cheater bar. But this method can be dangerous and could damage the wrench. Torque multipliers safely provide extra torque. These are small gearboxes that fit between the wrench and the fastener. They multiply the torque by a factor between 4 and 70. Torque multipliers also work with ordinary wrenches for loosening tight fasteners.

When torquing fasteners, never forget to take the multiplication factor into account when setting or reading the torque wrench. Since the gears waste some of the tightening torque, you must also compensate by increasing the target torque slightly.

Figure Credit: Courtesy of ENERPAC

1.0.0 Section Review

1. What *must* you do to a threaded fastener to ensure that it holds properly?
   a. Over-torque it.
   b. Cross-thread it.
   c. Torque it to the correct value.
   d. Apply anti-seize compound.

2. Which kind of torque tool beeps or vibrates when it reaches the target torque?
   a. An electronic torque wrench
   b. A beam-type torque wrench
   c. A click-type torque wrench
   d. A pneumatic impact wrench

3. You must torque a metric fastener with a target torque of 36 Nm. Unfortunately, all you have is a click torque wrench rated in inch-pounds. Using *Table 1* and rounding to the nearest 1 in-lb, to what value should you set the wrench?
   a. 4 in-lb
   b. 27 in-lb
   c. 49 in-lb
   d. 319 in-lb

Section Two

2.0.0 Threaded Fasteners

Objective

Identify threaded fasteners and summarize their applications.

a. List common threaded fastener types and standards.
b. Explain how to install threaded fasteners.
c. List common concrete anchors and describe their installation.

Performance Tasks

1. Install instructor-selected threaded fasteners.
2. Tighten a group of bolts to the proper torque and in the proper sequence.
3. Install instructor-selected fasteners in hardened concrete.

Trade Terms

Clearance hole: A fastener hole large enough for everything but the fastener's head to pass through.

Countersunk: A fastener installation method in which the head is flush with or below the surface.

Grade: A fastener-ranking system that identifies quality and strength.

Handedness: The direction that the threads wrap around a fastener, determining how it tightens or loosens.

Nominal size: The typical diameter of a bolt or screw.

Thread class: A code indicating the amount of looseness between the threads of matching bolts and nuts.

Thread pitch: The distance from the crest (top) of one thread to the next.

Thread series: Combinations of fastener diameter and threads-per-inch associated with specific thread diameters.

Thread standard: An established set of specifications for machining threads.

Thread-locking fluid: A hardening liquid placed on threaded fasteners to make them more resistant to vibration loosening.

Threads-per-inch (tpi): The number of threads in one inch of fastener length.

Threaded fasteners appear in almost all applications. They assemble and mount numerous kinds of equipment and parts. Bolts, nuts, screws, threaded rods, and anchors all fall under this fastener category. Selecting threaded fasteners requires skill since they come in many styles and materials. Installing them correctly requires another equally important set of skills.

2.1.0 Threads and Fasteners

As their name suggests, all threaded fasteners rely on spiral ridges, called *threads*, to operate, attach, and grip. *External threads*, or male threads, wrap around the cylindrical or conical fastener. *Internal threads*, or female threads, wrap around the inside of a hole. Matching external and internal threads fit into each other, drawing parts together through tension.

Threaded fasteners come in many styles and work in several ways. Bolts can pair with nuts or just screw directly into a threaded hole. Screws pull their way into soft materials like wood or grip the edges of a hole in sheet metal. Some fasteners work with specialized hardware like washers, locknuts, or anchors. This section explores a wide variety of fasteners.

Since threads are the heart of all threaded fasteners, craftworkers must understand them thoroughly. They must recognize the terms, types, and the standards associated with them. To choose the best fastener for the job, craftworkers must also understand fastener materials and ratings.

2.1.1 Threaded Fastener Standards

Manufacturers fabricate fasteners to a particular **thread standard**. This published specification defines many of the fastener's properties. These include the following:

- **Nominal size** — The typical fastener diameter.
- **Threads-per-inch (tpi)** — The number of threads in one inch of fastener length.

- **Thread series** — The number of threads per inch for each fastener of a certain diameter.
- **Thread class** — The permissible looseness between the external and internal threads of matching fasteners.
- **Handedness** — The direction that the threads wrap around the fastener, which affects how the fastener tightens or loosens.
- Length — The fastener's length in inches and/or fractions of an inch. For most bolts, length is the distance from the underside of the head to the end of the threads. For bolts recessed below the surface, length is the distance from the top of the head to the end of the threads.

In the United States, the Unified Thread Standard (UTS) governs threaded fasteners. Fasteners produced under this standard follow a well-defined identification scheme, as *Figure 12* shows.

2.1.2 UTS Thread Series and Thread Class

The UTS defines three thread series. Each specifies the threads-per-inch for fasteners of certain diameters. Essentially, the series indicates how coarse or fine the threads will be. The three series are as follows:

- *Unified National Coarse (UNC) Thread* — Used for general-purpose applications. Fasteners with UNC threads assemble and disassemble rapidly even when slightly corroded or damaged.
- *Unified National Fine (UNF) Thread* — Used for applications that require a finer thread than UNC.
- *Unified National Extra Fine (UNEF) Thread* — Used for thin-walled tubes, thin nuts, ferrules, and couplings. Finer threads grip better in thin materials.

Thread class determines how tightly the threads of matching fasteners hold each other. There are three classes. Classes 1A, 2A, and 3A apply to external threads. Classes 1B, 2B, and 3B apply to internal threads. Higher numbers indicate a tighter fit.

Few applications use Class 1 fasteners since they separate easily and have limited strength. Applications requiring quick disassembly sometimes use them. Class 2 fasteners work well for many applications and are quite common. Class 3 fasteners provide maximum strength and resist separation. However, they are more difficult to assemble due to their tight fit.

Thread Pitch Gauges

If you have a threaded fastener but do not know the threads-per-inch value, a *thread pitch gauge* can help. Simply place the fastener against the tool's different leaves until you find a match. Always use a tool that matches the fastener's standard (US or metric). You may find a close match with the wrong tool, but it will not be quite correct.

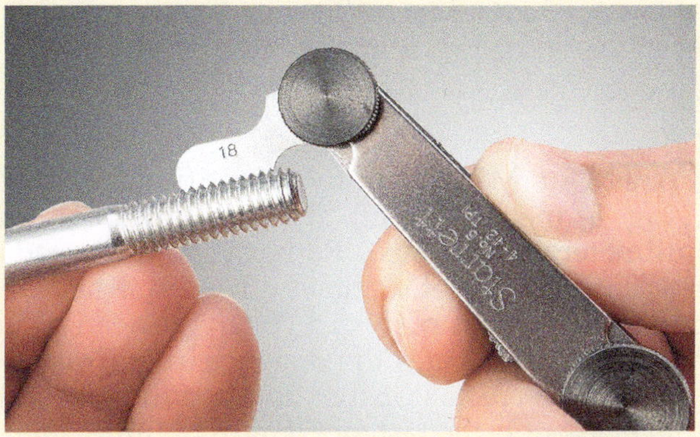

Figure Credit: The L.S. Starrett Company

2.1.3 Handedness

Fastener threads spiral in one of two directions. Thread direction, or handedness, determines the correct way to tighten or loosen the fastener. Right-hand (RH) threads tighten clockwise and loosen counterclockwise. Left-hand (LH) threads tighten counterclockwise and loosen clockwise.

Most applications use right-hand threads. Left-hand threads appear in rotating applications where the motion could unscrew a right-hand thread. Gas welding equipment uses fittings with left-hand threads for fuel cylinders. Oxygen cylinders use right-hand thread fittings. This strategy prevents accidentally connecting hoses backwards, a potentially lethal situation.

Fasteners with left-hand threads include an "LH" code in the fastener identification (*Figure 12*). Right-hand thread fasteners do not include a code.

2.1.4 Metric Threads

Metric-speaking countries follow the ISO metric screw thread standard. This standard is similar to, yet different from, the UTS. *Figure 13* shows a metric fastener identification.

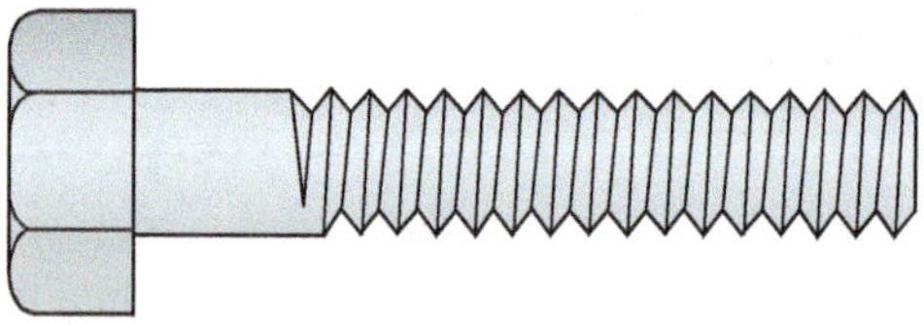

Figure 12 UTS fastener identification.

An M or MJ begins the identification, specifying a metric thread. The nominal size in millimeters comes next. Rather than a "threads-per-centimeter" value, however, metric fastener identifications use **thread pitch** instead. The thread pitch indicates the distance in millimeters between the crests (tops) of adjacent threads. The final element is the fastener length in millimeters.

Metric fasteners come in varying degrees of thread fineness. Unlike US standard fasteners, however, the terms "fine" and "coarse" don't apply to metric fasteners. Instead, the thread pitch value indicates the thread's fineness or coarseness. The smaller the thread pitch value, the finer the threads.

Metric fasteners beginning with an M have standard threads. Fasteners beginning with MJ have a modified thread style that is more resistant to fatigue. Some aerospace applications use the MJ style for this reason. An MJ bolt requires an MJ nut (or a hole with MJ internal threads). An M bolt, however, works with both M and MJ nuts (or holes with either type of internal threads).

Thread Lore

Why do threaded fasteners come in coarse- and fine-thread versions? It has to do with the application. Coarse-thread fasteners (or metric ones with a large thread pitch) assemble more easily and require less effort to turn. Large coarse-thread fasteners resist thread stripping. For most general-purpose applications, coarse-thread fasteners are the right choice.

Fine-thread fasteners (or metric ones with a small thread pitch) take more effort to turn. But they also resist loosening, so they are a good choice for vibrating applications. They work especially well with thin nuts or in thin metal. Since their threads are closer together, they grip the thin surface better.

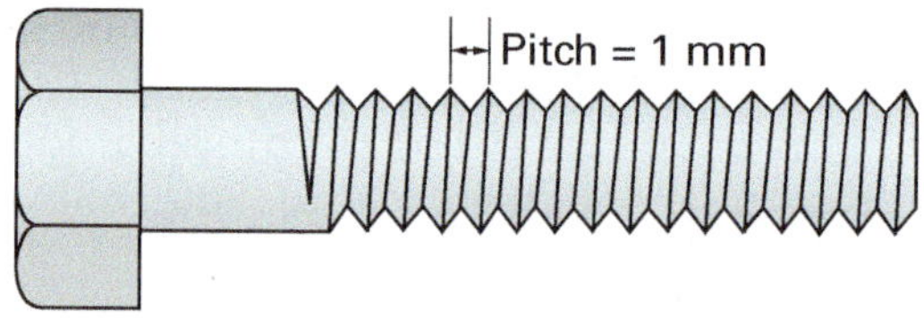

Figure 13 Metric fastener identification.

2.1.5 Threaded Fastener Grades

Fasteners may perform differently even if they look identical. Their material and the way that the manufacturer treated it can make a huge difference. A fastener made from low-quality steel is not very strong. It may work well in an undemanding application. But put under significant stress, it will fail. A high-quality steel that the manufacturer heat treats is much stronger and performs better under stress.

Manufacturers identify a fastener's quality and strength by assigning it a **grade**. Special markings on the fastener's head identify the grade. Standards organizations like SAE International and ASTM International specify a common set of markings, sometimes called *line markings*. *Figure 14* shows the most common ones. Notice that the lowest grades have no markings. Always assume that a fastener with a plain head is low-grade.

Grades 2, 5, and 8 are the most common. Made from low-carbon steel, Grade 2 fasteners work well for low-stress applications. A bracket supporting a sheet metal duct is an example. Made from high-quality, heat-treated steel, Grade 5 and Grade 8 bolts can handle more stress. For example, they might secure the cylinder head of an air compressor.

Higher-grade fasteners can be torqued to higher values. They also tend to stay torqued. Low-grade fasteners deform more easily and may not hold a torque value as well. For important or demanding applications, always use high-grade fasteners. Never forget that each grade has a different set of target torque values. Always follow equipment or manufacturer guidelines when torquing fasteners.

WARNING!

Never replace a higher-grade fastener with a lower-grade one. Doing this can cause equipment failure and injury to innocent bystanders.

ASTM and SAE Grade Markings for Steel Bolts & Screws

Grade Marking	Specification	Material
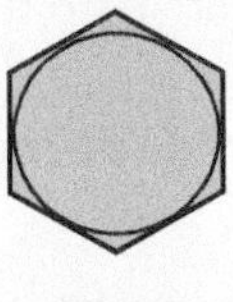	SAE-Grade 0	Steel
	SAE-grade 1 ASTM-A 307	Low-Carbon Steel
	SAE-Grade 2	Low-Carbon Steel
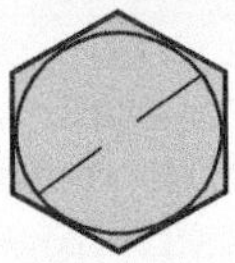	SAE-Grade 3	Medium-Carbon Steel, Cold Worked
	SAE-Grade 5	Medium-Carbon Steel, Quenched and Tempered
	ASTM-A 449	
	ASTM-A 325	Medium-Carbon Steel, Quenched and Tempered
	ASTM-A 354 Grade BB	Low-Alloy Steel, Quenched and Tempered
	ASTM-A 354 Grade BC	Low-Alloy Steel, Quenched and Tempered
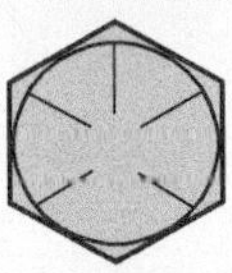	SAE-Grade 7	Medium-Carbon Alloy Steel, Quenched and Tempered Roll Threaded After Heat Treatment
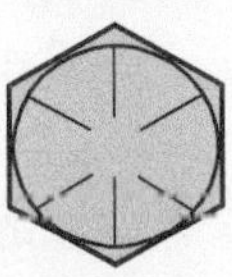	SAE-Grade 8	Medium-Carbon Alloy Steel, Quenched and Tempered
	ASTM-Aa 354 Grade BD	Alloy Steel, Quenched and Tempered
	ASTM-A 490	Alloy Steel, Quenched and Tempered

ASTM Specifications

A 307 – Low-Carbon Steel Externally and Internally Threaded Standard Fasteners.
A 325 – High-Strength Steel Bolts for Structural Steel Joints, Including Suitable Nuts and Plain Hardened Washers.
A 449 – Quenched and Tempered Steel Bolts and Studs.
A 354 – Quenched and Tempered Alloy Steel Bolts and Studs With Suitable Nuts.
A 490 – High-Strength Alloy Steel Bolts for Structural Steel Joints, Including Suitable Nuts and Plain Hardened Washers.

SAE Specification

J 429 – Mechanical and Quality Requirements for Threaded Fasteners.

Figure 14 Grade markings for US standard bolts.

Under Pressure

If you look closely at *Figure 15*, you'll notice that metric fastener strength measurements use a unit called the *pascal* (Pa). The pascal is the metric unit of pressure. A force of 1 N pressing down on a 1 m^2 surface generates 1 Pa of pressure. Metric fastener manufacturers perform a variety of tests on fasteners. These stretch and crush them to establish their strength. The manufacturer reports the results in pascals.

The pascal is named after Blaise Pascal (1623–1662), a French mathematician and physicist who researched pressure.

Metric fasteners use their own grading system (*Figure 15*). Instead of markings, numbers on the head identify the grade. Unmarked heads indicate the lowest grade.

2.1.6 Machine Bolts, Machine Screws, and Cap Screws

Craftworkers use machine bolts (*Figure 16*) to assemble parts and attach equipment. They come in a variety of sizes and grades. Diameters range from ¼" to 3" and lengths from ½" to 30". Metric machine bolts have a similar range. A machine bolt can screw directly into a threaded hole or a nut can secure it.

Carriage bolts are specialized bolts with unusual heads. Instead of fitting a wrench, the head is dome shaped with a square section underneath. This section fits into a square hole that keeps the bolt from turning while you torque the nut.

Grade	Head Marking	Size Range (mm)	Minimum Proof Strength (10^6 Pa)	Minimum Tensile Strength (10^6 Pa)	Material
4.6	4.6	M5–M36	225	400	Low or Medium Carbon Steel
4.8	4.8	M1.6–M16	310	420	Low or Medium Carbon Steel
5.8	5.8	M5–M24	380	520	Low or Medium Carbon Steel
8.8	8.8	M16–M36	600	830	Medium Carbon Steel, Q & T
9.8	9.8	M1.6–M16	650	900	Medium Carbon Steel, Q & T
10.9	10.9	M5–M36	830	1,040	Low Carbon Martensite Steel, Q & T
12.9	12.9	M1.6–M36	970	1,220	Alloy Steel, Q & T

Figure 15 Metric bolt grades.

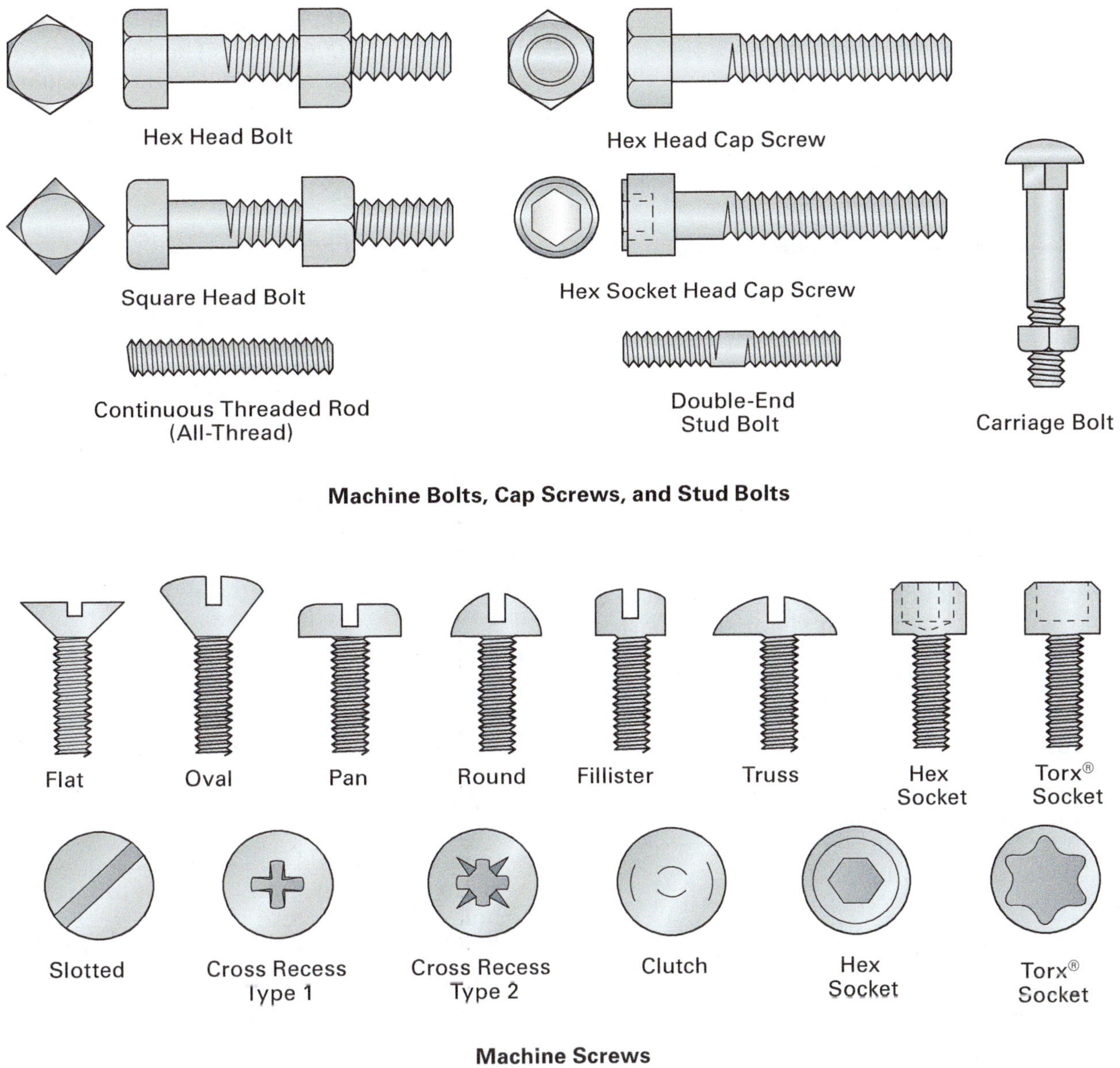

Figure 16 A variety of bolt-like fasteners.

Machine screws resemble small machine bolts but have different heads. Slotted heads accept a screwdriver. Specialized driver heads require a matching tool. Like machine bolts, machine screws perform a wide variety of assembly and attachment tasks.

Smaller US standard machine screws come in sizes #6 through #12 (0.060" to 0.2160"). Larger machine screws have diameters between $\frac{1}{4}$" and $\frac{1}{2}$". Lengths range from $\frac{1}{8}$" to 3". Metric machine screws follow the metric bolt identification system.

Cap screws give assemblies a finished appearance. They pass through a **clearance hole** in one part and screw into a threaded hole in the other. Manufacturers machine them to precise dimensions. They come in coarse- and fine-thread versions. Diameters range from $\frac{1}{4}$" to 2" and lengths from $\frac{3}{8}$" to 10". Metric cap screws come in similar sizes.

2.1.7 Stud Bolts and Threaded Rod

Threaded rod (all-thread) is a metal rod threaded over its entire length. It comes standard in 3', 6', and 10' lengths. Custom lengths are available too. Made from low-carbon steel, threaded rod works best with low-stress loads. Pipe hangers and trapeze assemblies for supporting multiple pipes or cable trays are common applications (*Figure 17*).

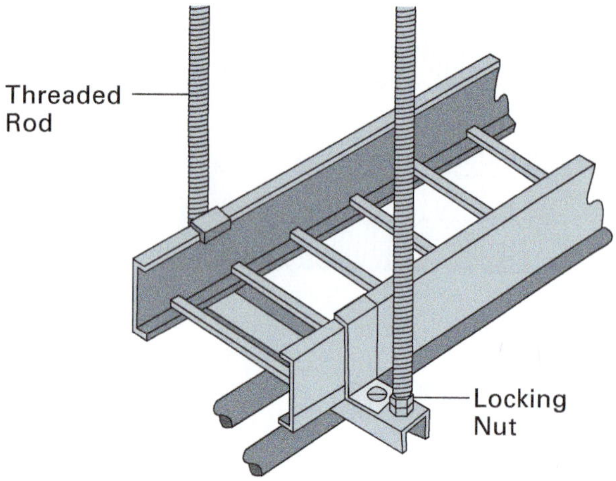

Figure 17 Threaded rods supporting a cable tray.

Stud bolts are headless bolts. They resemble a threaded rod with an unthreaded section in the middle. One end screws into a threaded hole. A part with a hole fits over the protruding stud. A nut and washer combination clamps the part in place. Some stud bolts have different threads on each end (for example, UNC and UNF).

2.1.8 Nuts

Unless a threaded fastener screws directly into a matching hole, it probably has a nut on the other end. General-purpose nuts come in several shapes. When used with bolts, choose a style that matches the bolt head. Nuts also come in a variety of special styles. *Figure 18* shows many different nut types.

Special-purpose nuts include the following:

- *Acorn* — These provide a finished appearance and cover exposed threads.
- *Castellated (castle)* and *slotted* — These nuts accept a pin in one of the slots. The pin keeps the nut from loosening.
- *Self-locking* — These nuts have nylon inserts that deform during tightening. The insert keeps the nut from working loose. Self-locking nuts cannot be reused and must be replaced after removal.
- *Wing* — Wing nuts allow for loosening and tightening without a wrench. They are useful for equipment that requires frequent assembly and disassembly.
- *Jam* — These thin nuts help lock a standard nut into place. Always install them under the regular nut, never on top (*Figure 18*).
- *U and J* — These nuts are simply threaded holes in a U- or J-shaped piece of metal (*Figure 19*). The nut slips over the edge of drilled sheet metal. A bolt then threads into the nut.
- *Cage* — A sheet metal cage encloses these square nuts (*Figure 19*). The cage has a spring clip at the bottom that fits into a square hole in an equipment rack. The nut can move in the cage to accommodate some misalignment.

2.1.9 Washers

Flat washers (*Figure 20*) distribute the pressure created by bolt heads and nuts. When used with softer materials, they keep the fastener from digging in and damaging it. Flat washers come in light, medium, heavy-duty, and extra heavy-duty versions. Fender washers are large-diameter flat washers that bridge oversized holes.

Lock washers come in many styles. When used properly, they keep fasteners from vibrating loose. They fit between a flat washer and the bolt head or nut. In general, they work better with softer, cheaper fasteners and materials rather than with high-grade ones. Lock washers include the following types:

- *Split ring* — Commonly used with bolts and cap screws.
- *External* — Provides the greatest resistance to vibration.
- *Internal* — Used with small screws.
- *Internal-external* — Used with oversized mounting holes.
- *Countersunk* — Used with flat or oval-head screws that fit into **countersunk** (recessed) holes.
- *NORD-LOCK®* — Uses a pair of washers equipped with wedges to hold the fastener in place.

CAUTION

Improperly using a lock washer can be worse than not using one. They must be torqued to a specific value and never over-torqued. An over-torqued lock washer behaves more like an ordinary flat washer.

Unless the manufacturer recommends it, never add lock washers to fasteners holding equipment together. Some precision machinery manufacturers avoid them, relying instead on high-grade fasteners torqued to a precise value. Adding a lock washer could defeat this feature.

2.1.10 Thread-Repair Inserts

Thread-repair inserts (*Figure 21*), often known by the brand name *HeliCoil®*, are a special kind of fastener. Resembling a coil spring, they fit tightly inside a hole, giving it internal threads. As the name implies, they replace damaged or stripped threads in tapped holes.

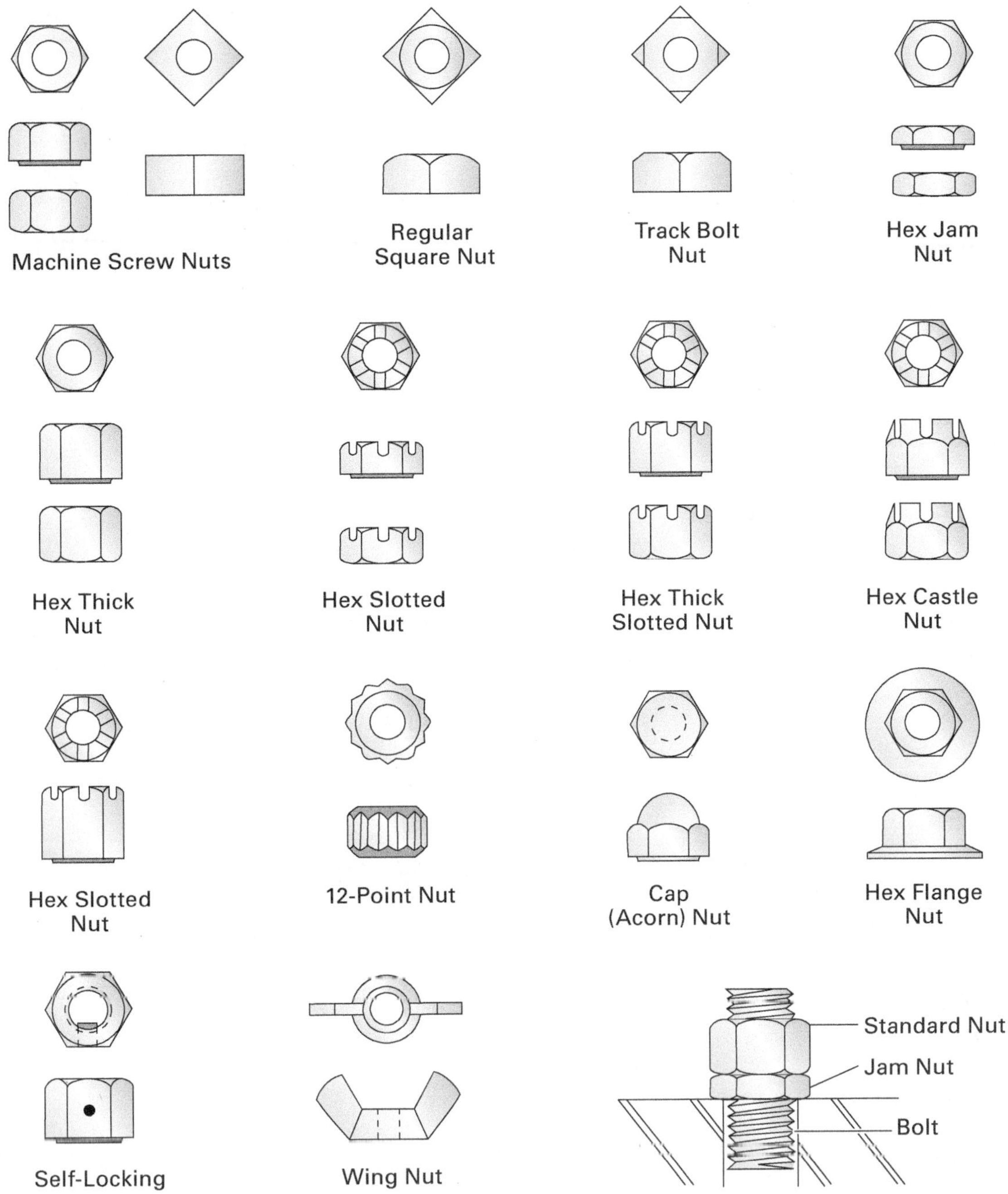

Figure 18 Nuts of all types.

They also offer an alternative to tapping holes in soft metals or plastics. Threads in these materials are very weak. Placing a thread-repair insert into a plain hole gives stronger threads.

Inserts come in standard sizes and forms, as well as metric versions. A special tool compresses the insert to fit it into the hole. It then expands, tightly anchoring itself in place.

For demanding applications, solid-body thread-repair inserts, like KEENSERTS®, work better. These permit multiple fastener removal cycles, which many other insert types discourage.

2.1.11 Eye Bolts

Eye bolts (*Figure 22*) have a loop, or eye, at one end and a threaded section at the other. Some eyes are formed and welded. Others are forged. Eye bolts never fasten parts together. Instead, they provide attachment points. Some manufacturers insert shoulder-forged eye bolts into equipment to act as lifting points.

Figure 19 Specialty nuts.

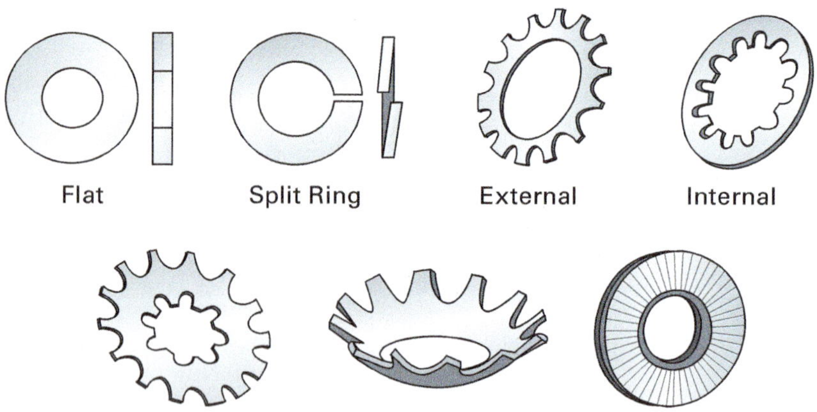

Figure 20 Flat and lock washers.

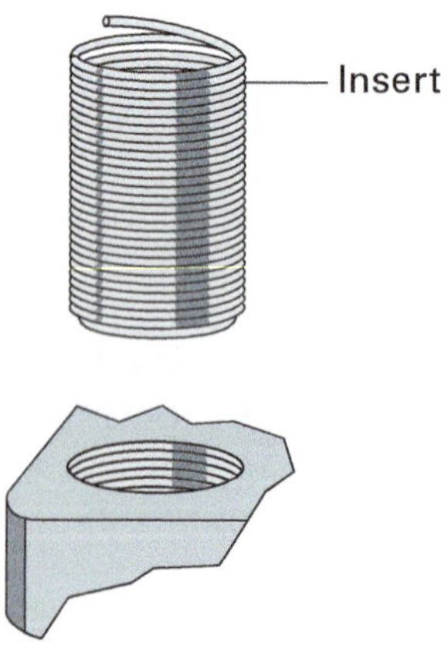

Figure 21 Thread-repair insert.

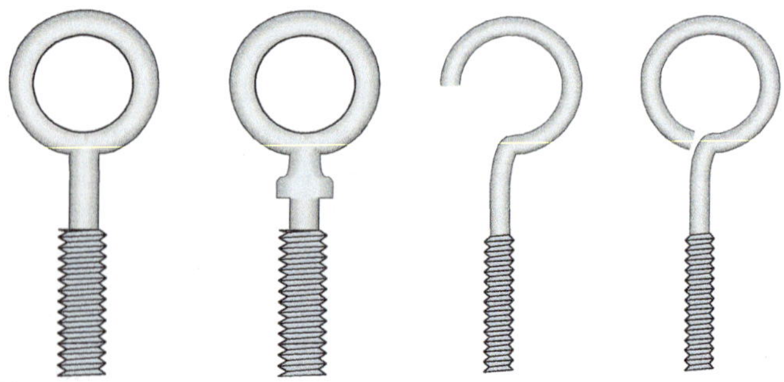

Figure 22 Eye bolts.

2.1.12 Toggle Bolts

Toggle bolts (*Figure 23*) secure light loads to concrete block or sheetrock walls. Regardless of the bolt's strength, they are never more secure than the wall material. Overloading one could pull it right through the wall.

2.1.13 Screws

Screws resemble bolts but often have tapered bodies with pointed ends. They do not require a threaded hole or a nut to hold them. Many produce their own matching threads when twisted into soft materials. Others grip the edges of holes in thin materials. Screws securely fasten parts together and may be used with parts requiring disassembly. They come with many head styles. Some accept ordinary screwdrivers. Others require special driver bits. *Figure 24* shows typical screw head types.

Some US standard screws come in numbered sizes, ranging from #2 to #24. Others have fractional diameter designations like ½". Lengths range from ¼" to 6". Metric screws follow the same general pattern as metric bolts.

Screws often come with specialized coatings or are plated with certain metals. These features improve corrosion resistance or make the screws more attractive. Screws made from brass or nylon handle specialized needs.

Lag screws attach heavy loads to wood structures. When used with a lead lag shield, lag screws attach objects to concrete or masonry (*Figure 25*). The shield fits into a pre-drilled hole. Threading the screw into the shield expands it, causing it to grip the hole. Overtightening can damage the shield, causing it to lose its grip.

Thread-forming screws fasten light-gauge metal parts. They form a thread as they enter the material, eliminating the need for tapping. Since they have a very sharp point, they can drill their own holes in thin metals. For thicker materials, self-drilling thread-forming screws work better. They have a special tip shaped like a drill bit that allows them to drill their own hole.

Thread-cutting (self-tapping) screws are more aggressive, cutting threads into heavier metal as they enter a pilot hole. *Figure 26* shows examples of both screw types.

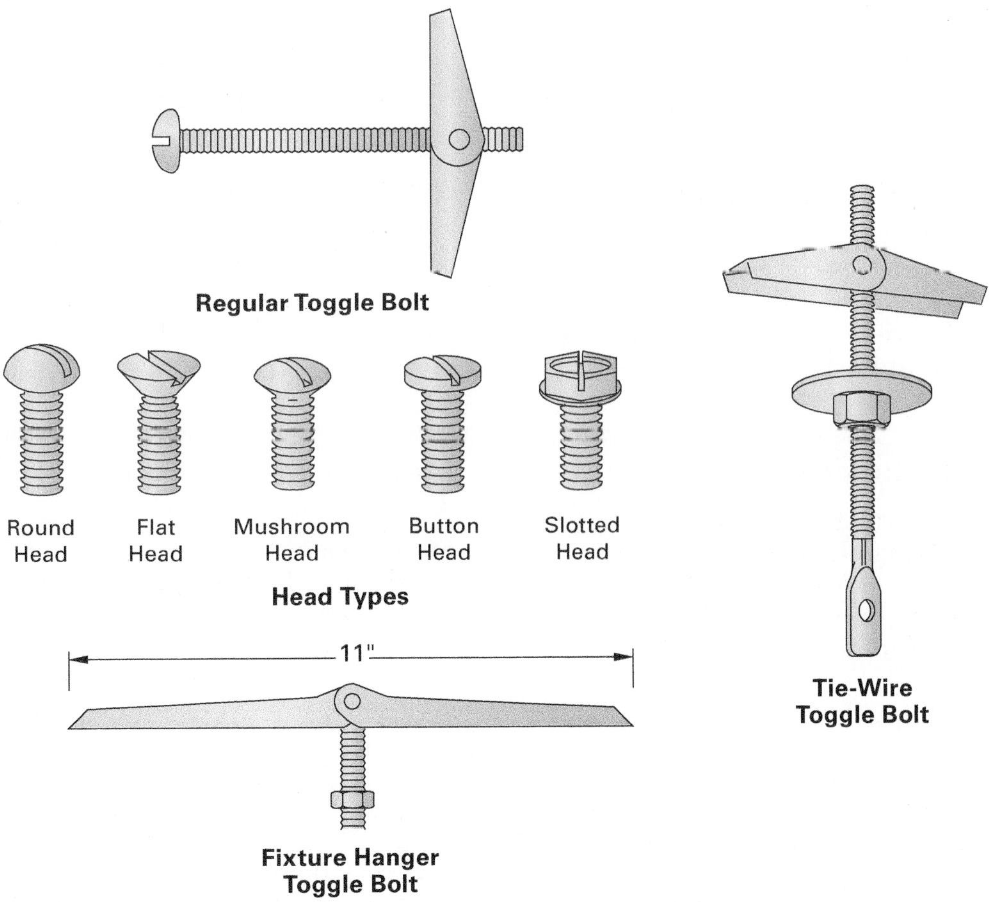

Figure 23 Toggle bolts.

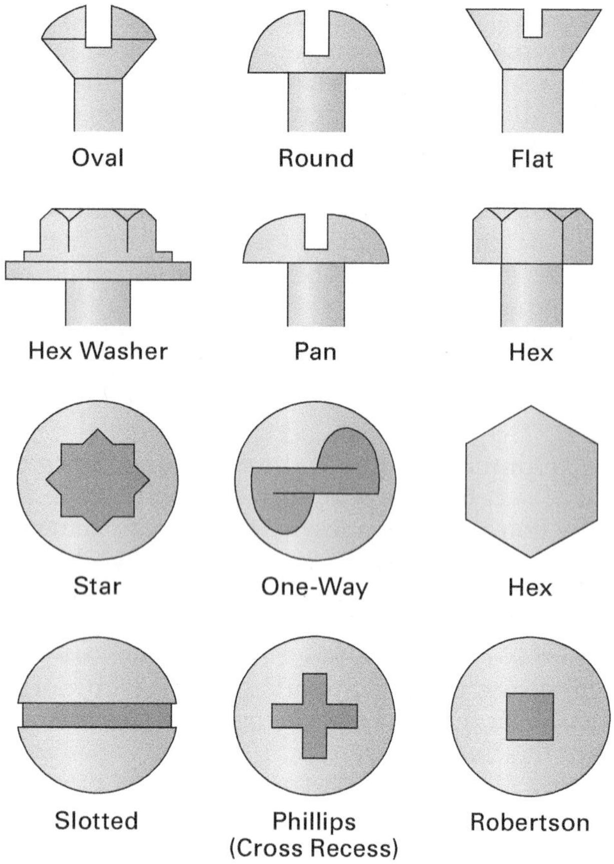

Figure 24 Screw head types.

2.1.14 Setscrews

Setscrews attach gears, pulleys, and couplings to shafts (*Figure 27*). Most do not have heads. Instead, they fully thread into the object and press against the shaft. Some press against other fastening devices. Made from hardened steel, setscrews have a variety of head and point styles (*Figure 27*). Most require a driver bit to install and tighten.

2.2.0 Installing Threaded Fasteners

Threaded fasteners do their job only when installed correctly. Since there are so many varieties, no one method works for everything. But each major fastener group has a general procedure. This section outlines these.

2.2.1 Installing Bolts, Machine Screws, and Cap Screws

All bolt-like fasteners follow a similar installation procedure. The most important factor is remembering to torque them properly. Most fastener failures result from under- or over-torquing the fastener. Follow these steps to install bolts, machine screws, and cap screws:

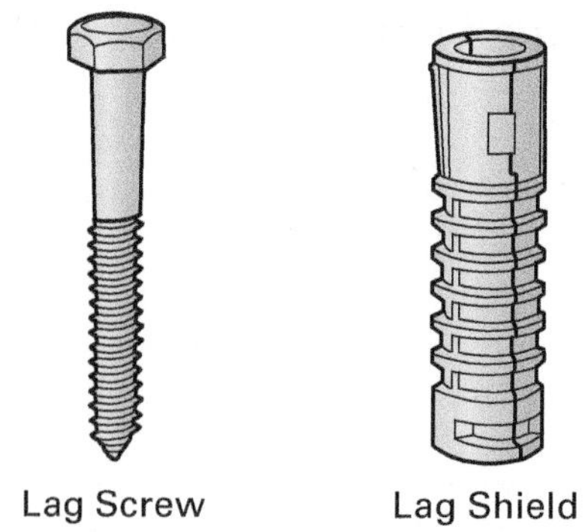

Figure 25 Lag screw and shield.

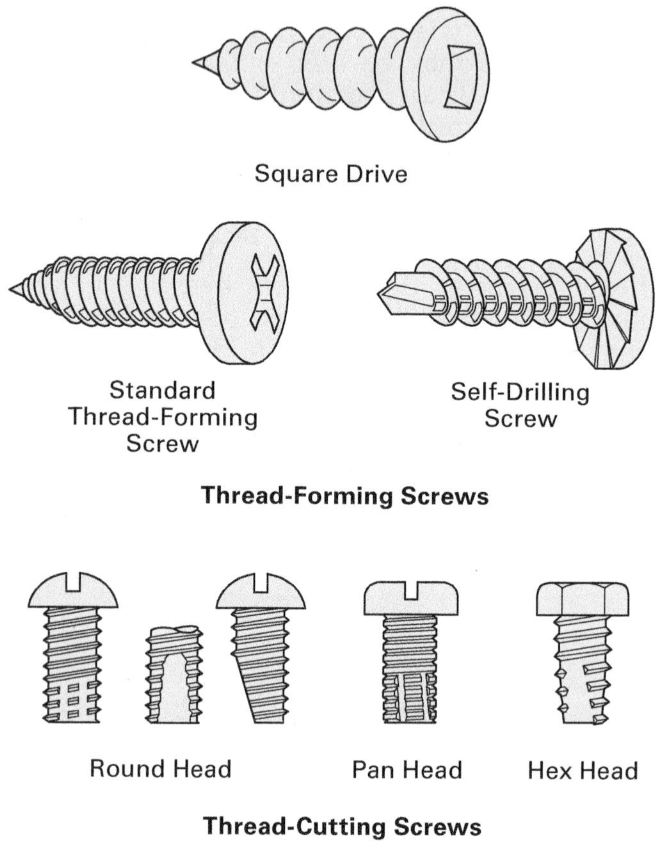

Figure 26 Thread-forming and thread-cutting screws.

Step 1 Determine the proper target torque for the fastener. When possible, follow the equipment manufacturer's guidelines. When this is not available, consult the fastener manufacturer.

Step 2 Select appropriate tools for turning and holding the fastener. Select a torque tool appropriate to the fastener's target torque value.

Step 3 Check the fastener threads for burrs and replace, if necessary. Burrs make accurate torquing difficult.

Step 4 Insert the fastener along with any required hardware, such as washers. If the fastener screws directly into a threaded hole, start it by hand. If the fastener passes through the hole, start the nut by hand.

CAUTION

Always confirm that threaded fasteners start properly to avoid damaging the threads (cross-threading). Once damaged, threads are difficult, if not impossible, to repair. Turning the fastener backwards a partial turn can help you find the correct starting point.

Step 5 Use an appropriate tool to tighten the fastener until it is just snug. When the fastener includes a nut, use a second tool to keep it from rotating.

Step 6 Once the fastener is snug, use the torque tool to tighten the fastener to its target torque value. Remember, tightening in stages of 33 percent, 66 percent, and 100 percent of the target torque is usually the best practice.

NOTE

When using a jam nut, place the thin nut on the fastener first. Torque it to between 25 and 50 percent of the target torque. Then place the standard nut on top and torque to the final value. While torquing, prevent the jam nut from turning by holding it with a thin wrench. If the jam nut turns, it will not do its job.

2.2.2 Installing Toggle Bolts

Follow these steps to install toggle bolts (*Figure 28*):

Step 1 Select an appropriate toggle bolt for the application. Confirm that the wall material is strong enough to support the expected load.

Step 2 Drill a hole in the wall large enough to accept the folded toggle bolt.

Step 3 Remove the bolt and insert it through the object being attached. Reassemble the bolt.

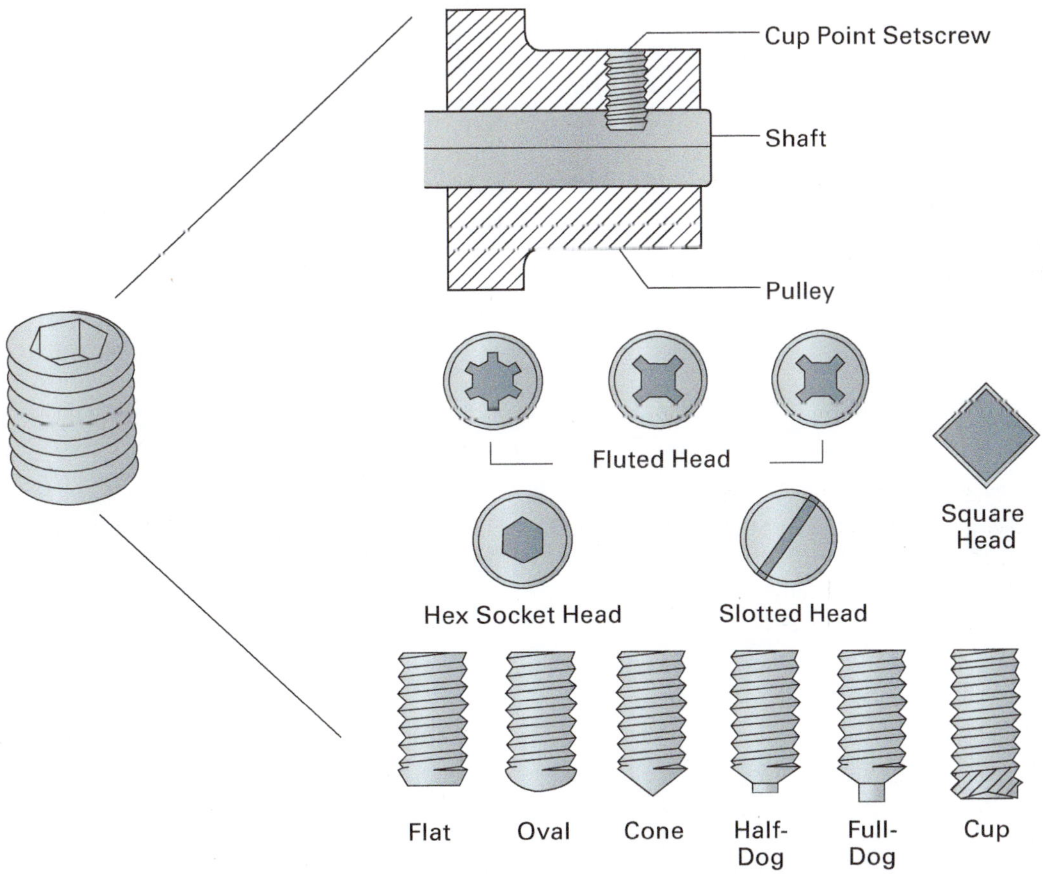

Figure 27 Setscrew heads and points.

Step 4 Fold the toggle bolt wings toward the bolt head. Insert the folded wings into the hole. They should spring open inside the wall.

Step 5 Tighten the bolt using an appropriate tool. Putting some tension on the bolt as you tighten it makes the process easier.

2.2.3 Installing Screws

Follow these steps to install screws:

Step 1 Select a tool that fits the fastener head correctly.

Step 2 If necessary, drill any required holes.

Step 3 Insert the screw and tighten it. When using thread-forming or thread-cutting screws, slowly and firmly turn the screw as it works its way through the material. **Be sure it enters the material straight.**

> **NOTE**
> When tightening screws into thin materials like sheet metal, do not overtighten or you may damage the hole.

Step 4 If the application requires a specific torque, use a torque tool to tighten the fastener to the target value.

2.2.4 Anti-Vibration Strategies

Vibrating equipment can loosen fasteners. Properly torquing them reduces this problem, but fasteners still can come loose. For this reason, some manufacturers recommend using extra measures to reduce vibration loosening.

Applying a **thread-locking fluid** can help keep fasteners in place. These are liquids applied to the fastener threads. Over the next hour or so, the liquid thickens inside the threads, making them more resistant to loosening.

The procedure for removing a fastener treated in this way depends on the fluid. Some release with a slow, steady pressure. Others require heating before they release. Manufacturers color code their fluids to identify them. The LOCTITE® brand includes a wide range of thread-locking products.

Some fasteners come with a small nylon patch melted onto one side of the threads. The nylon patch makes the fastener harder to turn. But once torqued to the correct value, the fastener resists coming loose.

2.3.0 Anchors and Anchor Bolts

Craftworkers frequently fasten equipment to concrete or masonry. Concrete cannot be threaded easily, nor can it hold regular threaded fasteners. Anchoring devices inserted into the concrete perform this function. They come in many types and styles. Some act like nails or screws. Others expand inside a hole, gripping it firmly once tightened. This section examines the most popular examples.

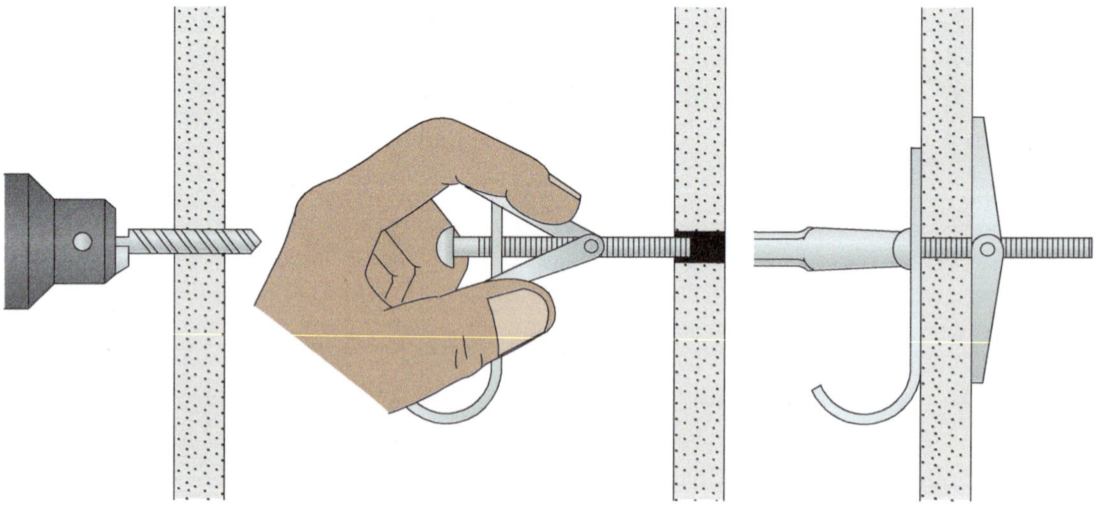

Figure 28 Toggle bolt installation.

CAUTION

Many concrete fasteners require drilled holes. Always follow the manufacturer's guidelines for drill size and hole depth. Incorrect holes are a major cause of anchor failure. Clear dust and debris from the hole before installing the fastener. Some anchors work correctly without clearing the hole, but they usually require deeper holes.

WARNING!

Drilling concrete or masonry produces a lot of dust, which can be a serious health hazard. Always wear appropriate respiratory protection when drilling concrete. Some organizations use water sprays to control dust. Follow all relevant safety standards and company safety policies.

2.3.1 Concrete and Masonry Screws

Concrete and masonry screws, commonly known as *Tapcon® screws* (*Figure 29*), have very coarse, sharp threads. They require a pilot hole of a specific diameter to hold properly. If the pilot hole is too large, they can pull out. If the hole is too small, they are difficult to drive and may break. Manufacturers often include a masonry drill bit with each box of fasteners. Using this bit ensures the correct pilot hole diameter.

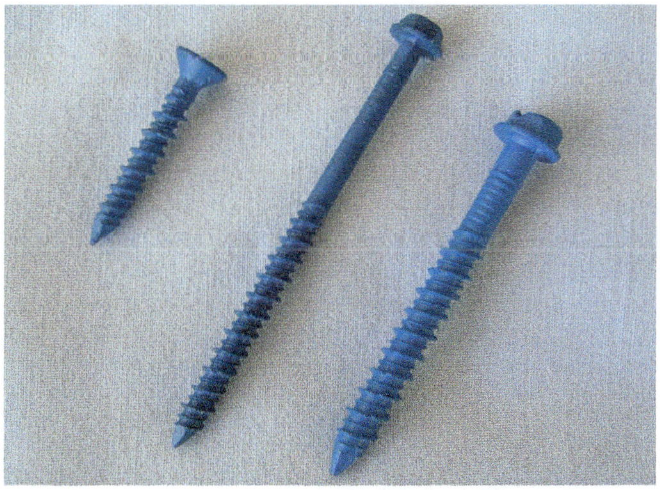

Figure 29 Tapcon® screws.

2.3.2 Hammer-Drive and Powder-Drive Pins and Studs

Hammer-drive pins or studs fasten wood or steel to concrete. They come in several styles. Each type requires a specific tool and installation procedure.

One variety has a tubular anchor surrounding the pin, which resembles a nail (*Figure 30*). Drill an appropriate hole in the concrete. Insert the anchor through the secured object into the hole. Using an ordinary hammer, drive the pin into the fastener. This action spreads the anchor in the hole, securing it. A version with a screw rather than a nail is also available.

Another type requires a special tool (*Figure 30*) and a heavy hammer. Simply position the fastener and drive it directly into the concrete by striking the tool with powerful blows. This variety requires the most brute strength to install but does not require drilling a hole.

Powder-drive fasteners use a small explosion to drive the fastener home. A special tool (*Figure 31*) holds the fastener and a bullet-like load. After positioning the tool, press it tightly against the surface and tap the end with a hammer or pull the trigger. This fires the charge, which drives the fastener into the concrete. Manufacturers color code the loads to identify their driving power. Picking the correct size ensures that the fastener reaches the correct depth.

In the United States, OSHA requires powder-drive tool operators to have a valid license. Tool manufacturers offer certification courses and tests, oftentimes online. Never operate one of these tools without a license.

All three systems require a specific concrete thickness and strength. Always follow the manufacturer's guidelines or the fasteners may not hold correctly.

WARNING!

Hammer-drive and powder-drive fasteners require eye protection. Wear ear protection when using powder-drive fasteners since the explosion can be quite loud. Treat the charges with the same care as ammunition. Never use powder-drive technology without proper training and a valid license.

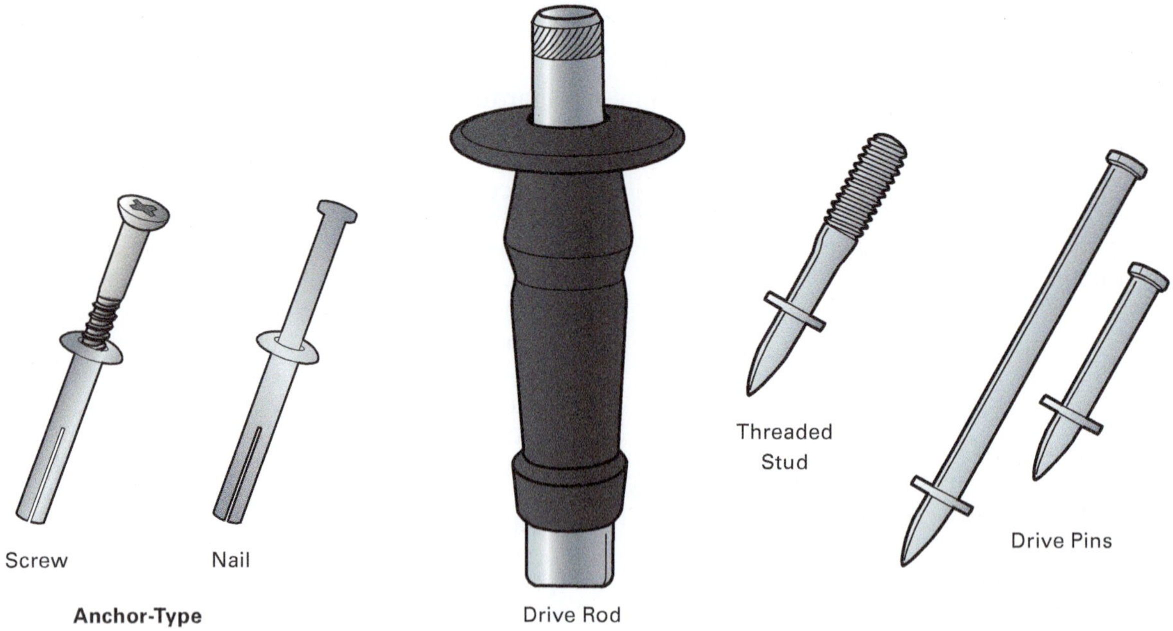

Figure 30 Hammer-drive pins.

2.3.3 One-Step Anchors

One-step anchors require a pre-drilled hole matching the anchor's diameter. When tightened, the anchor expands to grip the hole tightly. They come in many styles (*Figure 32*). Diameters range from $\frac{1}{4}$" to $1\frac{1}{4}$", with lengths from $1\frac{3}{4}$" to 12". Metric versions are similar.

- *Wedge anchors* — These are heavy-duty anchors supplied with nuts and washers. Insert the anchor into the hole. Drive it with a hammer so at least six threads are below the surface. Tighten the nut to expand the anchor and secure it in the hole.
- *Stud bolt anchors* — These are heavy-duty threaded anchors that take a matching nut. They require a clean hole of a specific depth. After insertion, repeated hammer blows expand the anchor, fixing it in the hole.
- *Sleeve anchors* — These are multipurpose anchors. The hole depth is not critical if it is deep enough for the anchor. Clean the hole and tap the anchor into place. Tightening the nut or screw expands the anchor in the hole.
- *One-piece anchors* — These are multipurpose anchors that work like springs. They require a clean hole deeper than the anchor length. Carefully follow the manufacturer's guidelines. Driving the anchor into the hole compresses it. Once seated, its spring-like sleeve expands, holding it in place. Some manufacturers sell special tools for inserting these anchors.
- *Sammy® anchors* — These install in concrete, steel, or wood. The Sammy® anchor supports a threaded rod, which screws into the anchor head. A special nut driver installs the anchor, which screws into the concrete.

2.3.4 Bolt Anchors

Bolt anchors install flush with the concrete surface. Bolts or machine screws thread into them. Some can accept threaded rods. They come in several styles (*Figure 33*).

Drop-in anchors are for heavy-duty applications. One type works in solid concrete and masonry. It has an internally threaded expansion anchor and plug. Once tapped into the prepared hole, a special setting tool expands the anchor. The bolt or screw then installs into the anchor.

The second type works in hollow materials, such as concrete block (cinder block). Hollow-set drop-in anchors have a slotted, tapered expansion sleeve and a serrated expansion cone. The

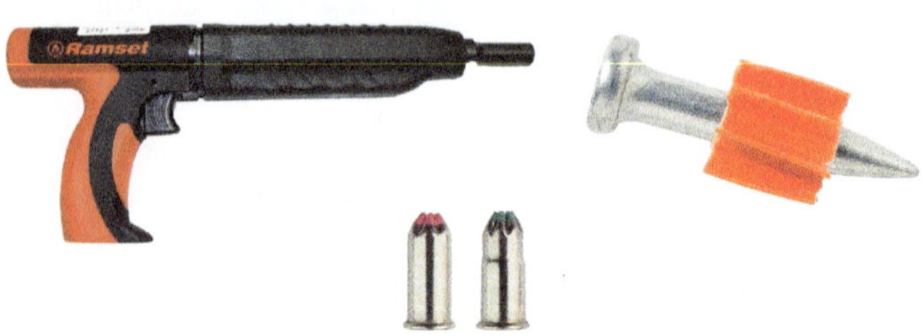

Figure 31 Powder-drive installation tool, pins, and charges.

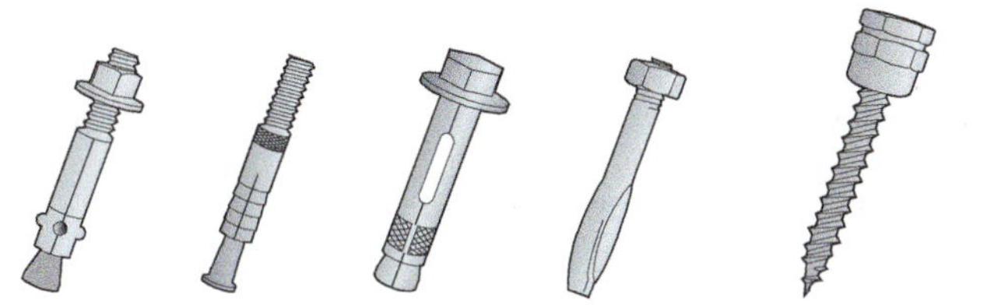

Figure 32 One-step anchors.

pre-drilled hole extends into the hollow space. After tapping the anchor into place, attach the mounted component. Insert the bolt or screw. Tightening the bolt expands the anchor, holding it in place.

Single- and double-expansion anchors install using a similar technique to drop-in anchors. Double-expansion anchors are good with weaker concrete. As with many anchors, they expand as the fastener is tightened.

2.3.5 Self-Drilling Anchors

Self-drilling anchors require a hammer drill to install (*Figure 34*). The fastener itself acts as the drill bit. After drilling the hole, remove the anchor and clean the hole. Insert the expander plug into the sleeve and drive it into the hole with the drill. This action expands the anchor, holding it in place. The top part then snaps off. Finally, a bolt threads into the anchor.

2.3.6 Epoxy Anchoring Systems

Epoxy resins come as two separate liquids. Once mixed, a chemical reaction hardens them into a strong plastic-like material. The hardening period, or *cure time*, takes anywhere from a few minutes to several hours. Some epoxies require mixing the two parts manually. Others come in dual-chambered cartridges that mix the parts as they emerge (*Figure 35*). A tool like a caulk gun extrudes the mixed epoxy.

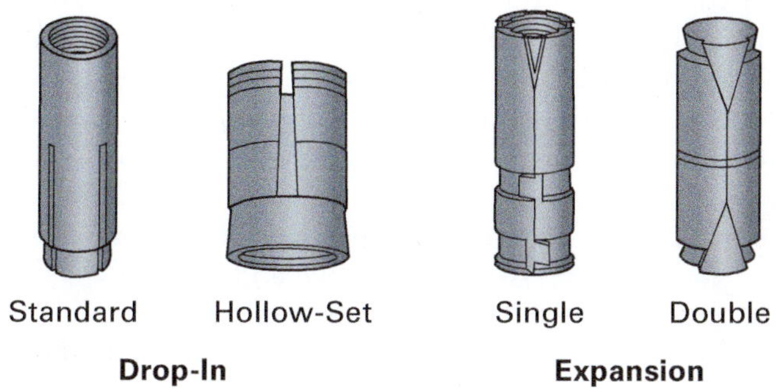

Figure 33 Bolt anchors.

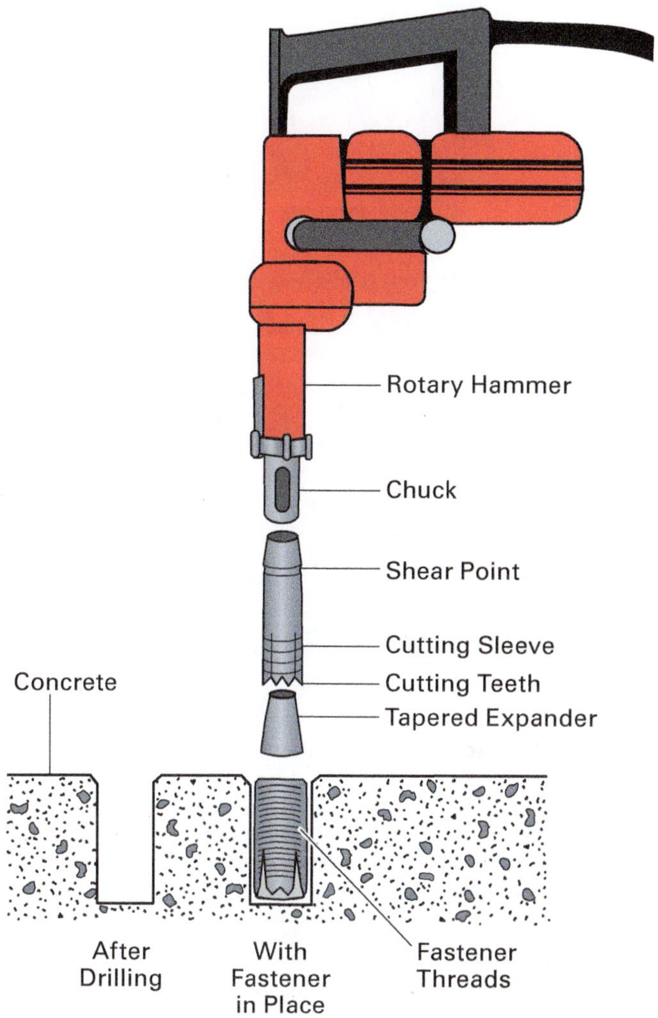

Figure 34 Self-drilling anchor.

Epoxies can hold a fastener in concrete. Simply drill a hole of the correct diameter and depth. Clean it thoroughly. Confirm that the epoxy system is mixing the liquids correctly. A uniform color indicates a good mix. Inject the mixed epoxy into the hole to the proper depth. Insert the fastener, twisting it slowly to eliminate bubbles or voids. Confirm that the anchor is level. Once the epoxy has cured, the fastener is ready for use (*Figure 36*).

Using Epoxy

Once mixed, epoxy has a limited working time. When manually mixing it, never mix more than you can use within the working time. Always allow epoxy to cure the full time specified by the manufacturer. Most epoxies look cured before they reach full strength. Loading them too soon can cause them to fail.

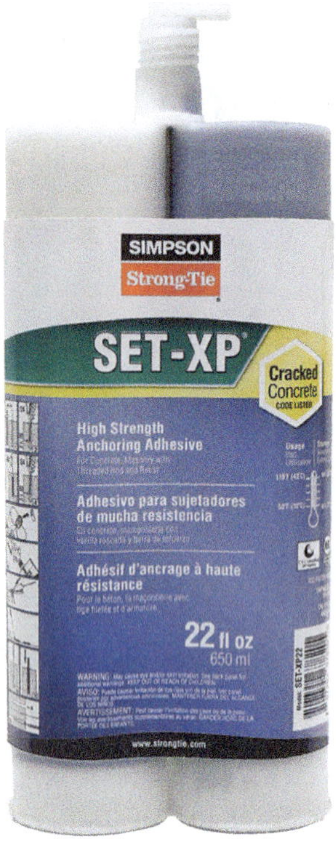

Figure 35 Two-part anchor epoxy.

Epoxy works with hollow concrete materials too but requires a different technique. A hollow metal or plastic mesh tube called an *anchor screen* holds the epoxy in the hole. After filling the screen with epoxy, insert it into the hole. Slowly insert the fastener, twisting it to eliminate voids and bubbles. Once the epoxy cures, the fastener is ready.

2.3.7 Using Anchors Successfully

When correctly selected and installed, anchors do an excellent job. The best path to success is following the manufacturer's guidelines. Confirm that the anchor itself is strong enough for the application. Given enough stress, most anchors will pull out of their holes. Anchors can also snap sideways (shear off).

Always install anchors in holes of the correct diameter and depth. Confirm that the material itself is appropriate for the anchor. Some materials are weaker than others and cannot hold certain anchor types. The material must be thick enough too.

Proper drilling technique is crucial. A hammer drill and appropriate masonry bit is the correct tool. If the manufacturer supplies a bit with the anchors, use it. Check the drill and bit before starting work. Replace damaged tools immediately.

Hammer Drills

A hammer drill (or rotary hammer) drills holes in masonry and concrete. It both spins and pounds as it operates. This combined motion goes through these materials far faster than an ordinary drill.

Figure Credit: Image property of Stanley Black & Decker. Used with permission.

Make sure that the surface is smooth and provides solid footing. Uneven footing could cause the installed equipment to twist, warp, not tighten properly, or vibrate. Set the drill depth gauge to the required hole depth. When drilling, let the drill do the work. Avoid pushing down too hard. For large holes, start with a smaller bit and then change to a larger one. Follow the manufacturer's instructions for installing the fastener (*Figure 37*).

WARNING!

Drilling concrete produces dust, flying particles, and noise. Always wear eye and hearing protection. Respiratory protection may be necessary too. Pay attention to nearby workers and verify that they have suitable protection as well.

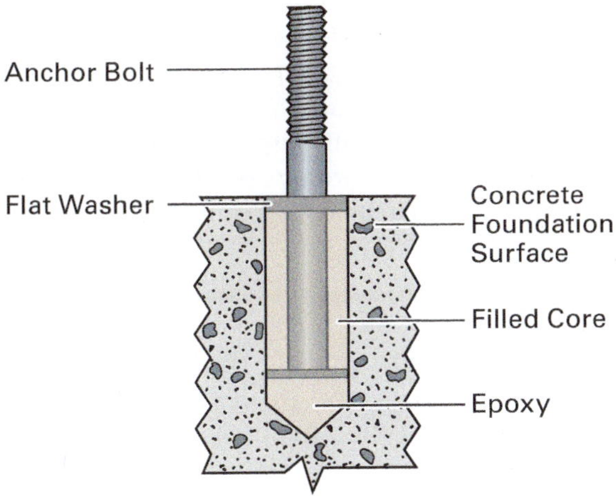

Figure 36 Fastener anchored in epoxy.

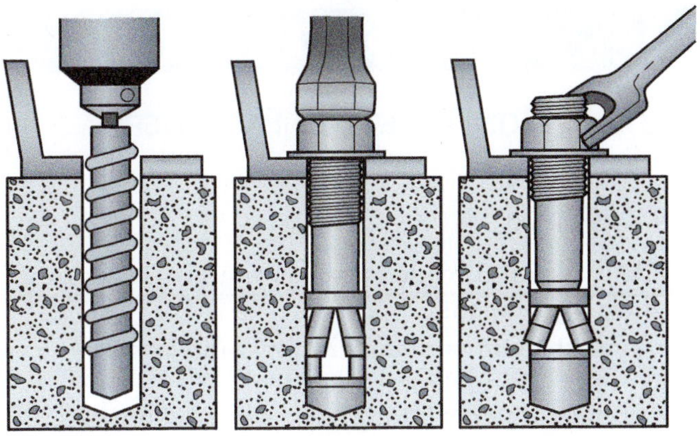

Figure 37 Installing an anchor bolt in concrete.

2.0.0 Section Review

1. US standard fastener identifications include a threads-per-inch value, while metric fastener identifications use _____.
   a. thread class
   b. thread pitch
   c. threads-per-centimeter
   d. thread handedness

2. What can you apply to a threaded fastener to keep it from vibrating loose?
   a. Graphite or molybdenum
   b. Anti-seize compound
   c. Thread lubricant
   d. Thread-locking fluid

3. One-step concrete anchors work by _____.
   a. exploding inside the hole
   b. hardening inside the hole
   c. expanding inside the hole
   d. melting inside the hole

Section Three

3.0.0 Non-Threaded Fasteners

Objective

Identify non-threaded fasteners and summarize their applications.

a. List common retainers, pins, and keys, describing their applications.
b. Explain how blind rivets work and how to install them.

Performance Tasks

4. Install instructor-selected non-threaded fasteners.
5. Install a snap ring.

Trade Term

Blind rivet: A fastener that installs from one side of the work surface and requires a special tool to expand and tighten it.

Non-threaded fasteners tend to be more specialized than threaded fasteners. Many keep parts from moving or coming apart. Others keep components aligned or transfer motion between them. This section explores common examples, briefly examining their applications.

3.1.0 Retainers, Pins, and Keys

This fastener category prevents components from coming apart. Some also transfer motion between parts. In each case, the fastener either grips the component or simply sits in a hole or slot.

3.1.1 Retainers and Lock Rings

Retainers and lock rings (*Figure 38*) prevent components from coming apart. Some snap into place without requiring special tools. Retainers (also called *snap rings* or *circlips*) are springy. They sit in a groove machined into the part. Special pliers stretch them during installation. Once in place, they snap back and provide a physical barrier against motion. Snap rings often hold gears and pulleys in place on shafts.

WARNING!

Large snap rings are powerful springs that store a lot of energy. If they get away from you during installation or removal, they can shoot across the room with surprising force. Always wear appropriate protective equipment and take great care when installing or removing them.

3.1.2 Pins

Pins (*Figure 39*) align mating parts, secure slotted nuts, and hold gears and pulleys on shafts. They come in several styles, each of which performs specific tasks.

Dowel pins are simple cylinders, often with beveled edges. They help keep components aligned or prevent them from moving.

Taper pins are conical. They fasten gears, pulleys, and collars on shafts. Their shape naturally grips when driven into a matching hole. Taper pins also help align motors and pumps on their mounting bases.

Spring pins (*roll pins*) fasten gears, pulleys, and collars on shafts (*Figure 40*). They can function as hinge pins too. C-shaped, springy, and hollow, the pin compresses when installed. Once in place, it expands, gripping tightly.

Clevis pins hold two pieces of equipment together but allow movement as well. They resemble an unthreaded bolt with a hole drilled through one end. Usually, a cotter pin fits through the hole, keeping the clevis pin from coming out. Clevis pins form part of simple hitch mechanisms used for towing. They also work with certain kinds of shackles.

Cotter pins come in several styles (*Figure 41*). They lock parts together but still allow movement. In general, they are best for low-stress applications. The standard version fits through a hole or slot in the part. Once installed, bend the ends apart to keep the pin from coming out.

Humped, cinch, and hitch-type cotter pins are self-locking. Most cotter pins should be discarded and replaced after removal. The hitch-type may be reused.

Using Clevis and Hitch Pins

Mobile hoists often use clevis and hitch pins for quick setup and disassembly. Scaffolding systems use them too for the same reason.

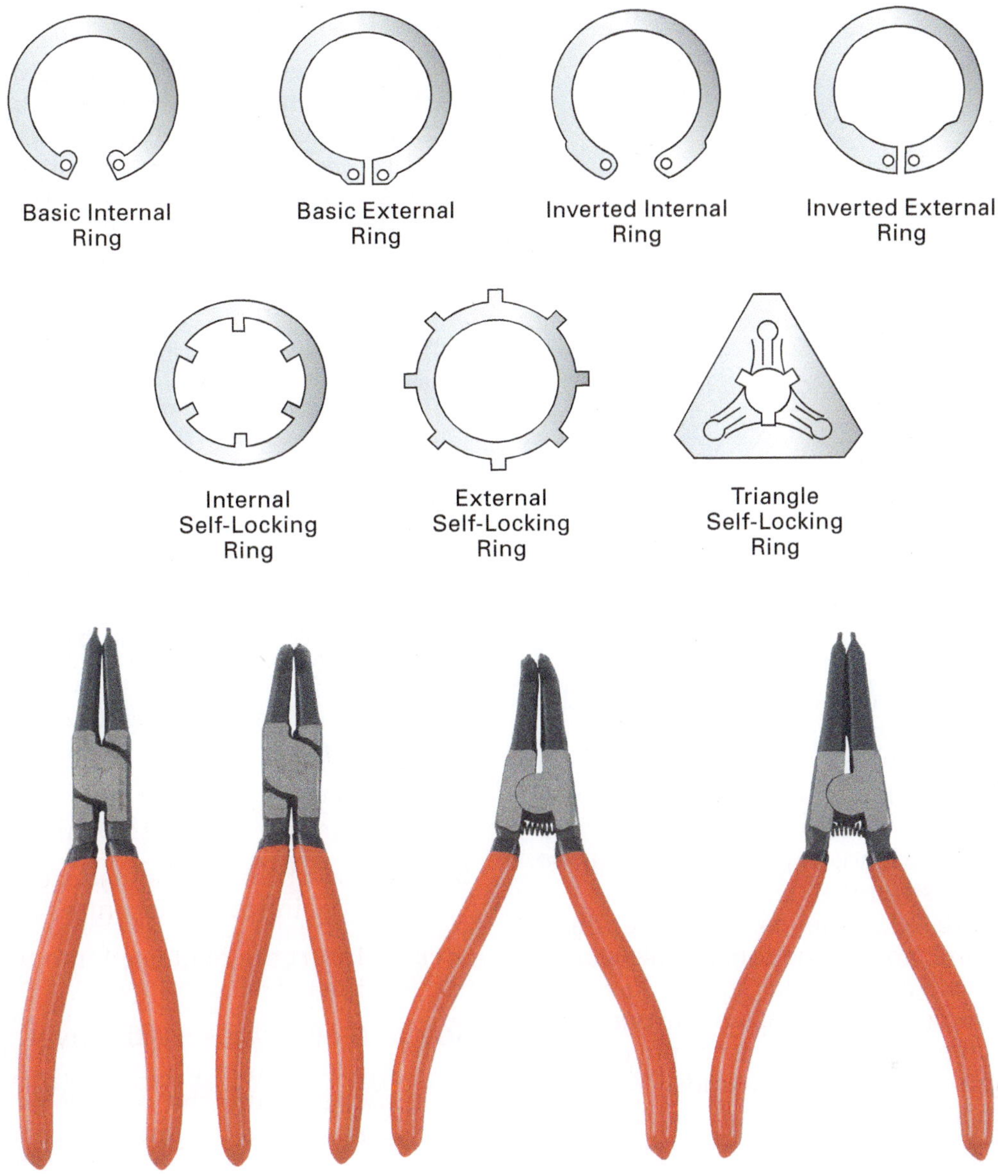

Figure 38 Retainers, lock rings, and pliers.

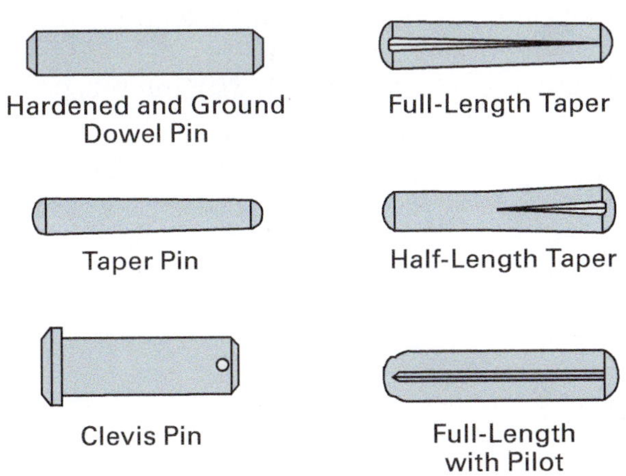

Figure 39 Common pins.

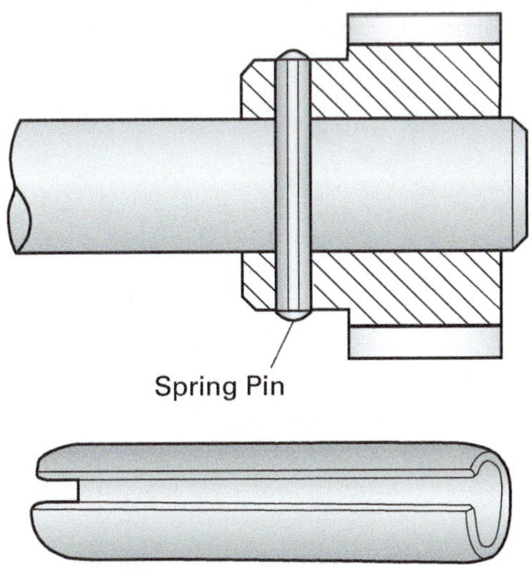

Figure 40 Spring pin.

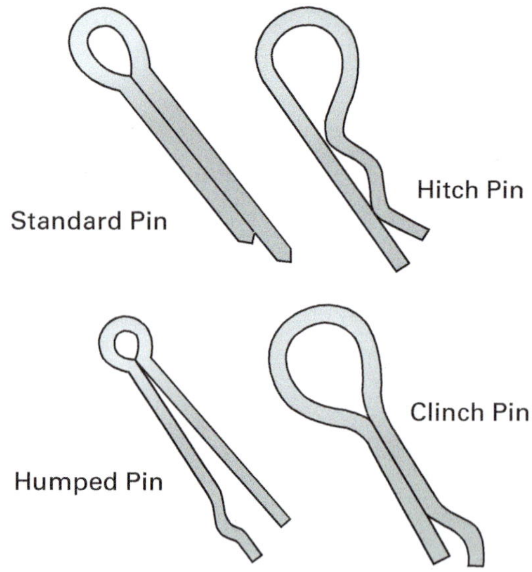

Figure 41 Cotter pins.

3.1.3 Keys

A *key* is a metal piece that fits into a matching slot called a *keyway*. Keys hold gears and pulleys on shafts. They also help link shafts together. Each keyway engages half of the key. As the parts rotate, the key holds them together. In fact, the key carries the entire load, making it the weakest link. Keys come in several styles (*Figure 42*).

Square keys are as wide as they are high. Typically, they are about one-fourth the shaft's diameter. *Gib head keys* are like square keys but have a small triangular protrusion on one end. This feature makes them easier to remove. *Pratt and Whitney keys* have rounded ends. *Woodruff keys* are half circles. One keyway must match this shape. The other is a standard square style.

3.2.0 Blind Rivets

A **blind rivet** (or *pop rivet*) fastens sheet metal parts together. Blind rivets also work with plastics, fiberglass, and belting material. Unlike standard rivets, blind rivets do not require a tool on both sides of the material during installation. A simple tool called a *rivet gun* does everything (*Figure 43*). Pop rivets come in several sizes and materials. Aluminum is the most common.

> **WARNING!**
>
> Always wear appropriate eye and face protection when riveting.

Use the following procedure to install blind rivets (*Figure 44*):

Step 1 Select a rivet of the proper diameter and length.

Step 2 Drill an appropriate hole through both parts.

Step 3 Inspect the rivet gun for any defects or damage that might make it unsafe.

Step 4 Insert the rivet's mandrel into the proper hole in the rivet gun.

Step 5 Insert the rivet itself into the hole through the parts.

Step 6 Slowly squeeze the rivet gun's handles. This action pulls the mandrel upwards, expanding the rivet. When the mandrel snaps off, the process is complete.

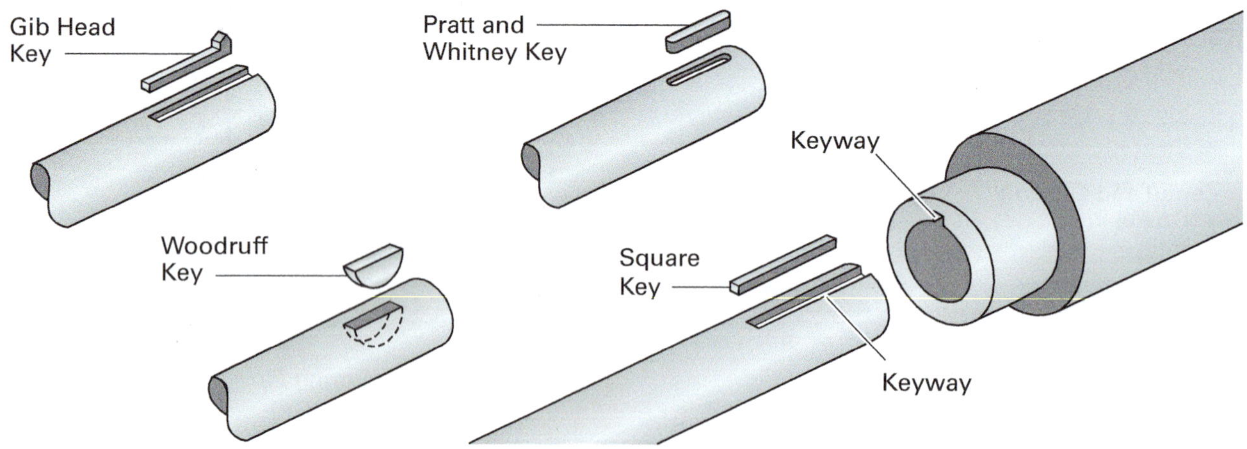

Figure 42 Keys and keyways.

Around the World

NEMA and IEC Keys

Craftworkers who use European equipment may discover that their keys look different. Motors and related equipment manufactured in the United States follow the NEMA (National Electrical Manufacturers Association) standard. Those manufactured in Europe follow the IEC (International Electrotechnical Commission) standard. NEMA equipment uses square keys. IEC equipment uses rectangular keys with rounded ends.

Some European manufacturers sell hybrid equipment in the United States. It may have metric dimensions and IEC-style flanges. But shafts may be specified in inches and have NEMA-style keyways.

Figure 43 Rivet gun.

Step 7 Inspect the rivet to make sure that it is firm and the joint is tight.

To remove blind rivets, chisel, grind off, or drill out the heads. Punch out the shaft using a hammer and punch. Perform all steps carefully to avoid damaging the parts. Always wear appropriate eye protection.

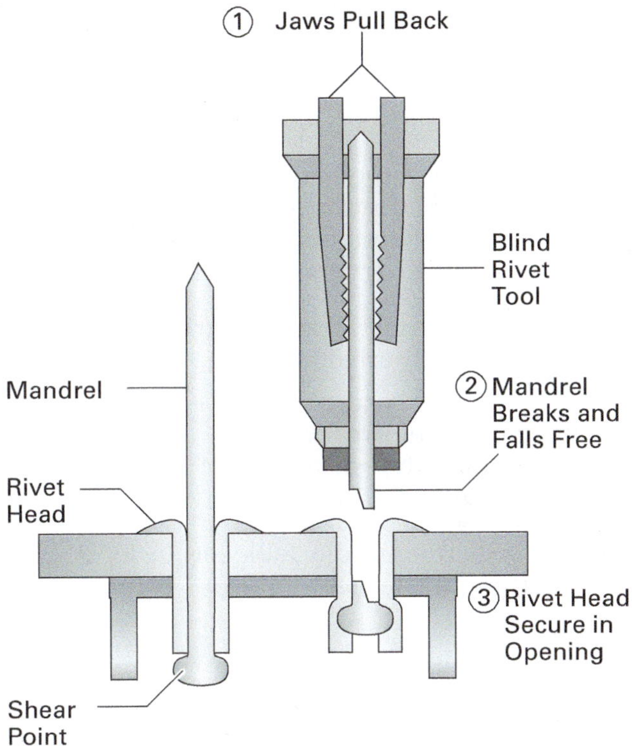

Figure 44 Blind rivet installation.

3.0.0 Section Review

1. A metal piece that sits inside a matching slot is a _____.
   a. keyway
   b. taper pin
   c. key
   d. spring pin

2. What kind of fastener requires a special gun to expand it by pulling on a mandrel?
   a. A blind rivet
   b. A thread-forming screw
   c. A toggle bolt
   d. A hammer-drive pin

Review Questions

1. What force do threaded fasteners use to hold things together?
   a. Adhesion
   b. Tensile strength
   c. Tension
   d. Elastic limit

2. What can happen if you over-torque a bolt?
   a. It can become hardened.
   b. It could permanently stretch.
   c. It resists vibration much better.
   d. It tends to dry torque itself.

3. A bolt suddenly locks up during tightening. It has just experienced _____.
   a. break-away torque
   b. elastic deformation
   c. run-down resistance
   d. set/seizure

4. Non-electronic, adjustable torque tools signal reaching the target torque by _____.
   a. clicking
   b. beeping
   c. vibrating
   d. releasing

5. A Class 3A / Class 3B fastener combination could come apart easily.
   a. True
   b. False

6. A bolt with "LH" in its identification loosens clockwise.
   a. True
   b. False

7. Which fastener would give the *most* finished appearance?
   a. Machine bolt
   b. Cap screw
   c. Machine screw
   d. Stud bolt

8. Which nut type *must* be replaced if it is removed?
   a. Wing
   b. Jam
   c. Cage
   d. Self-locking

9. To create threads in heavier pieces of metal without tapping, use a _____.
   a. lag screw
   b. thread-cutting screw
   c. coated screw
   d. thread-forming screw

10. A threaded fastener equipped with a nylon patch is good for applications with a lot of _____.
    a. torque
    b. heat
    c. vibration
    d. galling

11. A concrete fastener that requires a special tool and a heavy hammer to install is the _____.
    a. hammer-drill anchor
    b. powder-drive fastener
    c. hammer-drive fastener
    d. one-step anchor

12. Special pliers install a _____.
    a. snap ring
    b. blind rivet
    c. Woodruff key
    d. taper pin

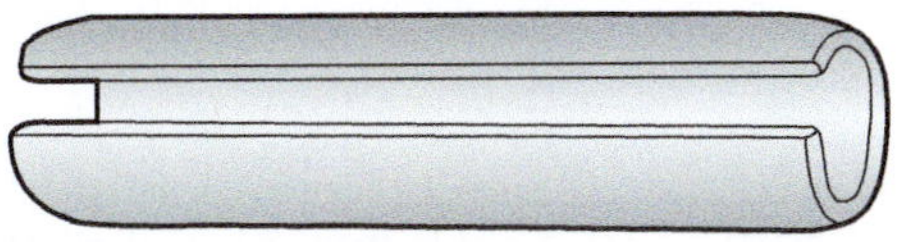

Figure RQ01

13. Which kind of pin is pictured in *Figure RQ01*?
    a. Spring pin
    b. Dowel pin
    c. Taper pin
    d. Cotter pin

14. Which type of cotter pin is self-locking?
    a. Gib head
    b. Standard
    c. Cinch
    d. Clevis

15. Blind rivets join parts accessible from one side only.
    a. True
    b. False

Trade Terms Quiz

Fill in the blank with the correct term learned from your study of this module.

1. A threaded fastener's _____________ determines the direction that it tightens and loosens.
2. _____________ is a twisting force that causes rotation.
3. A threaded fastener's _____________ indicates its quality and strength.
4. A fastener's _____________ value indicates how many threads it has in a one-inch section.
5. When objects rub together, they generate a force called _____________.
6. Threaded fastener manufacturers follow a set of specifications called a(n) _____________.
7. The _____________ is a large torque unit in the US standard system.
8. A fastener's _____________ identifies the coarseness or fineness of its threads.
9. Stretching a material beyond its _____________ permanently deforms it.
10. The distance between the tops of adjacent threads is a fastener's _____________.
11. A fastener head that is flush with or below the surface is _____________.
12. Applying _____________ to a threaded fastener helps keep it from vibrating loose.
13. A(n) _____________ is large enough for everything but a fastener's head to pass through.
14. The _____________ indicates how loosely or tightly the threads of matching fasteners fit.
15. Loosening a threaded fastener usually requires applying a large force called _____________.
16. Threaded fasteners hold things together through a pulling force called _____________.
17. A(n) _____________ holds sheet metal together after a special gun expands the fastener.
18. Stretching a material beyond its _____________ breaks it.
19. The _____________ is a small US standard torque unit.
20. A threaded fastener that locks up while being tightened has experienced _____________.
21. The _____________ is the metric torque unit.
22. Adding extra hardware to a threaded fastener can increase the _____________ felt when tightening it.
23. A fastener's _____________ is its typical diameter.

Trade Terms

Blind rivet	Friction	Run-down resistance	Thread series
Break-away torque	Grade	Set/seizure	Thread standard
Clearance hole	Handedness	Tensile strength	Thread-locking fluid
Countersunk	Inch-pound (in-lb)	Tension	Threads-per-inch (tpi)
Elastic limit	Newton-meter (Nm)	Thread class	Torque
Foot-pound (ft-lb)	Nominal size	Thread pitch	

Cornerstone of Craftsmanship

James Maloney
Resource Manager
FLSmidth, Inc.

How did you choose a career in the industry?

I'm a third-generation millwright, and it's kind of a strange story. I didn't choose this path right away. I served four years in the US Navy during Desert Storm, and then I started working at a computer company called Gateway 2000 in '92 as a computer service technician. I quickly worked my way up to IT, but I also did millwright work in my spare time during vacations and leaves of absence. Ultimately, I chose the millwright field because I enjoyed the constant mechanical challenge of the equipment we work on.

Who inspired you to enter the industry?

My father and grandfathers inspired me to enter the industry. Three of my five sons are also in the business, and I've trained my little brother, along with many other apprentices.

What types of training have you been through?

I learned from a master mechanic in the late 80s, and my technical school training came from NCCER as a Certified Industrial Millwright. I've also been OSHA/MSHA certified and laser-alignment certified with Ludeca since 2006.

How important is education and training in construction?

It's very important. Most industries are just about getting the job done, and not a lot of time is devoted to training. But I believe it should be. This can create loyal employees who want to stick around if we invest in them.

How important are NCCER credentials to your career?

They're very important, and they definitely can increase your pay with most companies. Knowledge and respect in the industry are your resume.

How has training/construction impacted your life?

It's given me the opportunity to advance in my career. As a millwright, you're exposed to a lot of other industries, and it broadens your ideas and possibilities of pursuing different avenues. Cross-training is huge, and a true millwright should be able to do just about any other craft.

What kinds of work have you done in your career?

I worked on my tools for many years and then as a foreman, supervisor, rotating equipment specialist, mechanical turbine specialist, millwright superintendent, project manager, and the list continues! Currently, I'm the Service Execution Department Resource Manager at FLSmidth, Inc., the largest mining equipment manufacturer in the world.

Tell us about your current job.

I manage a department of millwrights and welders who specialize in repairing mining equipment. This includes developing training programs, collaborating with stakeholders, and communicating daily with colleagues to ensure that everything runs smoothly.

What do you enjoy most about your job?

The constant challenge and training the next generation of millwrights.

What factors have contributed most to your success?

Ambition and the willingness to learn. My wife says I'm a sponge for information. You've got to put yourself out there and in positions where you may not be an expert, but you can learn. You must earn the respect of the people you work with and they'll get you the right training to excel in the industry.

Would you suggest construction as a career to others? Why?

Yes, absolutely. There's a lot of opportunity, and you can always find work when you're a journeyman. There are a million different jobs available.

What advice would you give to those new to the field?

Keep your mouth shut and your ears open. No, seriously. Learn all you can, and always pursue the next challenge when it's offered. Ask questions and make sure you understand what they're showing you in the field.

Interesting career-related fact or accomplishment?

The cool thing is, if you're a true millwright, you have the fundamentals to do most other crafts well enough to gain work anywhere.

How do you define craftsmanship?

It's having the knowledge, training, and experience to truly repair, build, or modify equipment in the industry while leaving a quality job behind you. Your reputation is everything in this industry. Never forget that.

Trade Terms Introduced in This Module

Blind rivet: A fastener that installs from one side of the work surface and requires a special tool to expand and tighten it.

Break-away torque: The torque required to loosen a fastener.

Clearance hole: A fastener hole large enough for everything but the fastener's head to pass through.

Countersunk: A fastener installation method in which the head is flush with or below the surface.

Elastic limit: The point after which stretching a material permanently deforms it.

Foot-pound (ft-lb): A US standard torque unit equal to the force of one pound applied to a lever one foot long. One foot-pound equals 12 inch-pounds. Also called the *pound-foot (lb-ft).*

Friction: A motion-resisting force produced when two surfaces rub together.

Grade: A fastener-ranking system that identifies quality and strength.

Handedness: The direction that the threads wrap around a fastener, determining how it tightens or loosens.

Inch-pound (in-lb): A small US standard torque unit equal to the force of one pound applied to a lever one inch long. One inch-pound equals $\frac{1}{12}$ foot-pound. Also called the *pound-inch (lb-in).*

Newton-meter (Nm): A metric torque unit equal to the force of one newton applied to a lever one meter long.

Nominal size: The typical diameter of a bolt or screw.

Run-down resistance: Additional torque required to overcome the resistance created by extra hardware, such as washers and locknuts.

Set/seizure: Fastener lockup that occurs when approaching the target torque.

Tensile strength: The point at which stretching a material breaks it.

Tension: A pulling force, such as that produced by some springs.

Thread class: A code indicating the amount of looseness between the threads of matching bolts and nuts.

Thread pitch: The distance from the crest (top) of one thread to the next.

Thread series: Combinations of fastener diameter and threads-per-inch associated with specific thread diameters.

Thread standard: An established set of specifications for machining threads.

Thread-locking fluid: A hardening liquid placed on threaded fasteners to make them more resistant to vibration loosening.

Threads-per-inch (tpi): The number of threads in one inch of fastener length.

Torque: A twisting force that produces rotary motion.

Additional Resources

This module presents thorough resources for task training. The following reference material is suggested for further study.

Concrete Fastening Systems. **www.confast.com**

Gesipa Fasteners USA Inc. **www.gesipausa.com**

Machinery's Handbook. Erik Oberg; Franklin Jones; Holbrook Horton; Henry Ryffel; Christopher McCauley. Latest Edition. South Norwalk, CT: Industrial Press.

Figure Credits

iStock@Three Spots, Module Opener

Image property of Stanley Black & Decker. Used with permission., Figures 6–7, 38, 43

Courtesy of The Eastwood Company, Figure 8

Courtesy of HYTORC/UNEX Corp, Figure 9

Courtesy of ITW Ramset, Figure 31

Simpson Strong-Tie Company Inc., Figure 35

Section Review Calculations

Section 1.0.0

Question 3

Torque (in-lb) = torque (Nm) × 8.851
Torque = 36 × 8.851
Torque = **319 in-lb** (rounded to the nearest 1 in-lb)

Section Review Answer Key

Section 1.0.0

Answer	Section Reference	Objective
1. c	1.1.2	1a
2. a	1.2.1	1b
3. d	1.2.2; *Table 1*	1b

Section 2.0.0

Answer	Section Reference	Objective
1. b	2.1.4	2a
2. d	2.2.4	2b
3. c	2.3.3	2c

Section 3.0.0

Answer	Section Reference	Objective
1. c	3.1.3	3a
2. a	3.2.0	3b

This page is intentionally left blank.

NCCER CURRICULA — USER UPDATE

NCCER makes every effort to keep its textbooks up-to-date and free of technical errors. We appreciate your help in this process. If you find an error, a typographical mistake, or an inaccuracy in NCCER's curricula, please fill out this form (or a photocopy), or complete the online form at **www.nccer.org/olf**. Be sure to include the exact module ID number, page number, a detailed description, and your recommended correction. Your input will be brought to the attention of the Authoring Team. Thank you for your assistance.

Instructors – If you have an idea for improving this textbook, or have found that additional materials were necessary to teach this module effectively, please let us know so that we may present your suggestions to the Authoring Team.

NCCER Product Development and Revision
13614 Progress Blvd., Alachua, FL 32615
Email: curriculum@nccer.org
Online: www.nccer.org/olf

❑ Trainee Guide ❑ Lesson Plans ❑ Exam ❑ PowerPoints Other ____________

Craft / Level: Copyright Date:

Module ID Number / Title:

Section Number(s):

Description:

Recommended Correction:

Your Name:

Address:

Email: Phone:

This page is intentionally left blank.

Basic Layout

OVERVIEW

Whenever new equipment arrives at a facility, millwrights and industrial mechanics help position it. Some machines connect to each other, so millwrights must align them carefully to prevent damage. To do these tasks, they interpret drawings, establishing locations based on key reference points.

Anchors attach most equipment to the floor or a foundation. Millwrights determine where to install these anchors. Positions must be precise, since even small errors can cause problems. Imagine lowering a large machine onto multiple mounting bolts, only to find one out of place.

The millwright and industrial mechanic crafts use many specialized tools for layout work. This module introduces them and explains the proper way to use them.

Module 15104

Trainees with successful module completions may be eligible for credentialing through the NCCER Registry. To learn more, go to **www.nccer.org** or contact us at 1.888.622.3720. Our website, **www.nccer.org**, has information on the latest product releases and training.

Your feedback is welcome. You may email your comments to **curriculum@nccer.org**, send general comments and inquiries to **info@nccer.org**, or fill in the User Update form at the back of this module.

This information is general in nature and intended for training purposes only. Actual performance of activities described in this manual requires compliance with all applicable operating, service, maintenance, and safety procedures under the direction of qualified personnel. References in this manual to patented or proprietary devices do not constitute a recommendation of their use.

15104 V4.0

15104

BASIC LAYOUT

Objectives

Successful completion of this module prepares you to do the following:

1. Identify basic layout tools and use them to construct layout elements.
   a. Name and describe the function of common layout tools.
   b. Construct the preliminary elements for an equipment layout.
   c. Use layout tools to perform layout tasks.

Performance Tasks

Under supervision, you should be able to do the following:

1. Determine column lines from a building drawing.
2. Use the 3-4-5 method to lay out perpendicular baselines.
3. Use the arc method to lay out perpendicular baselines.
4. Lay out equipment centerlines.
5. Use a framing square to scribe lines perpendicular to a baseline.
6. Use a combination square to scribe lines parallel to an edge.
7. Use a combination square to scribe parallel 45-degree lines.
8. Use a protractor to scribe angled lines.
9. Use dividers to scribe lines perpendicular to a baseline.
10. Use dividers and a reference point to scribe lines perpendicular to a baseline.

Trade Terms

Baselines
Benchmarks
Bisect
Intersection
Perpendicular
Ratio
Tangent

Industry Recognized Credentials

If you are training through an NCCER-accredited sponsor, you may be eligible for credentials from NCCER's Registry. The ID number for this module is 15104. Note that this module may have been used in other NCCER curricula and may apply to other level completions. Contact NCCER's Registry at 1.888.622.3720 or go to **www.nccer.org** for more information.

You can also show off your industry-recognized credentials online with NCCER's digital badges. Transform your knowledge, skills, and achievements into badges that you can share across social media platforms, send to your network, and add to your resume. For more information, visit **www.nccer.org**.

NOTE

This module uses US standard and metric units in up to three different ways. This note explains how to interpret them.

Exact Conversions
Exact metric equivalents of US standard units appear in parentheses after the US standard unit. For example: "Measure 18" (45.7 cm) from the end and make a mark."

Approximate Conversions
In some cases, exact metric conversions would be inappropriate or even absurd. In these situations, an approximate metric value appears in parentheses with the ~ symbol in front of the number. For example: "Grip the tool about 3" (~8 cm) from the end."

Parallel but not Equal Values
Certain scenarios include US standard and metric values that are parallel but not equal. In these situations, a slash (/) surrounded by spaces separates the US standard and metric values. For example: "Place the point on the steel rule's 1" / 1 cm mark."

Contents

Figures

Figures (continued)

Section One

1.0.0 Basic Layout Tools and Layouts

Objective

Identify basic layout tools and use them to construct layout elements.

a. Name and describe the function of common layout tools.
b. Construct the preliminary elements for an equipment layout.
c. Use layout tools to perform layout tasks.

Performance Tasks

1. Determine column lines from a building drawing.
2. Use the 3-4-5 method to lay out perpendicular baselines.
3. Use the arc method to lay out perpendicular baselines.
4. Lay out equipment centerlines.
5. Use a framing square to scribe lines perpendicular to a baseline.
6. Use a combination square to scribe lines parallel to an edge.
7. Use a combination square to scribe parallel 45-degree lines.
8. Use a protractor to scribe angled lines.
9. Use dividers to scribe lines perpendicular to a baseline.
10. Use dividers and a reference point to scribe lines perpendicular to a baseline.

Trade Terms

Baselines: Key reference lines around which all layout elements develop.

Benchmarks: Metal markers that precisely identify specific locations.

Bisect: To divide into two equal parts.

Intersection: The point where two lines or arcs cross.

Perpendicular: A line that is at a 90-degree angle to another line.

Ratio: A proportional relationship between two or more numbers.

Tangent: A line that touches a circle's circumference at exactly one point.

Almost all equipment requires mounting, either on a structure or another machine. For example, bolts hold a pump in place through holes in its baseplate. Locating those bolts requires a layout. A millwright creates a pump-mounting layout using special tools and techniques. This section introduces those tools and explains basic layout principles.

1.1.0 Layout Tools

Millwrights and industrial mechanics use specific tools to create layouts. Some take measurements or mark a surface. Others help create straight or angled lines. Learning to use the following tools is a key craft skill:

- Scribers
- Steel rules
- Straightedges
- Steel squares
- Protractors
- Combination sets
- Dividers
- Trammel points
- Prick punches
- Center punches
- Transfer punches

1.1.1 Scribers

A scriber's sharpened tip scratches (or *scribes*) thin lines on metal parts. Some have angled ends to reach inside cylindrical objects (*Figure 1*). Always keep scribers sharp so they make a clear mark.

Hold the scriber at an angle and pull it firmly towards you. In many cases, a pattern or other tool guides the scriber. To make the line easier to see, some millwrights first coat the metal surface with a blue dye. The shiny scratch shows clearly against the blue background.

1.1.2 Steel Rules

The steel rule (*Figure 2*), a flat metal strip marked in $\frac{1}{64}$" increments, makes accurate measuring simple. Some have $\frac{1}{8}$" and $\frac{1}{16}$" markings down one side, plus $\frac{1}{32}$" and $\frac{1}{64}$" markings down the other. Metric rules have 1 mm and 0.5 mm markings. Double-sided rules have inch markings on one side and millimeter markings on the other. Steel rules commonly come in lengths between 6" and 6'. Metric equivalents range from 150 mm to 2 m.

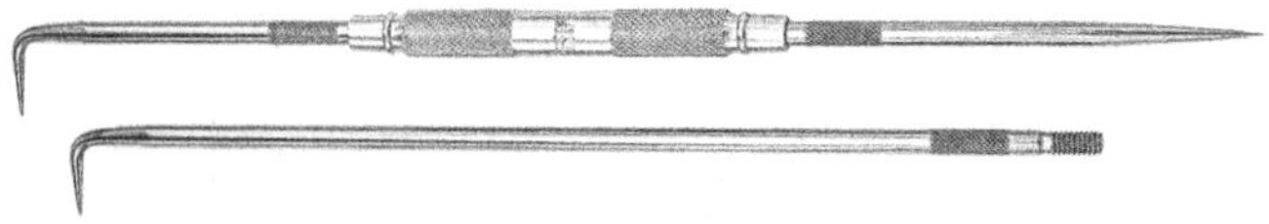

Figure 1 Scribers.

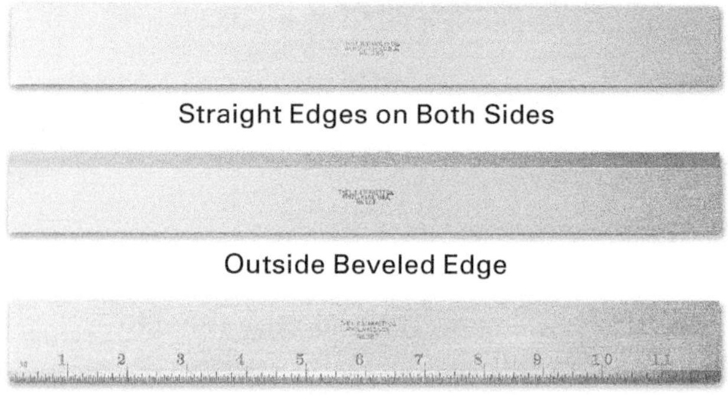

Figure 3 Straightedges.

1.1.3 Straightedges

A straightedge (*Figure 3*) looks like a steel rule but may not have any markings. It has very straight, precision-machined edges. Millwrights use them to scribe straight lines or clamp them to surfaces where they act as reference lines. You can check a surface for flatness by placing a straightedge on top of it. If light shines through, the surface isn't flat.

1.1.4 Steel Squares

The steel square (*Figure 4*) performs several tasks. It squares up stock and patterns. Millwrights lay out and check right angles with them. They also work as rules and straightedges. The framing square has a 24" blade and a 16" tongue (600 mm and 400 mm for the metric version). Machinist try squares are similar but smaller. Both types have markings that go down to $\frac{1}{64}$" and/or 0.5 mm.

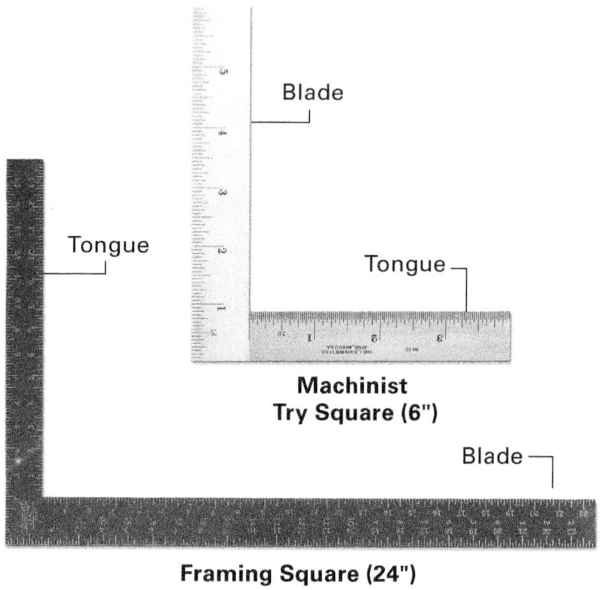

Figure 4 Steel squares.

1.1.5 Protractors

Protractors (*Figure 5*) help lay out and measure angles. Shaped like a half-circle, they have degree marks running from 0 to 180 around the curved edge. The straight bottom edge has a center mark and $\frac{1}{16}$" or 1 mm markings.

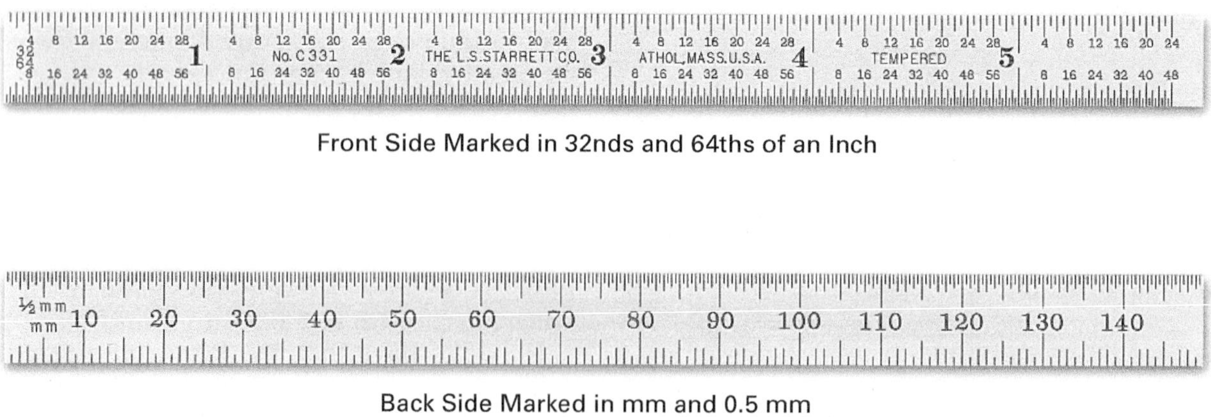

Figure 2 Steel rules.

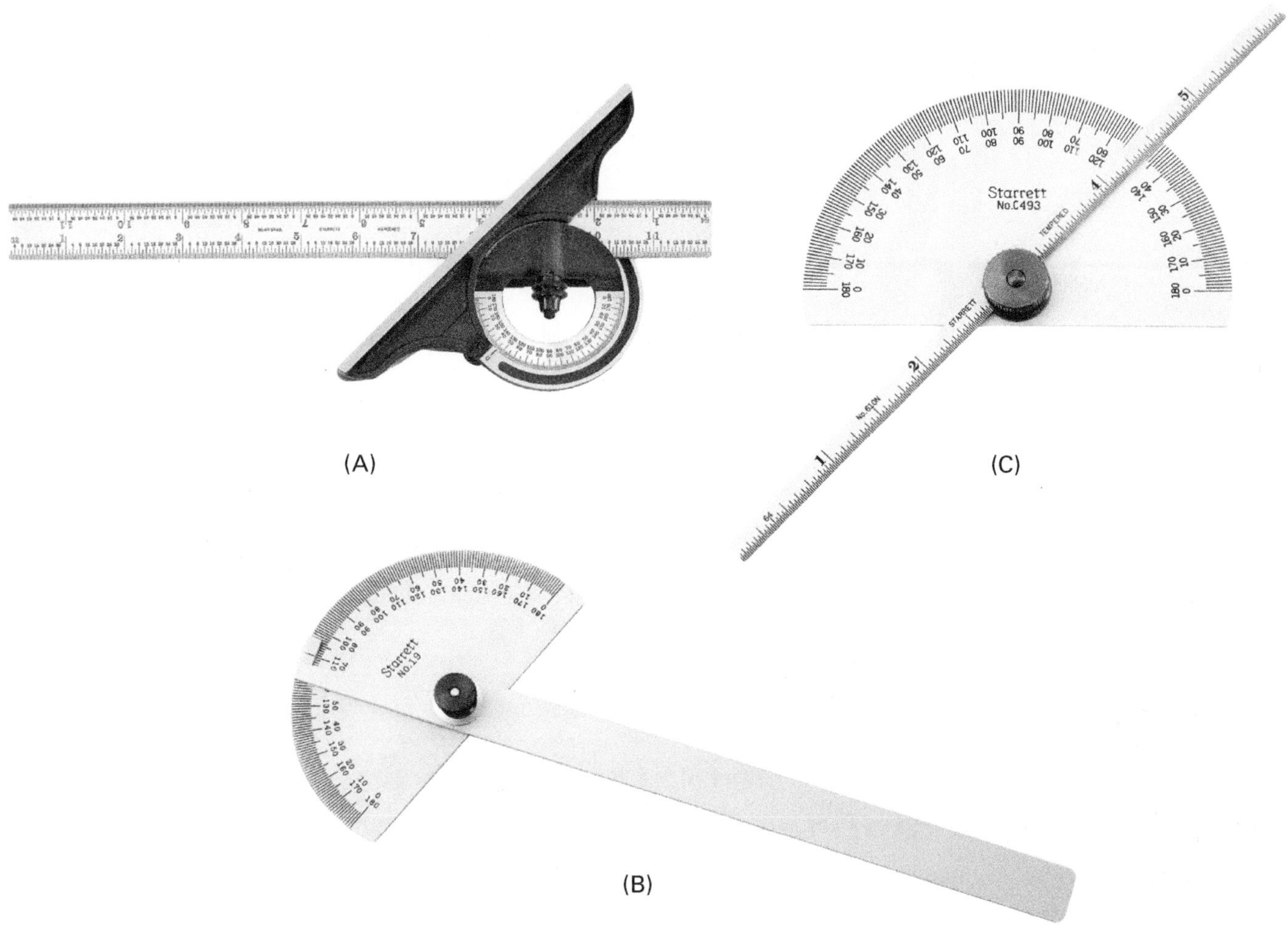

Figure 5 Protractors.

1.1.6 Combination Sets

The combination set (*Figure 6*) offers the millwright a lot of versatility. It includes a steel blade, a square head, a center head, and a protractor head. The square head and the protractor head often contain built-in spirit levels.

The tool functions as a combination square for laying out 45- and 90-degree angles, as well as parallel lines. It measures heights and angles. Millwrights and industrial mechanics also use it to locate the centers of round or square bars and as a steel rule, level, or marking gauge.

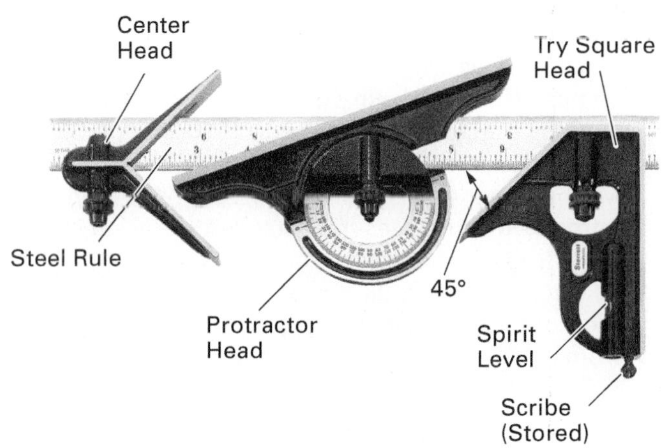

Figure 6 Combination set.

1.1.7 Dividers

Dividers (*Figure 7*), also known as *compasses*, perform several useful tasks. They transfer measurements and compare distances. Millwrights scribe circles and arcs with them. Most have a pair of steel points that must be kept sharp. Usually, a thumbscrew adjusts the distance between the legs.

Specialized dividers (*Figure 8*), called *calipers*, measure distances inside or outside objects. These may include a locking mechanism to hold the legs in a specific position.

Figure 7 Dividers.

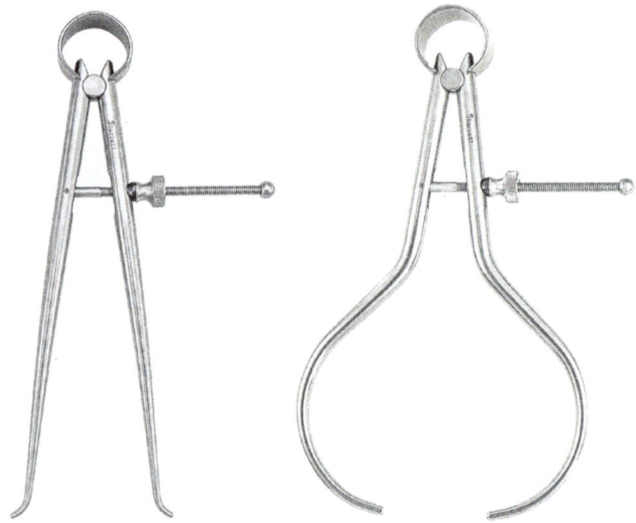

Figure 8 Inside and outside calipers.

Did You Know?

Measurement Standards

Suppose you were a carpenter who lived in a world without rulers or measuring tapes. How would you cut wood accurately? You might use your arms, hands, or fingers as a kind of body ruler. Many ancient cultures measured this way. Even now, a few of these old units survive. For example, some countries use *hands* (4") to express a horse's height.

But body part size varies from person to person, so this method isn't very good. Some ancient cultures started using standardized sticks and rods for measurements. Sargon of Akkad, who ruled in Mesopotamia from around 2334 BC to 2284 BC, introduced early measuring standards. The surveyor's chain (pictured here), developed by Edward Gunter in 1620, helped standardize and simplify land measurement. But measuring units varied widely from nation to nation, so consistent measurement remained challenging.

Figure Credit: Maurice Savage / Alamy Stock Photo

Today, the world recognizes two major standards—the inch-pound and metric systems. The United States, the United Kingdom, and Canada use the inch-pound system. It also goes by the names *Imperial, English, American,* or *US standard* system. Most of the world, however, uses the metric system. The United States has tried to "go metric" several times in its history but with only limited success.

1.1.8 Trammel Points

Trammel points (*Figure 9*) scribe large circles and arcs. They attach to a blade and slide along its length to set the spacing. If desired, a pen can replace one point. The term *beam compass* sometimes refers to a pair of trammel points attached to a blade.

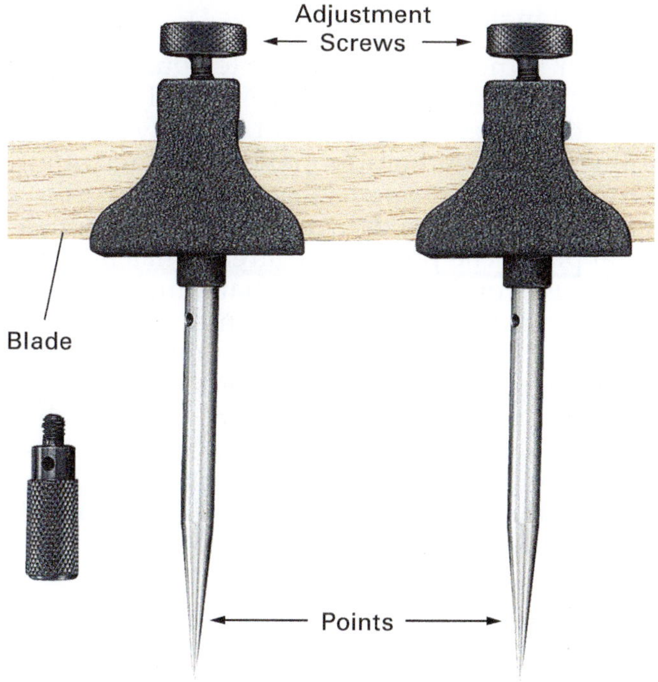

Figure 9 Trammel points.

1.1.9 Prick Punches

Prick punches (*Figure 10*) are pointed steel tools about 4" to 6" (~10 cm to 15 cm) long. The point's sides form a 60-degree angle. Striking the punch with a hammer produces an indentation. Millwrights and industrial mechanics use prick punches to mark centers and layout lines.

1.1.10 Center Punches

Center punches (*Figure 11*) look like prick punches but have blunter points. Their sides form a 90-degree angle, so they make a bigger indentation. Before drilling a hole, a millwright uses one to enlarge the mark left by a prick punch. The indentation keeps the drill bit from wandering.

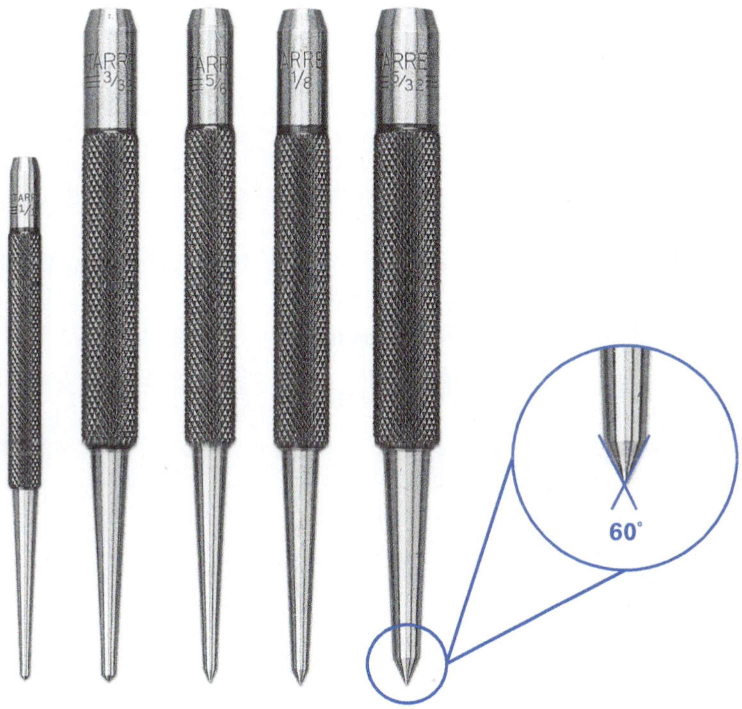

Figure 10 Prick punch set.

1.1.11 Transfer Punches

Transfer punches look like center punches but have a flat end tipped by a sharp point. They come in multiple sizes (*Figure 12*) to fit a variety of holes. Millwrights use them to mark the center of an existing hole onto material that they plan to drill.

Suppose you need to mount a pump on a metal plate. The pump base already has mounting holes. You need to drill matching holes in the plate. After placing the pump in position, you select a transfer punch that snugly fits the base's holes. Insert the punch into a hole and strike the end with a hammer. The punch's sharp point marks the metal plate precisely in the center of each pump baseplate hole. Drilling the indentations produces an identical hole pattern.

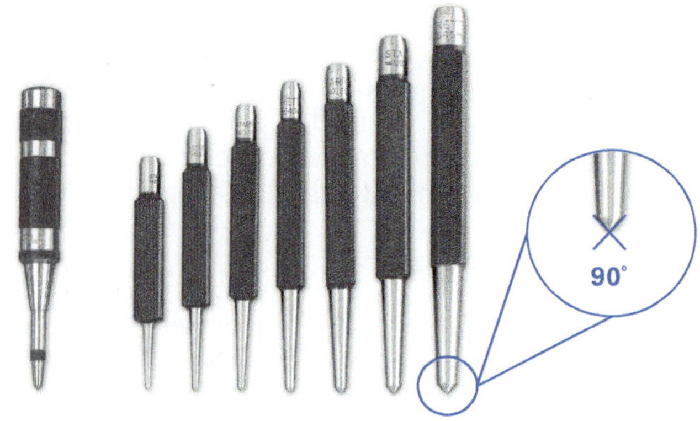

Figure 11 Center punch set.

1.2.0 Equipment Layout Preparations

Before installing new equipment, millwrights must establish its location using reference points. These may be building columns, walls, **benchmarks**, or other machines. The millwright constructs **baselines** using these references. Baselines are key reference lines for the layout. Equipment layout points depend on them. Most layouts have **perpendicular** baselines—two lines at a 90-degree angle.

Preparing to lay out equipment usually involves the following tasks:

- Locating column lines
- Laying out baselines
- Establishing equipment centerlines
- Installing permanent reference points
- Locating anchor bolts
- Establishing equipment elevation

Ball Point Attachment

Sometimes, millwrights need to place a trammel point in the middle of an existing hole. The ball point attachment solves this problem. The ball sits in the hole, and the trammel point fits into a small hole in the ball. Ball point attachments work with dividers too.

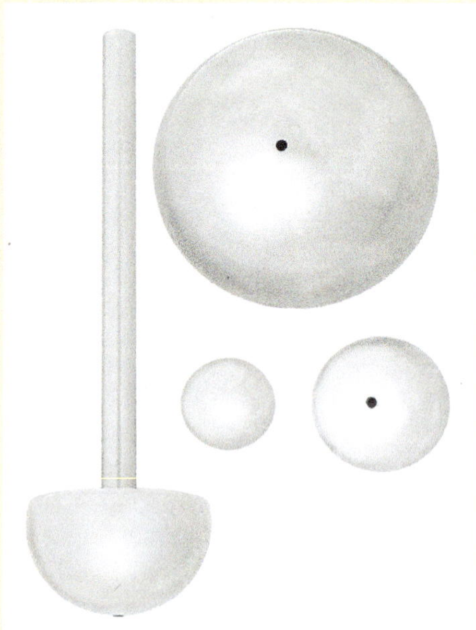

Figure Credit: The L.S. Starrett Company

Figure 12 Transfer punch set.

1.2.1 Locating Column Lines

Column lines provide handy references for millwrights to use when laying out baselines. To determine a column line, consult the building drawings.

Follow these steps to locate column lines on the drawing:

Step 1 Find the column nearest the equipment's planned location.

Step 2 Trace vertical and horizontal lines from the column out to the drawing's edges. One edge is lettered, while the other is numbered.

Step 3 Record the letter or number that each line touches.

Step 4 The column identifier is the letter plus the number. For example, *Figure 13* shows column B-3, which is at the **intersection** of lines B and 3.

1.2.2 Laying Out Baselines

Almost all other layout lines and measurements use the baselines for reference. Baselines may also establish centerlines or act as references for creating parallel lines (*Figure 14*). Laying out accurate baselines is crucial. Millwrights can choose from several methods when laying out baselines.

Column Lines Method

If available, column lines are handy references for constructing baselines. A millwright can lay out a pair of baselines from a pair of column lines.

Follow these steps to establish baselines from column lines, using *Figure 15* as an example:

Step 1 Measure to determine the exact center of columns D-2, D-3, and E-3. Mark their centers on the floor.

Step 2 Snap two chalk lines on the floor. One connects D-3 and E-3's center marks. The other connects D-3 and D-2's center marks.

> **NOTE**
> Keep the chalk line taut when snapping the lines. Baselines must be straight.

Step 3 Check the baselines to ensure that they are at right angles to each other. You can use a framing square or the 3-4-5 method (discussed later) to check the baselines.

3-4-5 Method

The second technique takes advantage of a quality of triangles. If their sides have lengths forming a 3-4-5 **ratio**, two sides will always form a right angle (*Figure 16*). Any multiple of 3-4-5, such as 6-8-10 or 9-12-15, will work.

Using Dividers and Trammels

To set a specific distance on dividers, put one leg's point on a steel rule at the 1" / 1 cm mark. Put the other point on the desired distance plus 1" / 1 cm. Set up trammel points similarly. To scribe an arc or a circle on sheet metal, mark the center with a prick punch. Place one point in the mark, hold the metal firmly in place, and scribe the line.

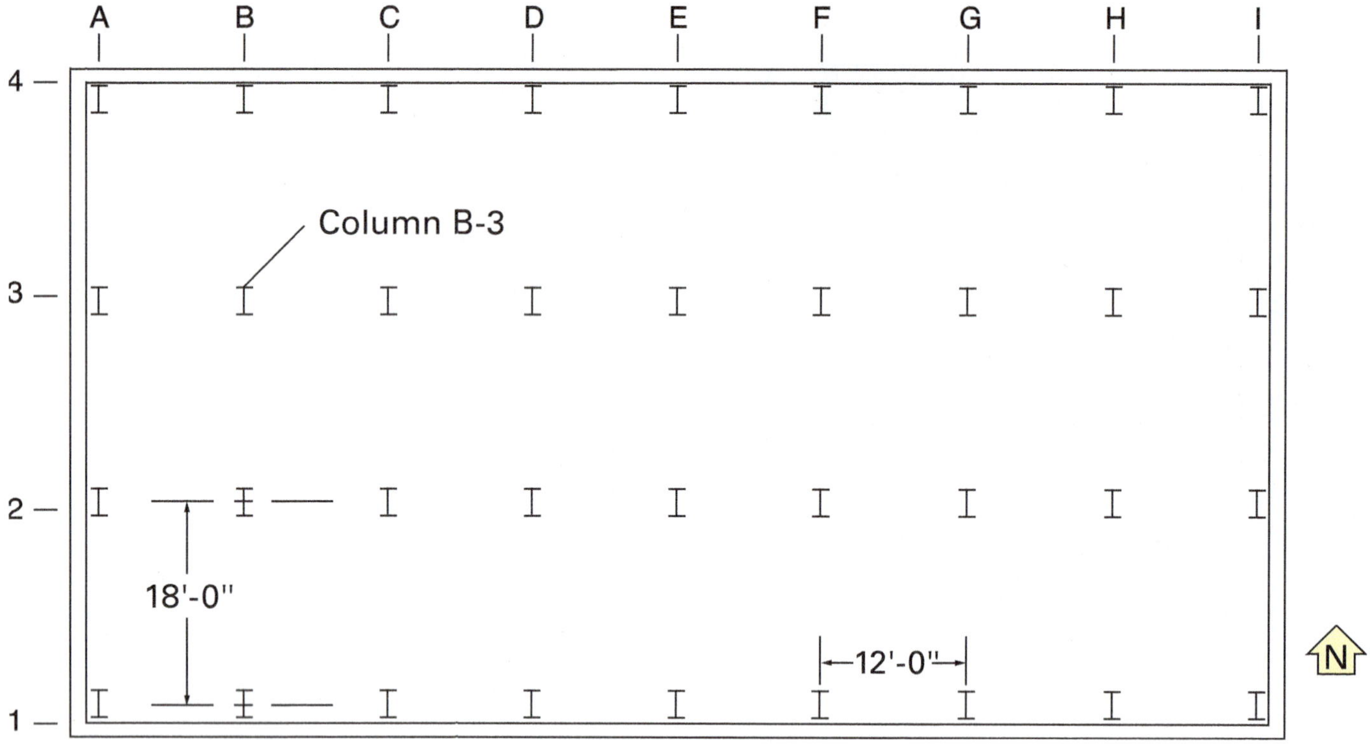

Figure 13 Column line designations.

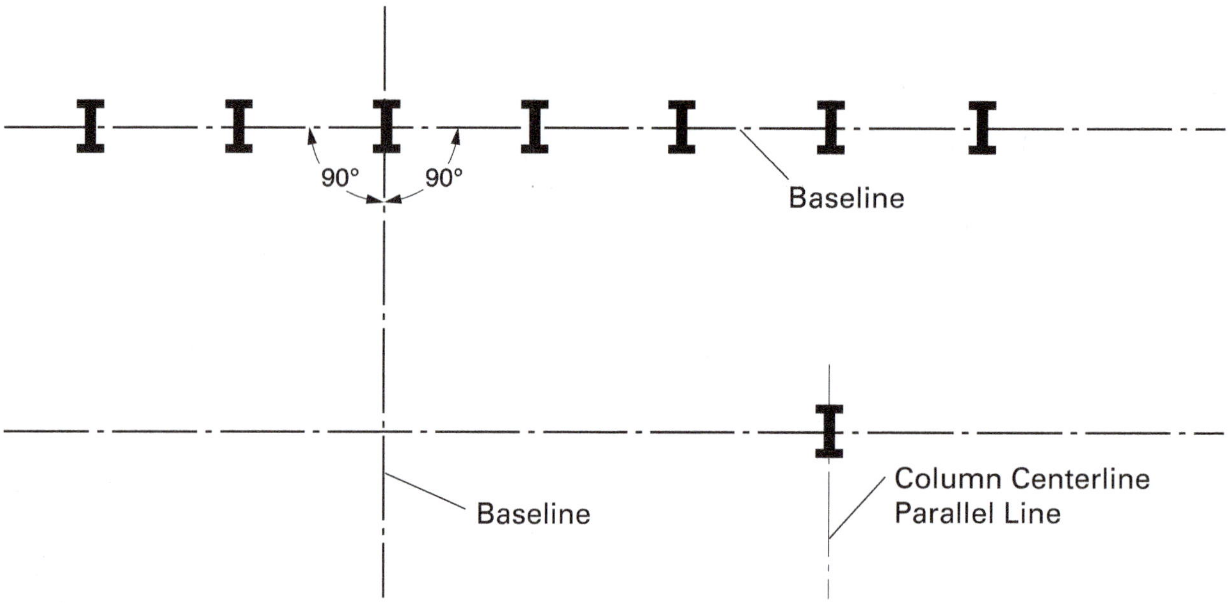

Figure 14 Baselines and parallel lines.

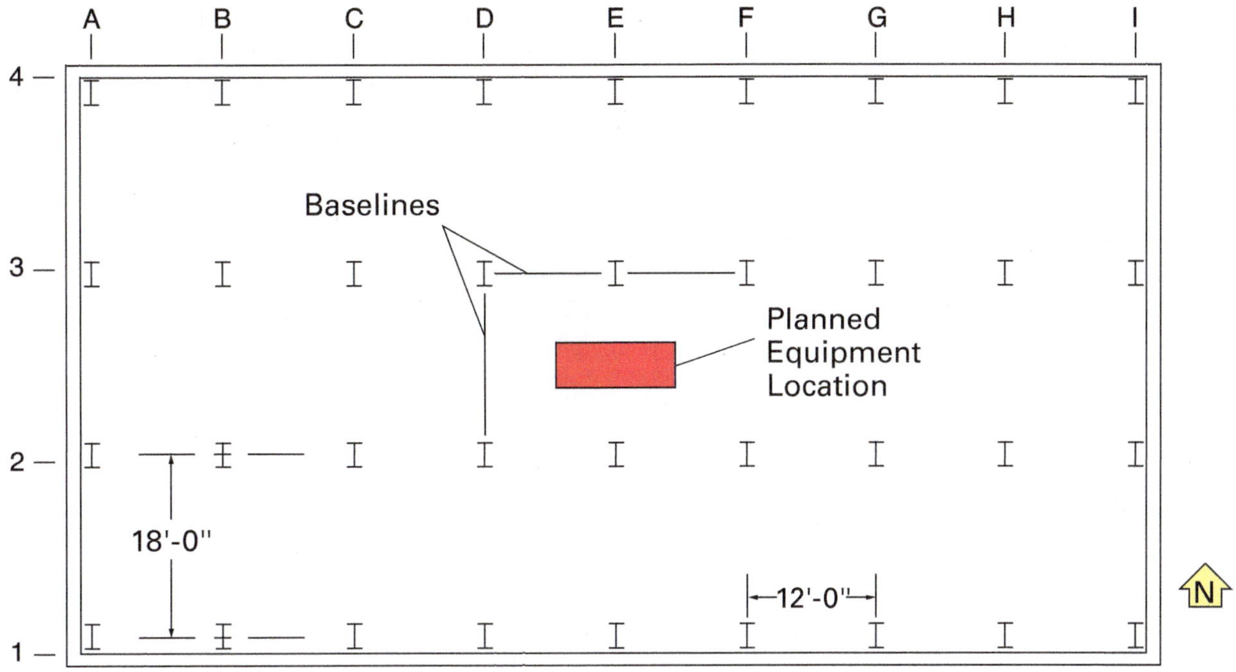

Figure 15 Establishing baselines from column lines.

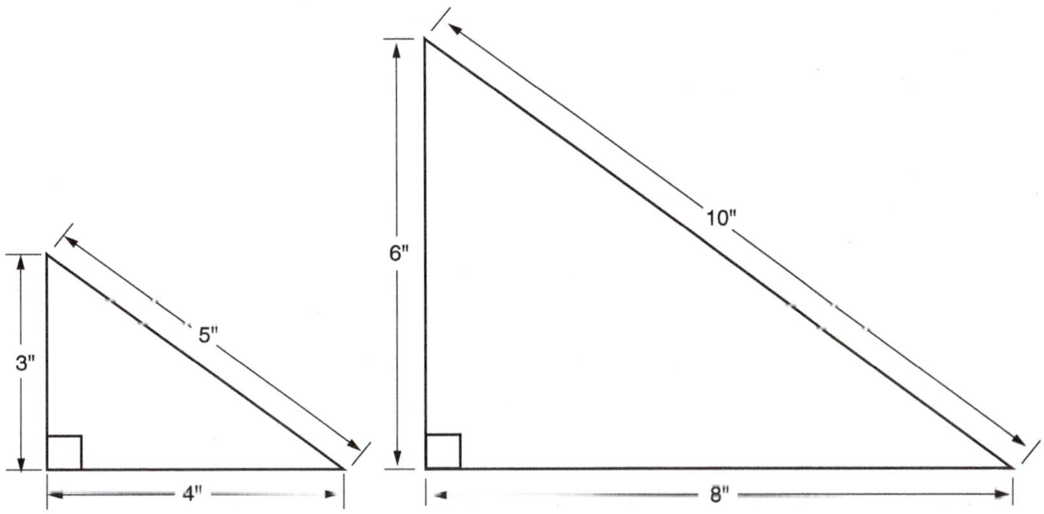

Figure 16 The 3-4-5 ratio.

Follow these steps to lay out baselines with the 3-4-5 method:

Step 1 Choose a measuring unit that when multiplied by 4 fits within the workspace. For example, if the workspace is 22' × 22', a unit of 5' is good.

> **NOTE**
> Larger units give more accurate layouts.

Step 2 Multiply the unit by 3, 4, and 5. For example, with a 5' unit, the results will be 15', 20', and 25'.

Step 3 Snap a chalk line on the floor to mark the first baseline. If available, a column line is a good reference. Mark the point on the first baseline through which the second baseline will pass.

> **NOTE**
> For the next steps, do not use a string or line that can stretch. Use a steel tape or wire.

Step 4 Measure 3 units (15' in this example) from the second baseline point at approximately 90 degrees to the first baseline. Swing arc A (*Figure 17*).

Step 5 Measure along the first baseline 4 units (20' in this example) from the second baseline point. Mark point B (*Figure 18*).

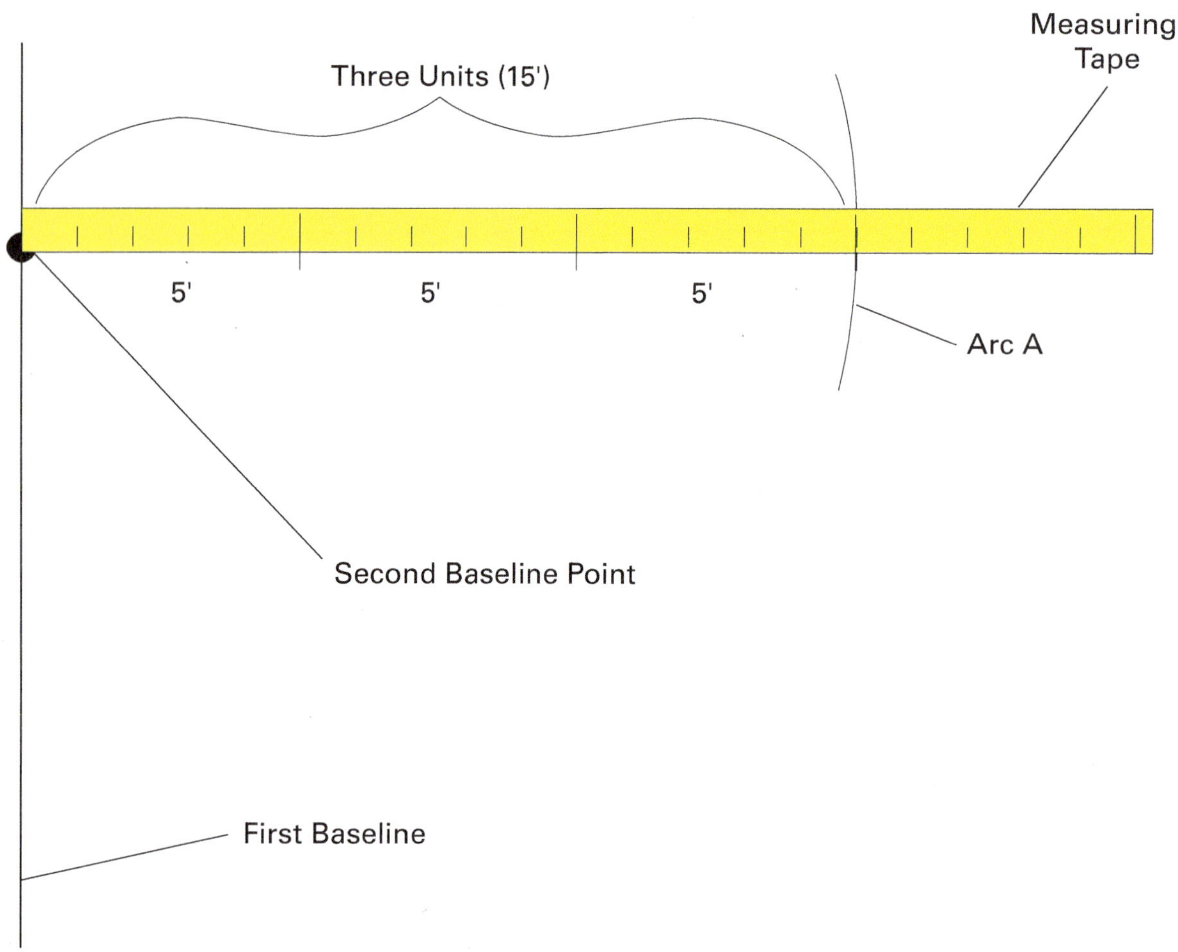

Figure 17 Swinging arc A.

Step 6 Measure 5 units (25' in this example) from point B to arc A. Mark point C on the arc (*Figure 19*).

Step 7 Draw a line from point C to the second baseline point (*Figure 20*). If you did all the steps correctly, the second baseline will be perpendicular to the first.

Checking Baselines with the 3-4-5 Method

You can also use the 3-4-5 method to confirm that two existing baselines are perpendicular. For example, it's always wise to check baselines created with the column lines method.

Follow these steps to check two existing baselines with the 3-4-5 method (*Figure 21*):

Step 1 Choose an appropriate unit.

Step 2 Measure along one baseline 3 units and mark a point.

Step 3 Measure along the other baseline 4 units and mark a point.

Step 4 Measure the distance between the two points.

Step 5 If the distance is exactly 5 units, the baselines are perpendicular.

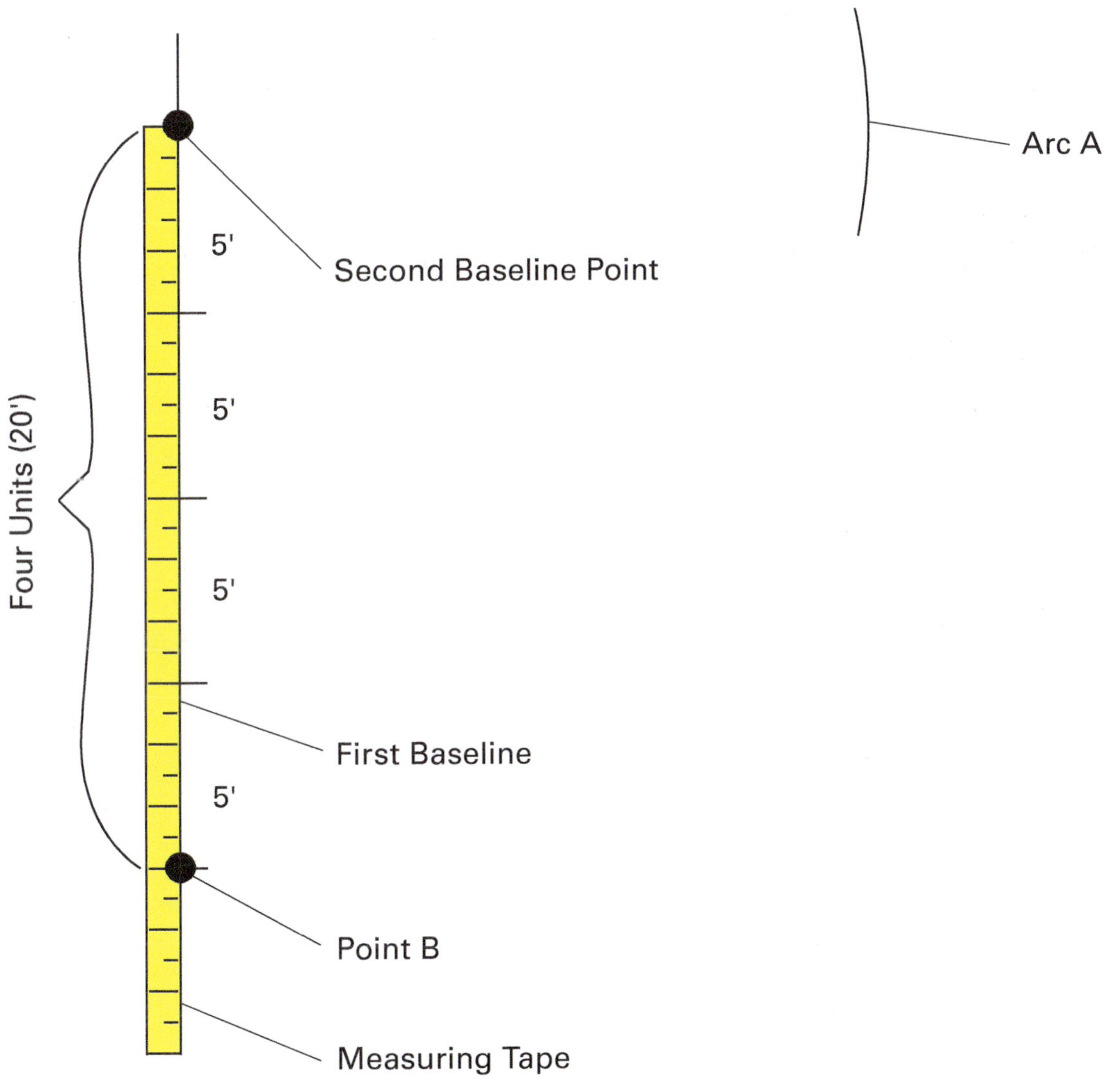

Figure 18 Locating point B.

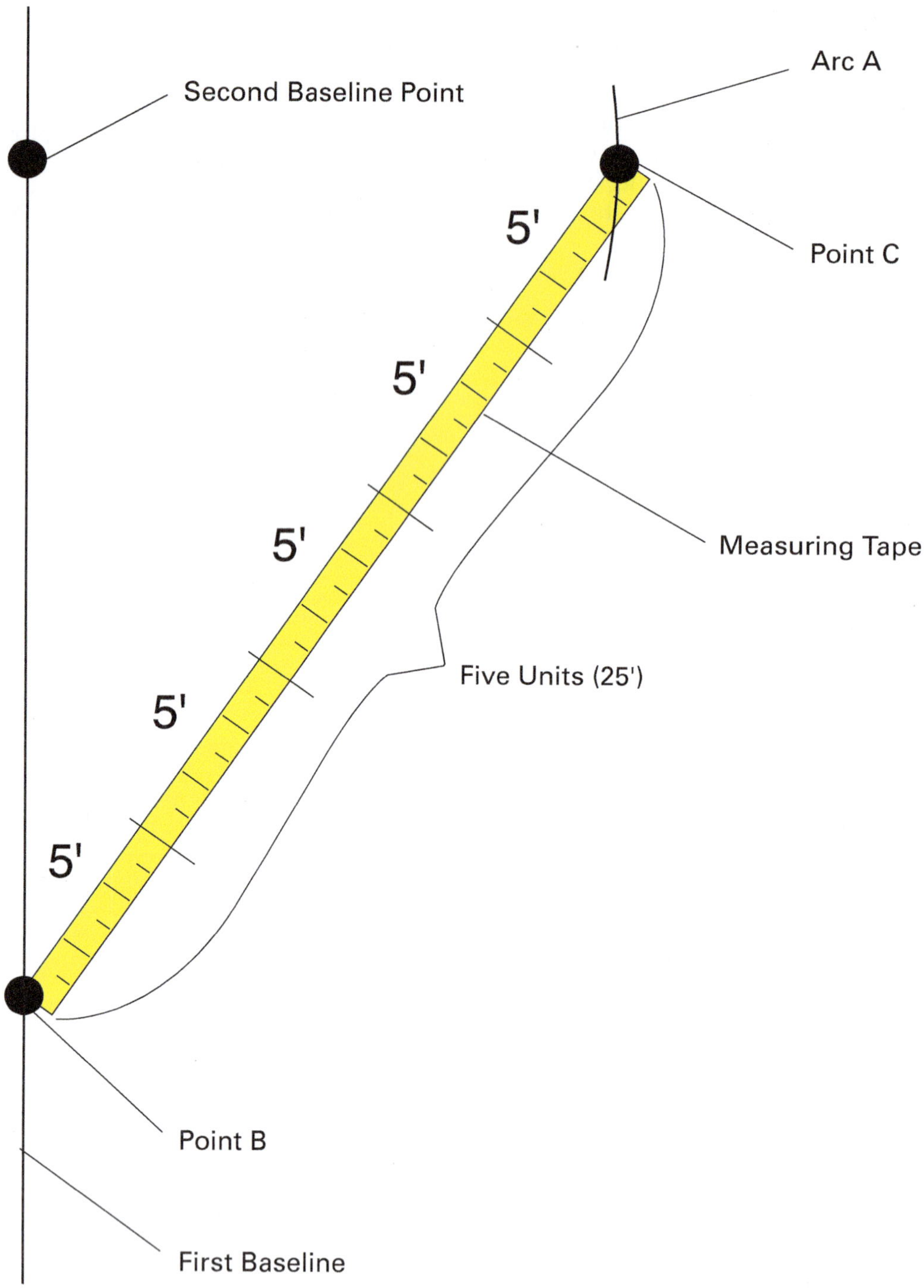

Figure 19 Locating point C.

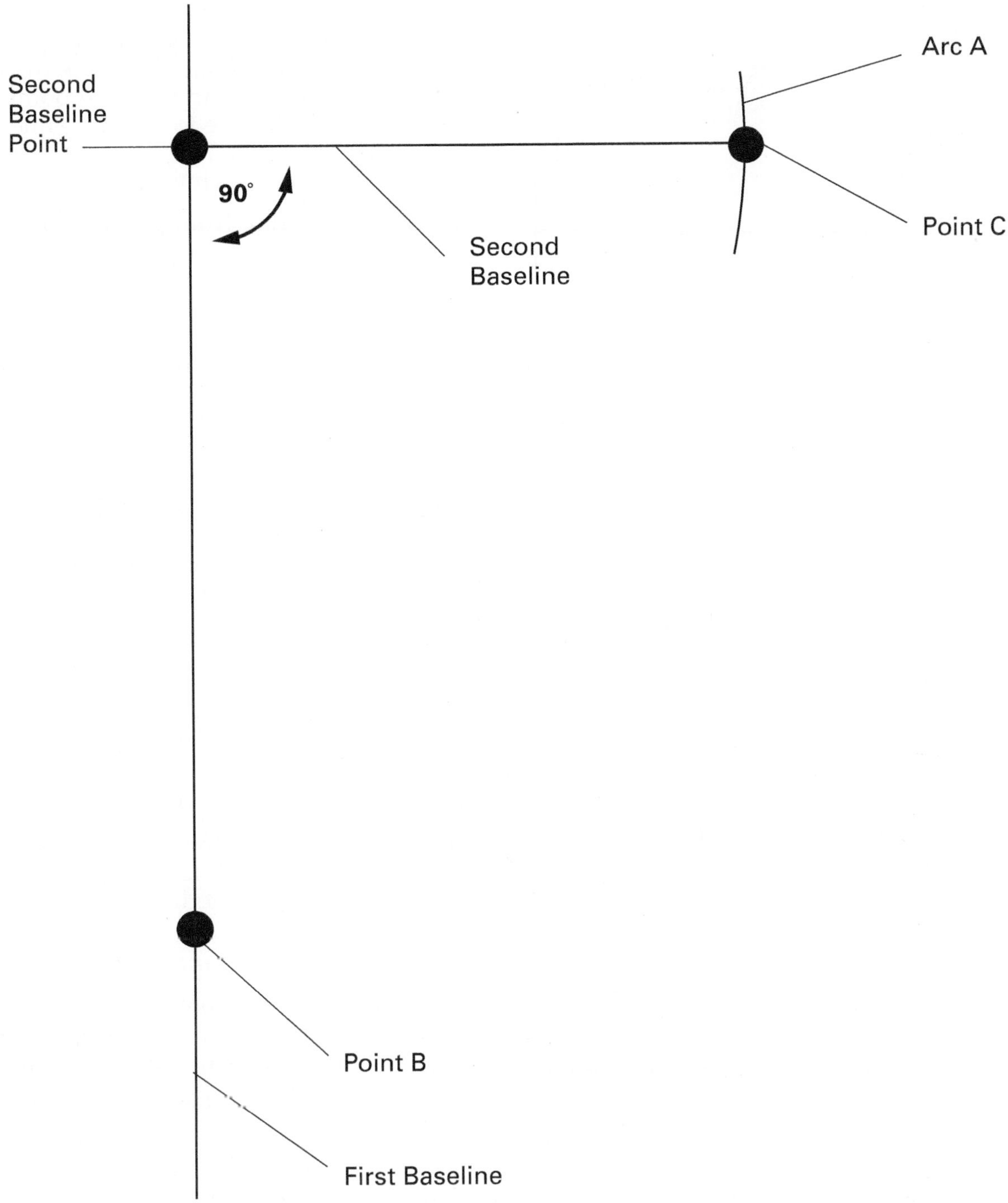

Figure 20 Perpendicular second baseline.

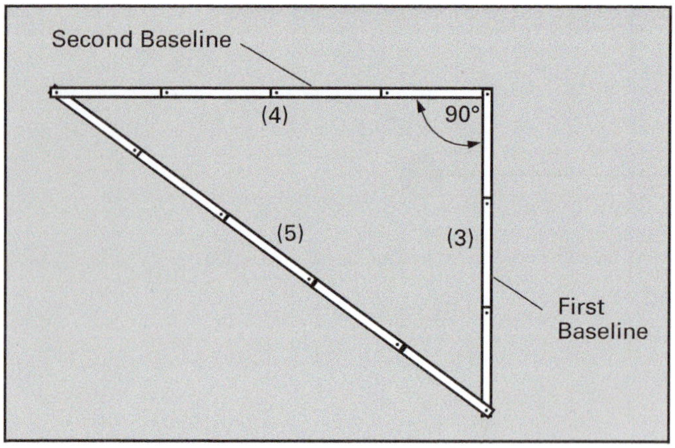

Figure 21 Checking baselines with the 3-4-5 method.

Arc Method

The third technique uses some basic geometry and simple tools to lay out a pair of baselines.

Follow these steps to lay out baselines with the arc method:

Step 1 Snap a chalk line on the floor to mark the first baseline. If available, a column line is a good reference. Mark the point on the first baseline through which the second baseline will pass.

> **NOTE**
> For the next steps, do not use a string or line that can stretch. Use a steel tape or wire.

Step 2 Place the steel tape at the second baseline point and swing two arcs to locate points A and B on the first baseline (*Figure 22*). The exact distance isn't important, but the arcs must be the same distance from the second baseline point.

Step 3 Swing a pair of arcs from point A and point B to locate point C (*Figure 23*). Use a distance about $1\frac{1}{2}$ times the distance from the second baseline point to point A or B. Be sure to use the same distance for both arcs. Mark point C at the place where the arcs intersect.

Step 4 Draw a line from point C to the second baseline point (*Figure 24*). If you did all the steps correctly, the second baseline will be perpendicular to the first.

1.2.3 Establishing Equipment Centerlines

After laying out the baselines, you can measure to create the equipment centerlines (*Figure 25*). These are crucial for positioning equipment.

Follow these steps to establish equipment centerlines:

Step 1 Consult the equipment drawings to determine the centerlines' dimensions.

Step 2 Measure from one of the baselines in two places, one near each end of the equipment's planned location.

Step 3 Mark the first centerline's endpoints on the floor.

Step 4 Snap a chalk line between these marks.

Step 5 Repeat *Steps 2* through *4*, measuring from the other baseline to establish the second equipment centerline.

1.2.4 Installing Permanent Reference Points

After establishing the centerlines, they need to be marked with brass plugs set in the floor. These plugs become permanent reference points.

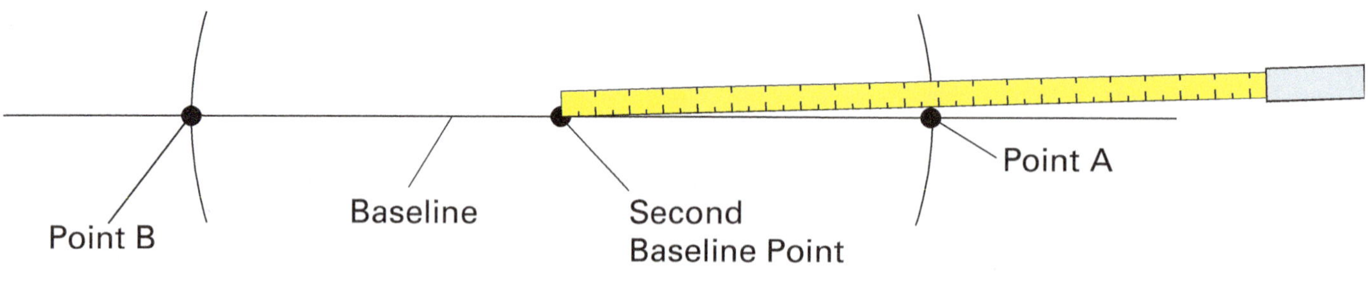

Figure 22 Locating points on the first baseline.

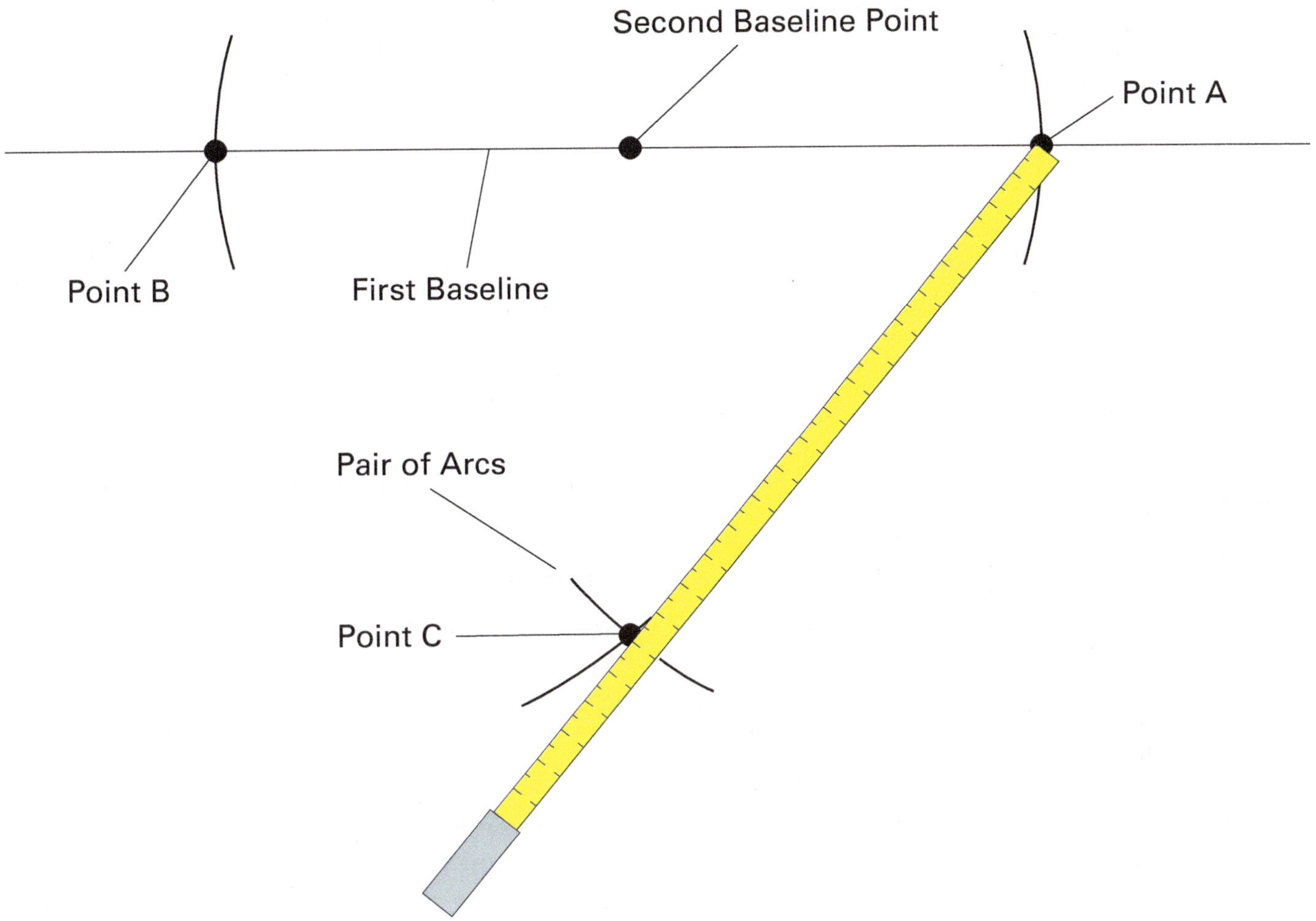

Figure 23 Locating point C.

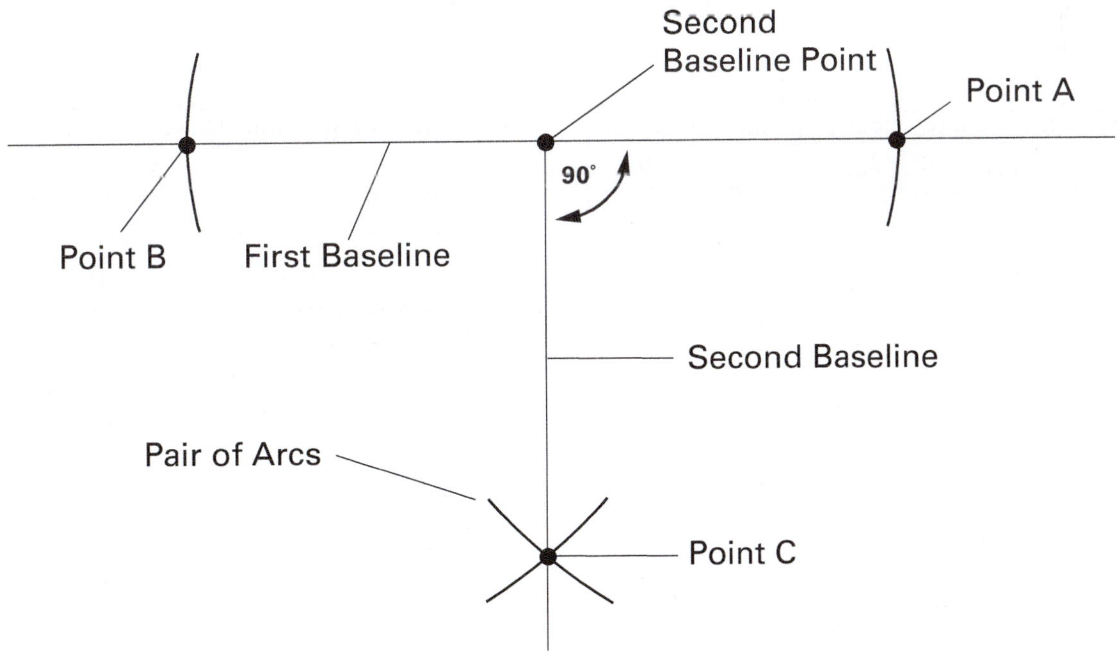

Figure 24 Perpendicular second baseline.

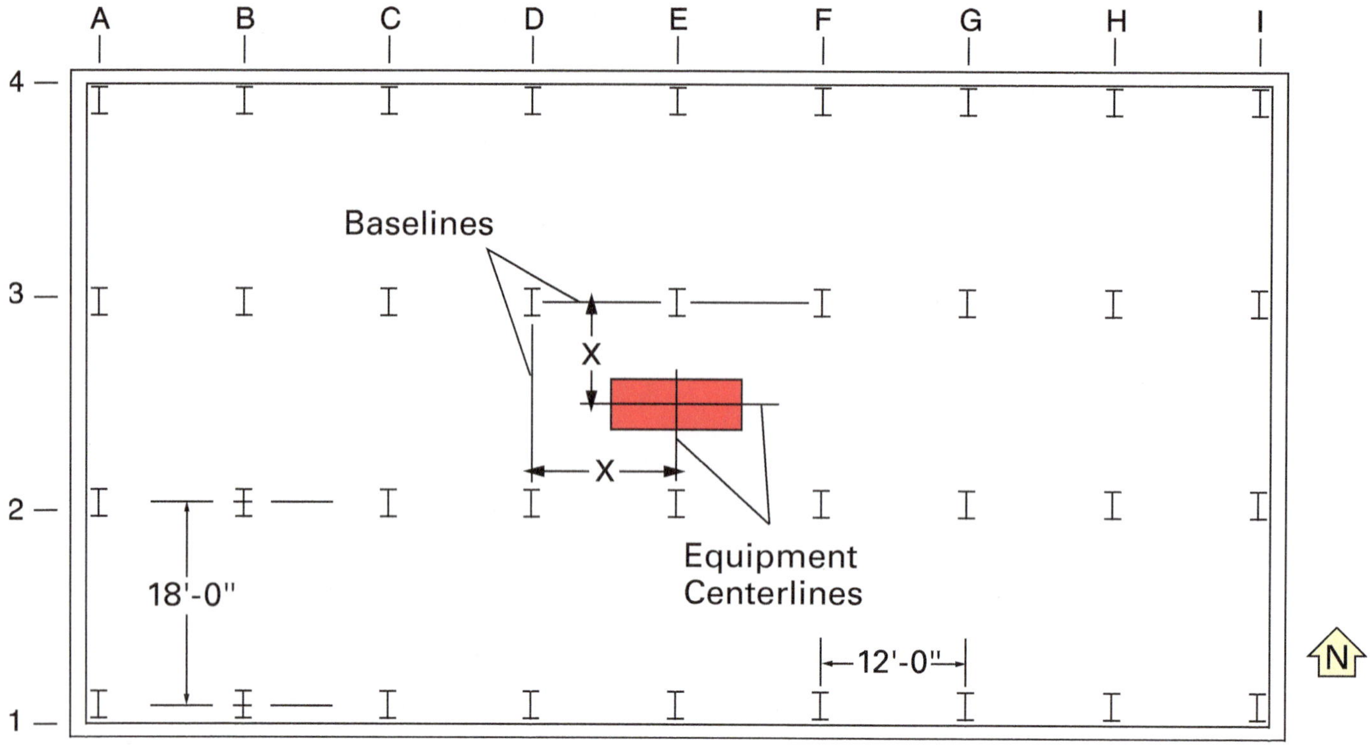

Figure 25 Equipment centerlines.

Follow these steps to install the permanent reference points:

Step 1 Confirm the centerlines' accuracy.

Step 2 Determine the best place on the centerlines for the plugs. The installed equipment should not hide them.

Step 3 Drill the plug holes using an appropriate core drill.

> **NOTE**
> The plugs should be about 2" in diameter and 3" long. Make the holes big enough for the plugs and the grout that holds them in place.

Step 4 Insert the plugs in the holes and grout them into place.

Step 5 Once the grout has cured, use a prick punch to mark the centerline position on the plugs.

1.2.5 Locating Anchor Bolts

Anchor bolts hold the equipment to the machine foundation or floor. Accurate placement prevents alignment problems during installation. Anchor bolt positions use the equipment centerlines as references (*Figure 26*).

Sometimes, the manufacturer provides a template for marking the anchor locations. Templates simplify the layout process since the millwright just transfers the pattern to the floor. Without a template, the millwright must measure from the equipment centerlines and mark each bolt location.

Follow these steps to locate the equipment anchor bolts by the measuring method:

Step 1 Measure carefully from the equipment centerlines to each mounting hole's centerline.

Step 2 Create a simple baseplate drawing showing all the measured dimensions.

> **NOTE**
> Some manufacturers provide a dimensioned drawing to aid in laying out the anchor bolts. If this is available, skip *Steps 1* and *2*.

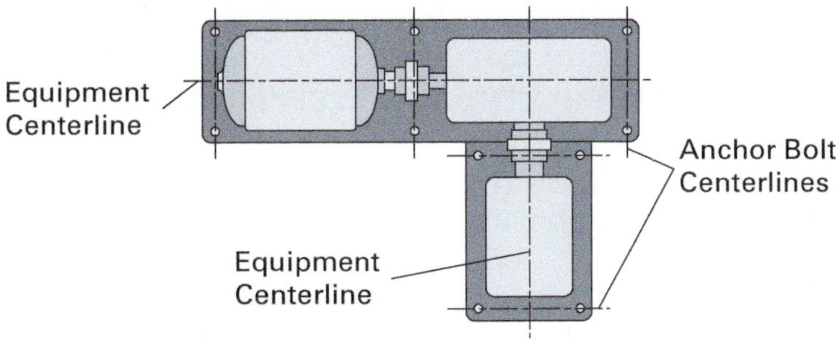

Figure 26 Equipment anchor bolt locations.

Step 3 Using the dimensions from the drawing, measure from the equipment centerlines on the floor. Mark each anchor bolt location.

Step 4 Double-check your measurements and confirm each mark.

Step 5 Drill the holes and install the anchor bolts.

1.2.6 Establishing Equipment Elevations

Many pieces of equipment sit directly on the floor. But some must sit on a raised base at a specific elevation. Manufacturers specify elevation in several ways. Sometimes, it is an above finished floor (AFF) value. In this case, simply measure from an appropriate reference point to the desired height.

In other cases, an elevation benchmark provides the reference. These may be inside or outside a building, with columns being popular locations. To determine a rough elevation from a column benchmark, measure from the benchmark to the floor. Then use the floor as the reference point.

For more accurate positioning, transfer the elevation from the benchmark to the equipment base with a string level, piano wire, or optical level. The optical level gives the best results.

Follow these steps to establish equipment elevation using an optical level:

Step 1 Set up the optical level where you can see the benchmark and the equipment location.

> **NOTE**
> The optical level must be set up on a firm, flat surface. It must be level in all directions and cannot move during the procedure.

Step 2 Point the level at the column above or below the benchmark. When the level faces the column, the horizontal crosshair will be above or below the benchmark.

Step 3 Determine the difference between the horizontal crosshair's elevation and the benchmark's elevation. Have an assistant hold a tape measure vertically against the column with the end at the benchmark (*Figure 27*). Read the tape through the level's telescope and record the positive or negative distance between the horizontal crosshair and the benchmark.

Step 4 Calculate the crosshair's elevation by adding or subtracting the measured value from the benchmark's elevation. For example, if the benchmark's elevation is 6' and the crosshair is 1" below the benchmark, the crosshair's elevation is 5'-11".

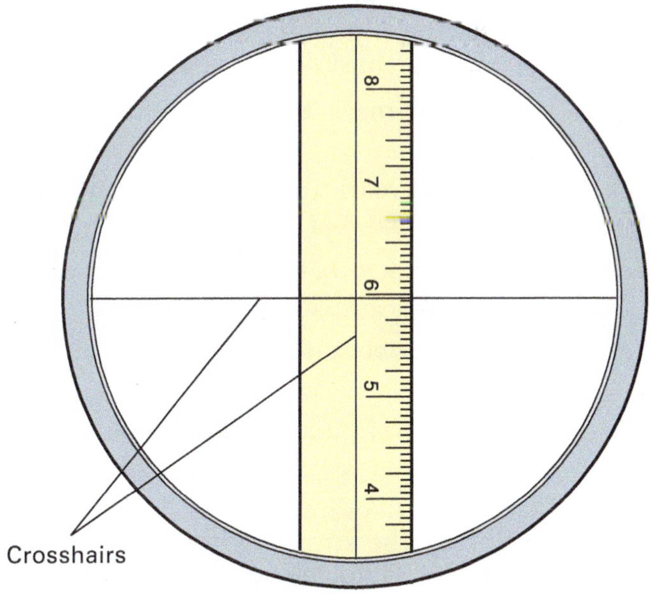

Figure 27 Taking a reading using an optical level.

Step 5 Carefully swing the level around to point at the equipment location. Confirm that the level has not moved and is still level in all directions.

Step 6 Have an assistant hold the tape measure vertically with the end on the equipment baseplate.

Step 7 Read the tape through the level's telescope where it intersects the horizontal crosshair.

Step 8 Use this reading and the crosshair's elevation to calculate the equipment base's current elevation.

Step 9 Adjust the equipment base as required.

1.3.0 Basic Layout Techniques

Millwrights and industrial mechanics create layouts for many different applications. These could be as small as a piece of sheet metal or as large as a factory floor. Layouts require drawing lines, arcs, and circles in specific positions. Distances, angles, orientations, and locations must be precise. The millwright cannot just guess or make casual measurements. When carried out correctly, the techniques introduced here produce accurate and precise layouts.

1.3.1 Scribing Straight Lines

When scribing short lines, use a steel rule or straightedge as a guide. If possible, choose one long enough for the entire line.

Follow these steps to scribe a straight line:

Step 1 Locate and mark the start point with a small V.

Step 2 Locate and mark the end point with another V.

Step 3 Using a prick punch, make marks at each V (*Figure 28*).

Step 4 Place the scriber point in one of the punch marks.

Step 5 Slide the straightedge into contact with the scriber.

Step 6 Place the scriber in the other punch mark and bring the straightedge into contact (*Figure 29*). Hold the straightedge firmly in place.

Step 7 Place the scriber against the straightedge at the start point. Angle it slightly.

Upright Measurements

This base holds a rule or combination square vertically. In this position, it performs a variety of tasks, including transferring measurements.

Figure Credit: The L.S. Starrett Company

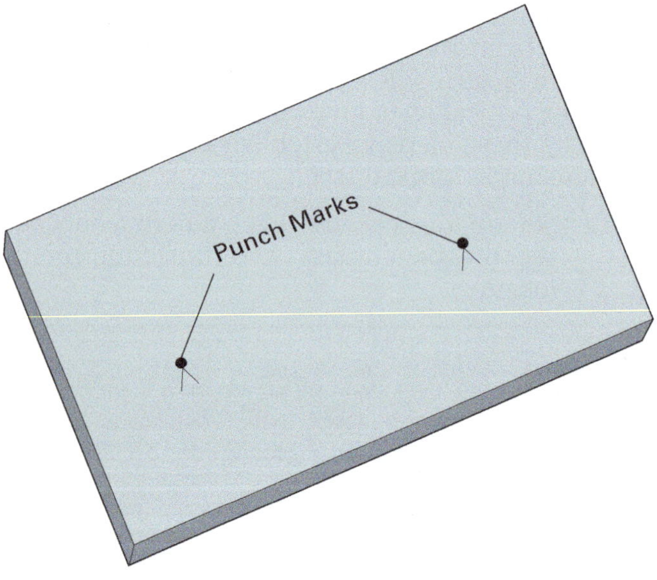

Figure 28 Punch marks.

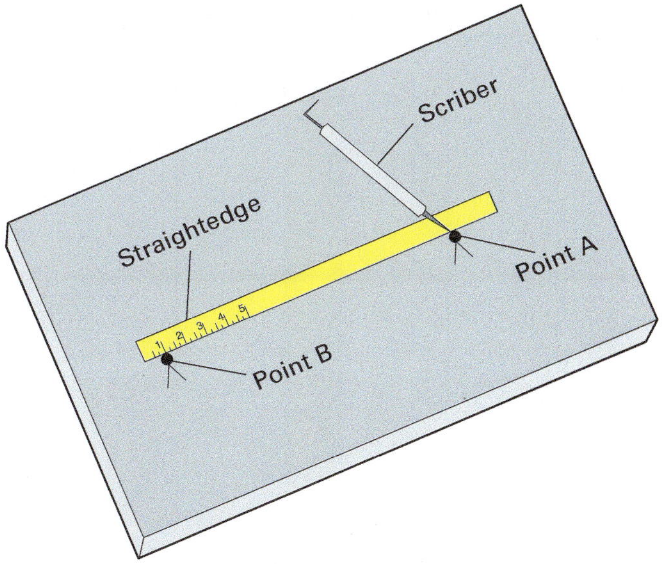

Figure 29 Aligning the straightedge.

Step 8 In a single, smooth motion, scribe a line to the end point. Do not make multiple passes.

1.3.2 Scribing Perpendicular Lines with a Framing Square

A framing square is ideal for creating a line perpendicular to a baseline. In fact, making multiple parallel perpendicular lines is just as simple.

Follow these steps to scribe parallel perpendicular lines using a framing square:

Step 1 Locate and scribe a baseline on the work surface.

Step 2 Align the straightedge with the baseline.

Step 3 Clamp the straightedge firmly in place.

Step 4 Place the framing square's blade against the straightedge.

Step 5 Scribe the perpendicular line along the square's tongue (*Figure 30*).

Step 6 Move the framing square the required distance and scribe another line. Repeat as many times as needed. These lines are parallel to each other and perpendicular to the baseline.

1.3.3 Scribing Lines with a Combination Square

Combination squares can also help you scribe perpendicular, parallel, or 45-degree lines to an edge. Maximum line length depends on the combination square's blade size.

Parallel Perpendicular Lines

Follow these steps to scribe parallel lines perpendicular to an edge:

Step 1 Select a smooth, straight edge to act as the reference edge.

Step 2 Place the square head of the combination square against the edge and scribe a line along the blade.

Step 3 Move the combination square the required distance and scribe another line. Repeat as many times as needed (*Figure 31*).

Single Parallel Line

Follow these steps to scribe a line parallel to an edge:

Step 1 Select a smooth, straight edge to act as the reference edge.

Step 2 Adjust the combination square blade to the required distance. Tighten the locknut.

Step 3 Place the square head against the reference edge. The blade end will be perpendicular to the edge.

Step 4 Hold the scriber point against the end of the blade.

Step 5 Slide the square and the scriber together, creating a line parallel to the edge (*Figure 32*).

Parallel 45-Degree Lines

Follow these steps to scribe parallel lines 45 degrees to an edge:

Step 1 Select a smooth, straight edge to act as the reference edge.

Step 2 Place the head's 45-degree angle against the reference edge.

Step 3 While holding the square head firmly, scribe a line along the blade.

Step 4 Move the combination square the required distance and scribe another line. Repeat as many times as needed (*Figure 33*).

1.3.4 Scribing Angled Lines with a Protractor

A protractor paired with a straightedge can assist you in scribing lines at any angle to a baseline.

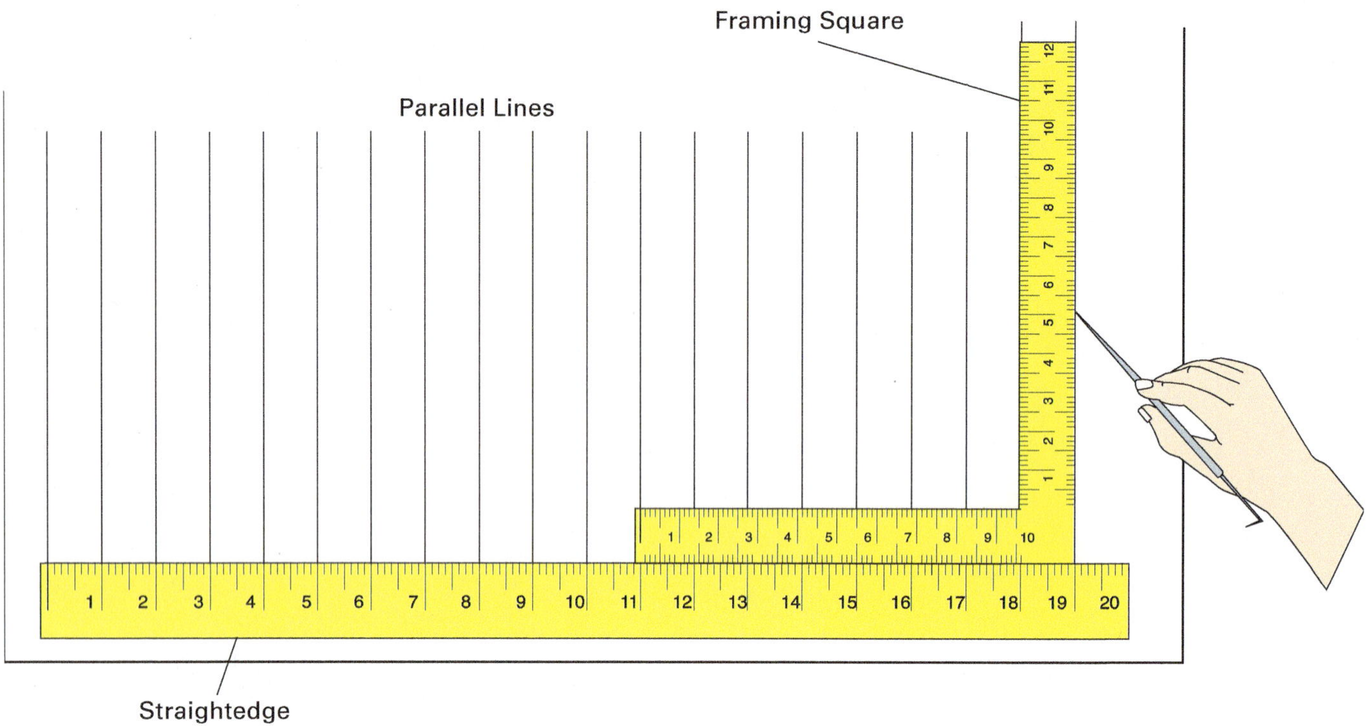

Figure 30 Scribing parallel lines.

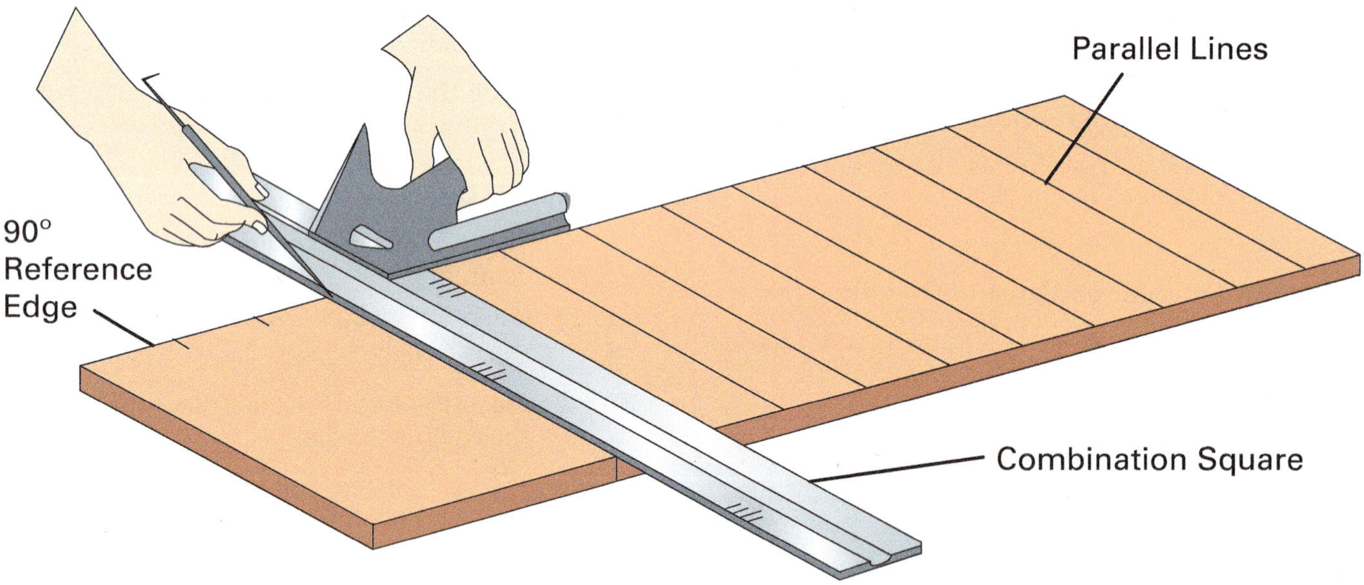

Figure 31 Scribing parallel lines perpendicular to an edge.

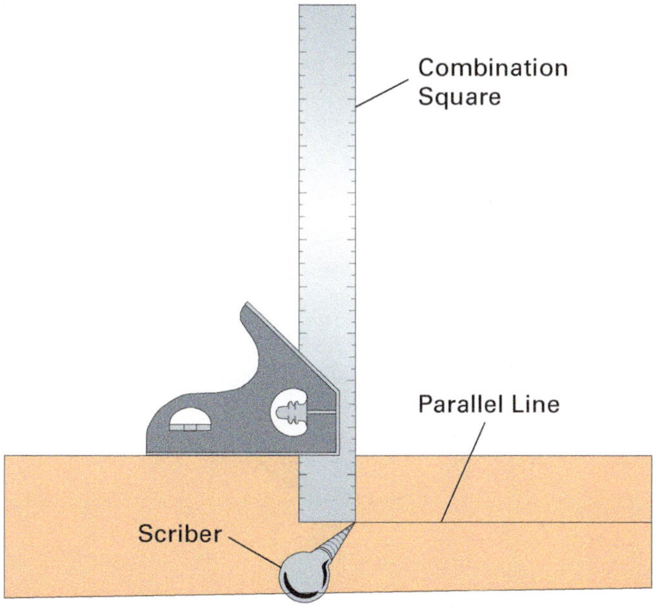

Figure 32 Scribing a line parallel to an edge.

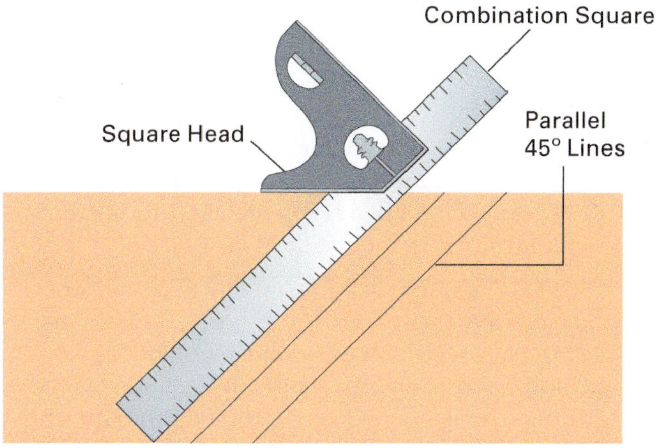

Figure 33 Scribing a line 45 degrees to an edge.

Follow these steps to scribe angled lines to a baseline:

Step 1 Locate and scribe a baseline on the work surface.

Step 2 Using a prick punch, make a mark at the proper location on the baseline.

Step 3 Place the protractor's center mark on the punch mark.

Step 4 Locate the correct angle on the protractor's scale and punch a mark there.

Step 5 Using the straightedge, scribe a line between the two punch marks (*Figure 34*).

1.3.5 Scribing Circles and Arcs with Dividers

Dividers are perfect for drawing accurate circles. Dividers draw arcs, which are just sections of circles, equally well.

Follow these steps to scribe a circle or an arc using dividers:

Step 1 Using a prick punch, mark the circle or arc's center point.

Step 2 Place one of the dividers' points on a steel rule's 1" / 1 cm mark.

Step 3 Turn the adjustment screw until the other point is at the correct radius (*Figure 35*).

Remember to add 1" / 1 cm to the value.

Step 4 Place one point in the punch mark.

Step 5 To scribe a full circle, tilt the dividers slightly forward and rotate 360 degrees using the thumb and index finger (*Figure 36*).

Step 6 To scribe an arc, tilt the dividers slightly forward and rotate enough to make the arc (*Figure 37*). For multiple arcs, repeat at another location.

Step 7 Using a steel rule, confirm the circle's diameter or the arc's radius.

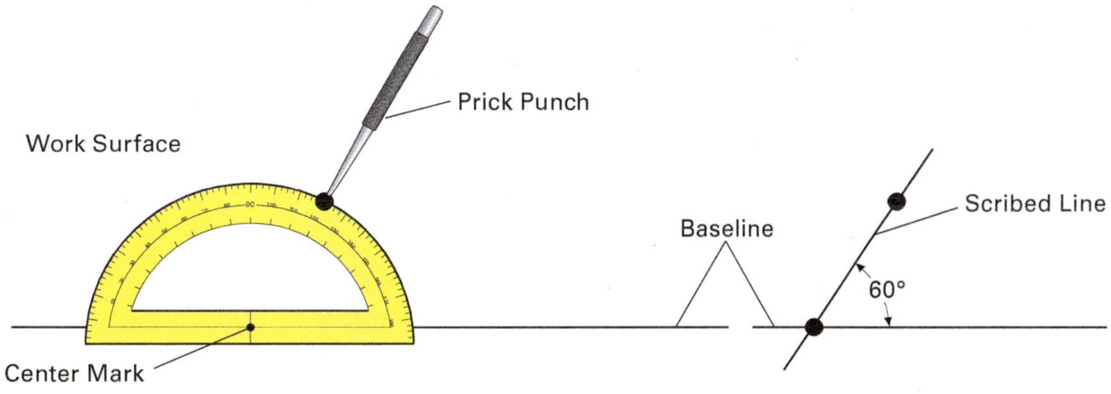

Figure 34 Scribing an angled line with a protractor.

Circles

From earlier training modules, you know that *radius, diameter,* and *circumference* describe parts of a circle. The following are a few more circle terms that you should know:

- *Chord* — A straight line joining any two points on the circumference.
- *Segment* — A section of the circle outlined by an arc and a chord.
- *Sector* — A pie-shaped slice outlined by an arc and two radius lines.
- *Tangent* — A line that touches a circle's circumference at just one point. Tangents are perpendicular to the radius that touches the point.

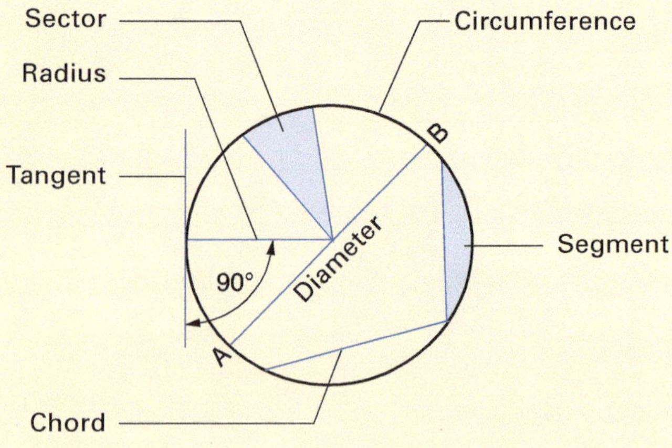

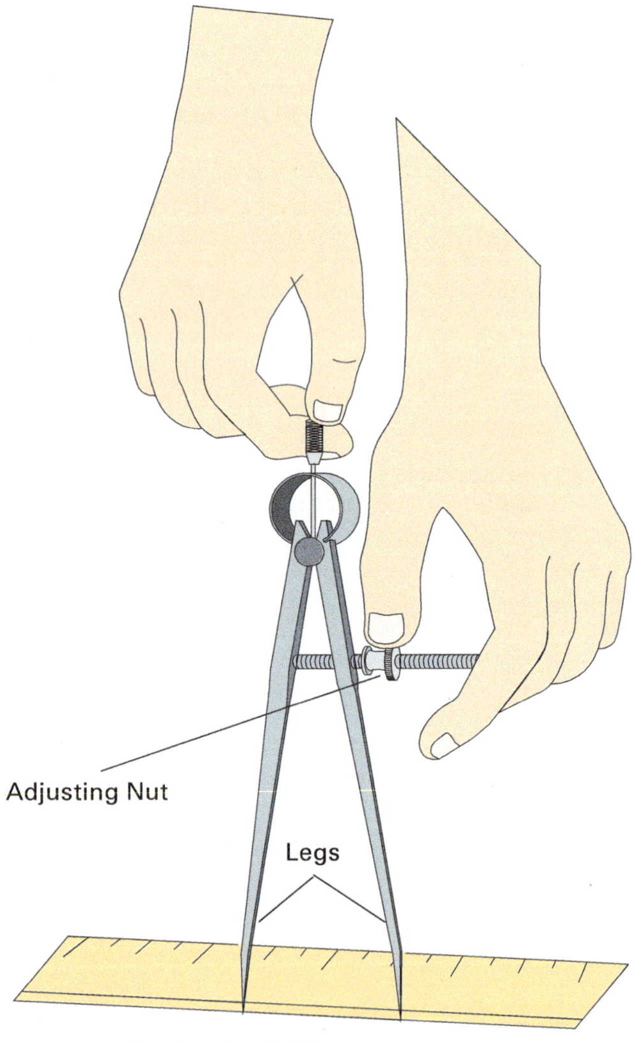

Figure 35 Adjusting the dividers.

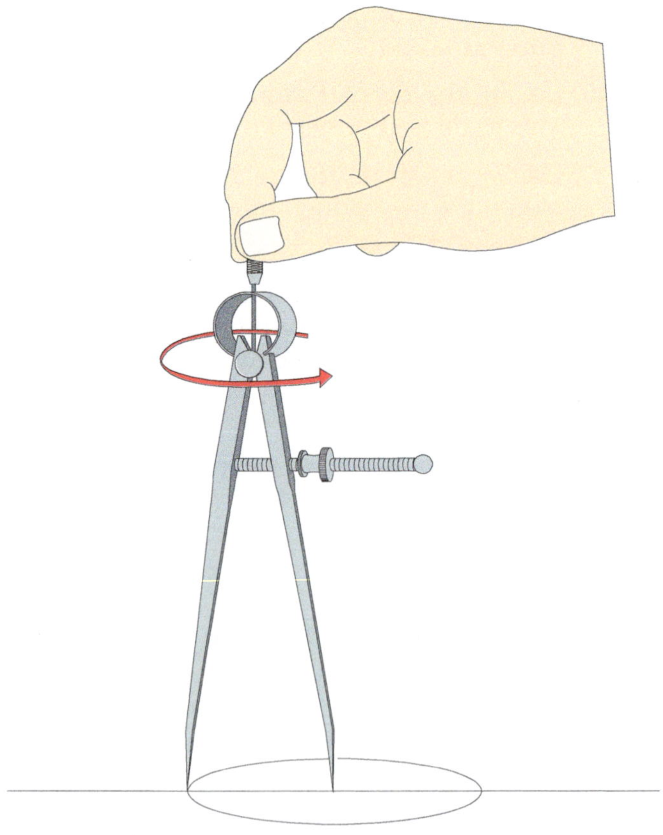

Figure 36 Scribing a circle.

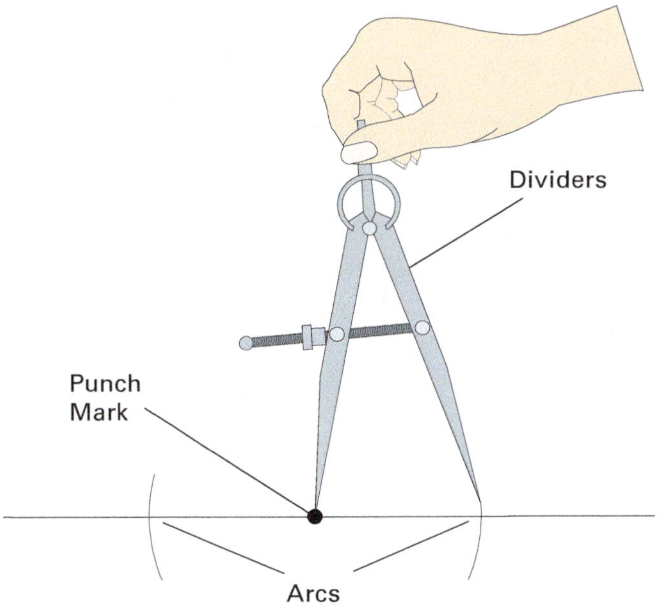

Figure 37 Scribing arcs.

1.3.6 Scribing Circles and Arcs with Trammel Points

Follow these steps to scribe a circle or an arc using trammel points:

Step 1 Using a prick punch, mark the circle or arc's center point.

Step 2 Place one trammel point on a steel rule's 1" / 1 cm mark.

Step 3 Slide the other trammel point to the proper distance plus 1" / 1 cm and tighten the thumbscrew.

Step 4 Place one trammel point in the punch mark. Hold it firmly.

Step 5 To scribe a circle, tilt the tool slightly forward and rotate it 360 degrees by pulling the scribing point (*Figure 38*).

Step 6 To scribe an arc, tilt the tool slightly forward and rotate enough to make the arc. If required, repeat at another location for a second arc.

Step 7 Using a steel rule, confirm the circle's diameter or the arc's radius.

1.3.7 Scribing Perpendicular Lines with Dividers

Dividers and a straightedge can help you scribe a line perpendicular to a baseline. There are two possible methods.

Method 1

Follow these steps to construct a line perpendicular to a baseline using dividers and a straightedge:

Step 1 Locate and scribe a baseline on the work surface.

Step 2 Using a prick punch, mark the location on the baseline where the perpendicular line will pass through.

Step 3 Adjust the dividers to the desired radius.

Step 4 Place one point in the punch mark.

Step 5 Scribe an arc through the baseline on each side of the punch mark (*Figure 39*).

Step 6 Punch a mark at each location on the baseline where the arc passes through.

Step 7 Adjust the dividers to a radius about $1\frac{1}{2}$ times larger.

Step 8 Place one point in one of the new punch marks.

Step 9 Scribe arcs above and below the center point (*Figure 40*).

> **NOTE**
> In some cases, there will not be enough space to scribe arcs on both sides of the baseline. If so, scribe them where they fit.

Step 10 Place one point in the other new punch mark.

Step 11 Scribe arcs above and below the center point. The arcs should cross the ones from *Step 9* (*Figure 40*).

Step 12 Scribe a straight line through the arc intersections and the center point (*Figure 41*). If you did all the steps correctly, the line will be perpendicular to the baseline.

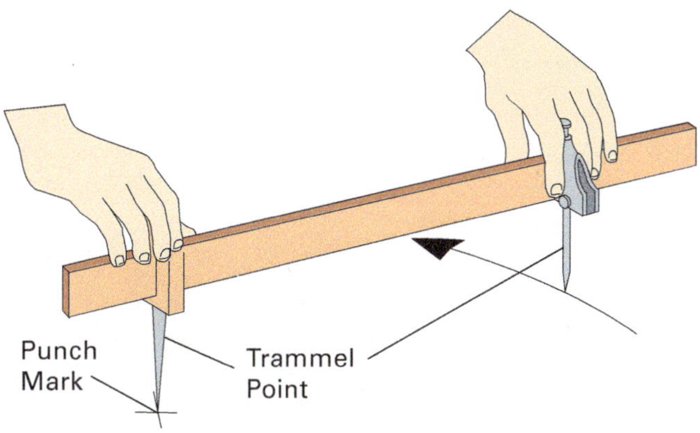

Figure 38 Scribing circles or arcs.

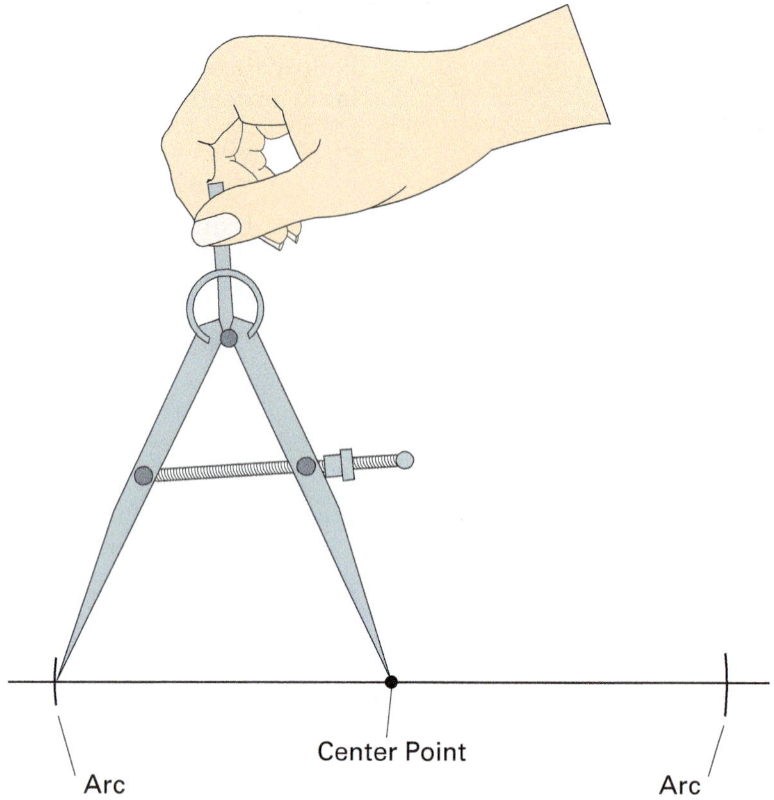

Figure 39 Scribing arcs on each side of the baseline.

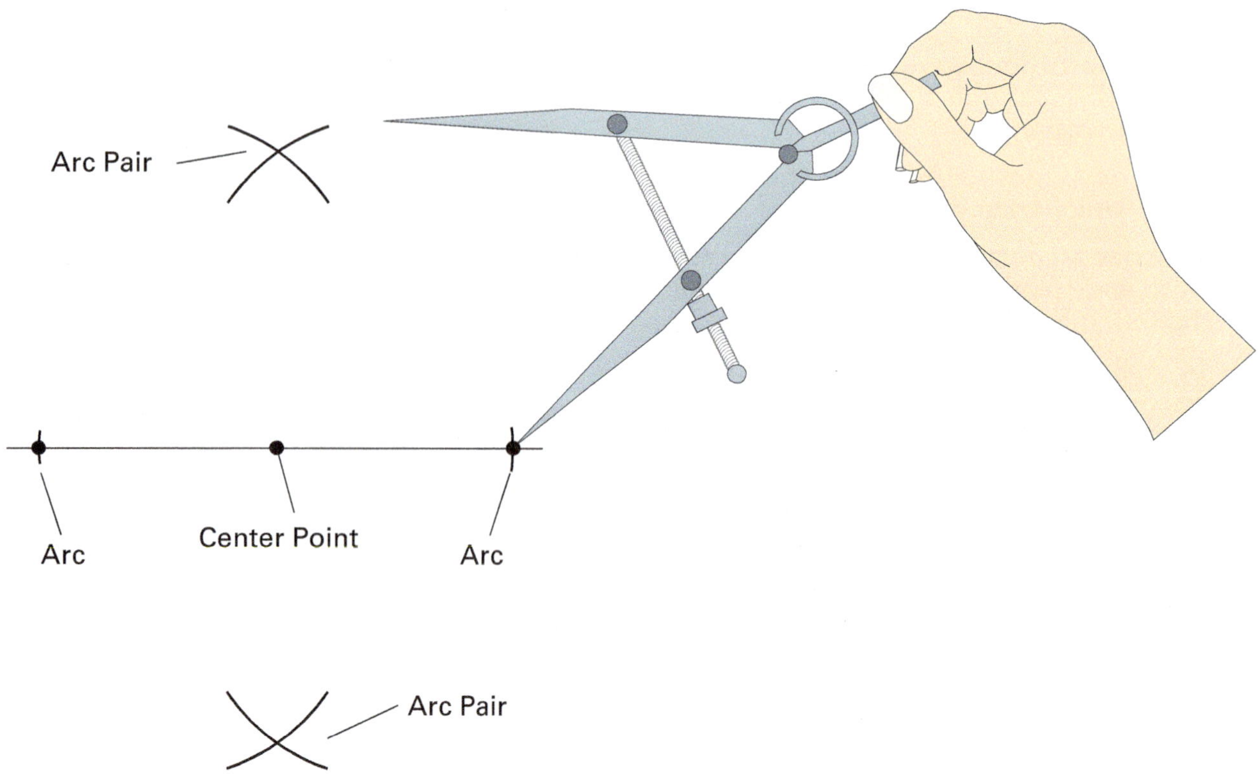

Figure 40 Scribing arcs above and below the center point.

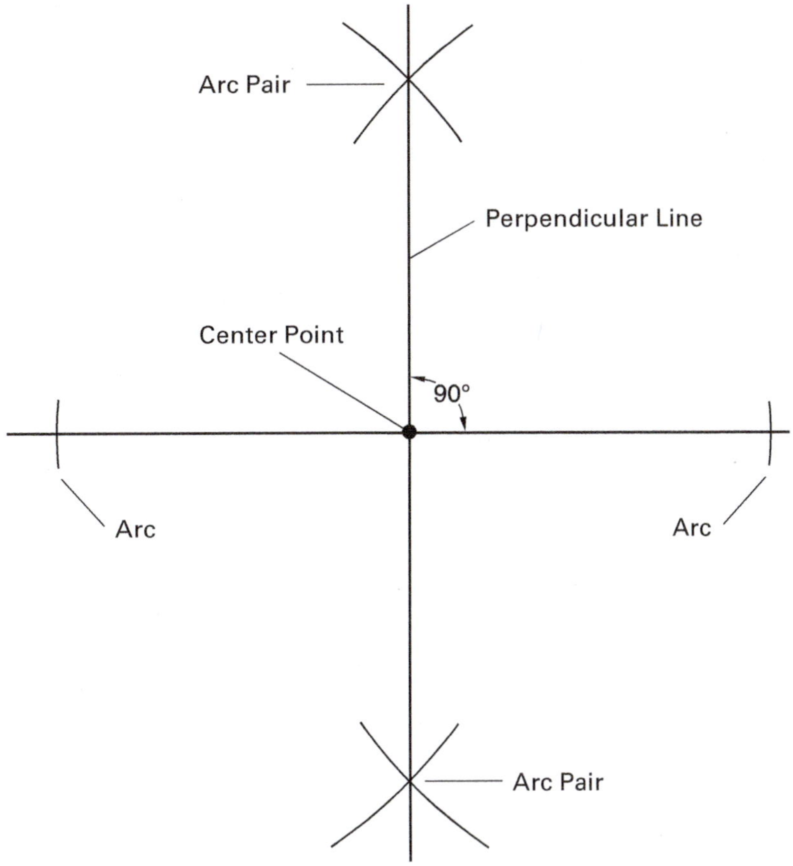

Figure 41 Perpendicular line.

Step 13 If one is available, use a protractor to confirm that the lines are perpendicular (90 degrees to each other).

Method 2

Follow these steps to construct a line perpendicular to a baseline using a reference point, dividers, and a straightedge:

Step 1 Identify the reference point and mark it using a prick punch.

Step 2 Locate and scribe a baseline on the work surface approximately 4" (~10 cm) below the reference point.

Step 3 Place one point on the reference mark. Adjust the dividers so it can sweep an arc that passes through the baseline at two points. Scribe the arc (*Figure 42*).

Step 4 Punch a mark at each location where the arc crosses the baseline.

Step 5 Adjust the dividers so the legs are slightly closer together.

Step 6 Set one point in one of the new punch marks.

Step 7 Scribe an arc above the baseline and below the reference point. Scribe a second arc below the baseline.

NOTE

In some cases, there will not be enough space to scribe arcs on both sides of the baseline. If so, scribe them where they fit.

Step 8 Set one point in the other new punch mark.

Step 9 Scribe an arc above the baseline and below the reference point. Scribe a second arc below the baseline. Both pairs of arcs should intersect (*Figure 43*).

Step 10 Scribe a line through the reference point, the arc intersections, and the baseline (*Figure 44*). If you did all the steps correctly, the line will be perpendicular to the baseline.

Step 11 If one is available, use a protractor to confirm that the lines are perpendicular (90 degrees to each other).

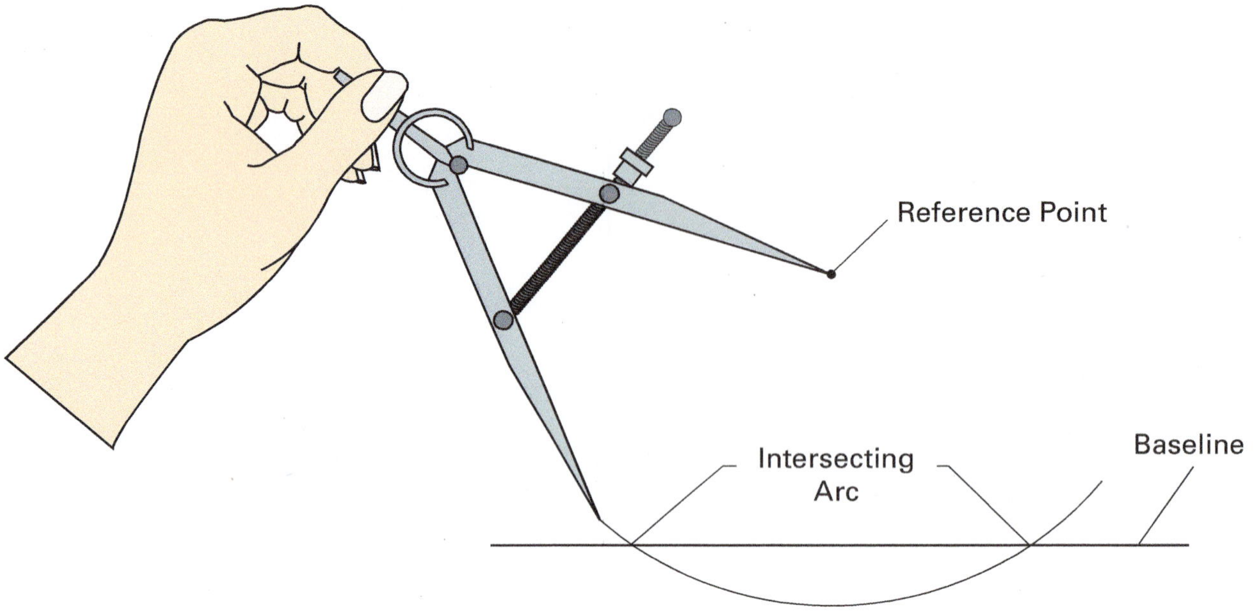

Figure 42 Arc intersecting the baseline.

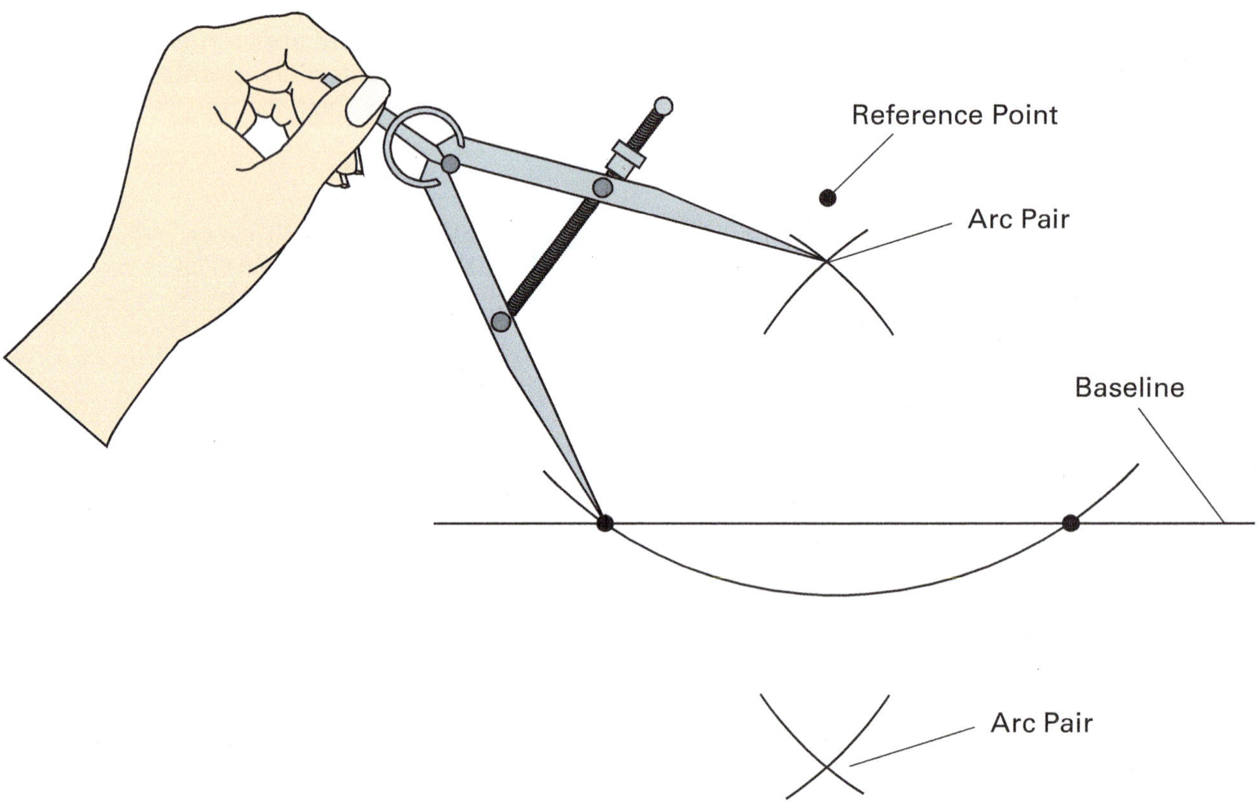

Figure 43 Arcs intersecting above and below the baseline.

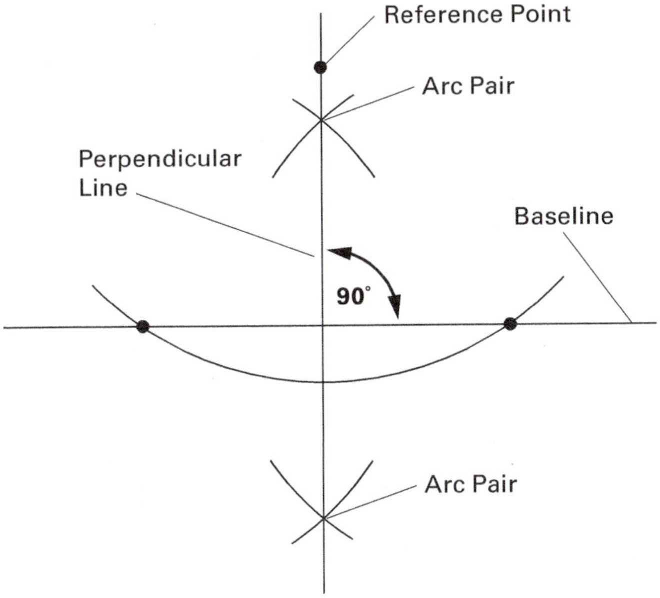

Figure 44 Scribing the perpendicular line.

1.3.8 Bisecting Angles with Dividers

Dividers offer a handy way to **bisect** an angle. This method gives accurate results and does not require an angle-measuring tool.

Follow these steps to cut an angle into two equal parts using dividers and a straightedge:

Step 1 Using a prick punch, mark the point where the angle's sides come together (the *vertex*).

Step 2 Set one point in the punch mark.

Step 3 Scribe an arc through the angle, intersecting both sides (*Figure 45*).

Step 4 Punch a mark at each location where the arc crosses the angle's sides.

Step 5 Set one point in one of the new punch marks.

Step 6 Scribe an arc between the angle's sides.

Step 7 Set one point in the other new punch mark.

Step 8 Scribe a second arc between the angle's sides. Both arcs should intersect (*Figure 46*).

Step 9 Scribe a line from the vertex through the point where the two arcs intersect (*Figure 47*). This line forms two equal angles.

Step 10 If one is available, check each angle using a protractor.

1.3.9 Dividing Lines into Equal Parts

If you had to divide a line into 5, 8, or 11 identical sections, how would you begin? For everyday tasks, you would probably use a tape measure or ruler. But millwrights sometimes use a special technique that gives accurate results without measuring.

Follow these steps to divide a line into equal parts using dividers and a straightedge:

> **NOTE**
> This technique can divide a line into any number of identical sections. This example uses five sections.

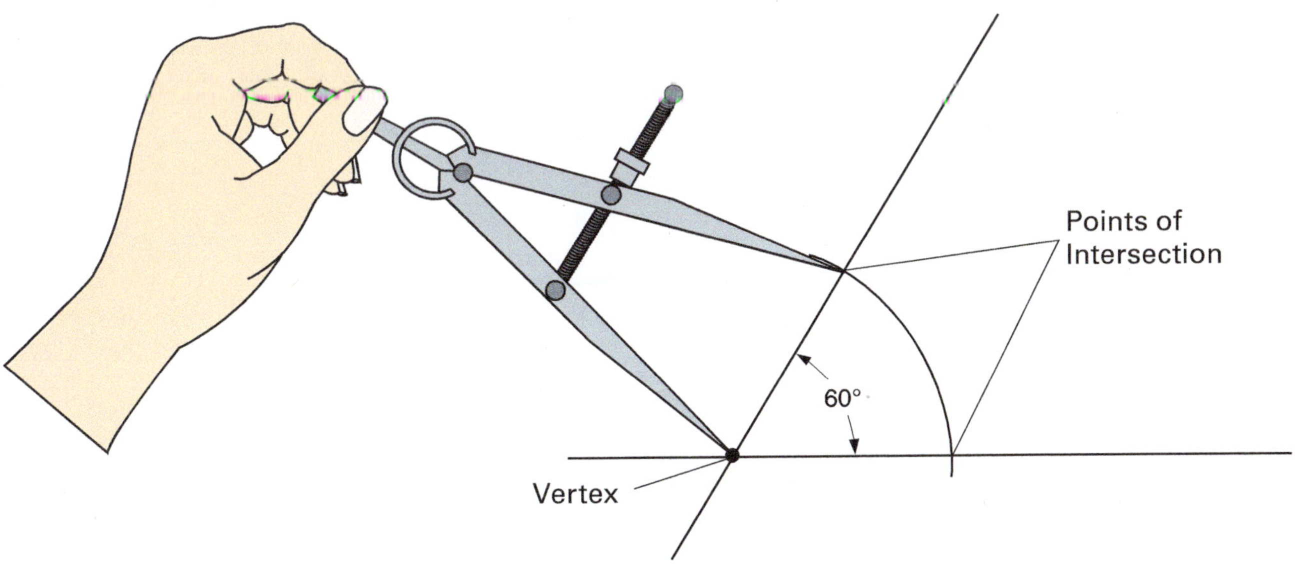

Figure 45 Scribed arc.

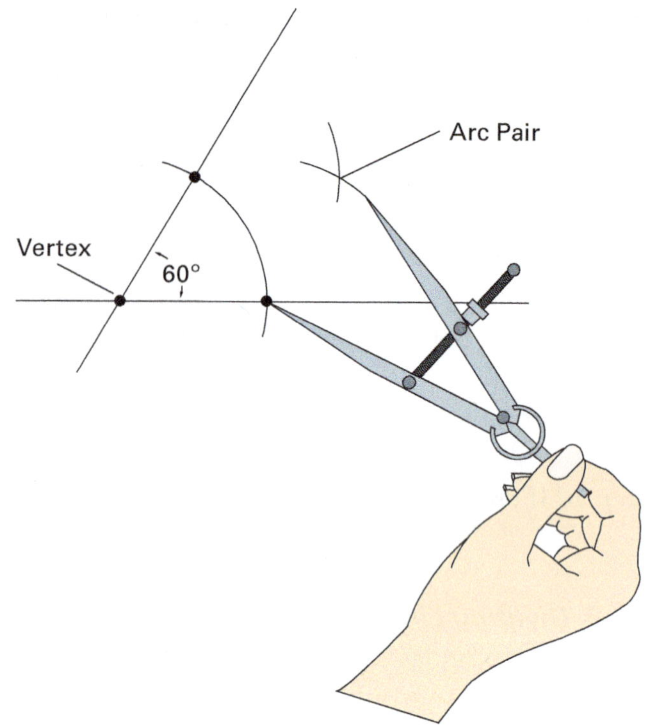

Figure 46 Intersecting arcs.

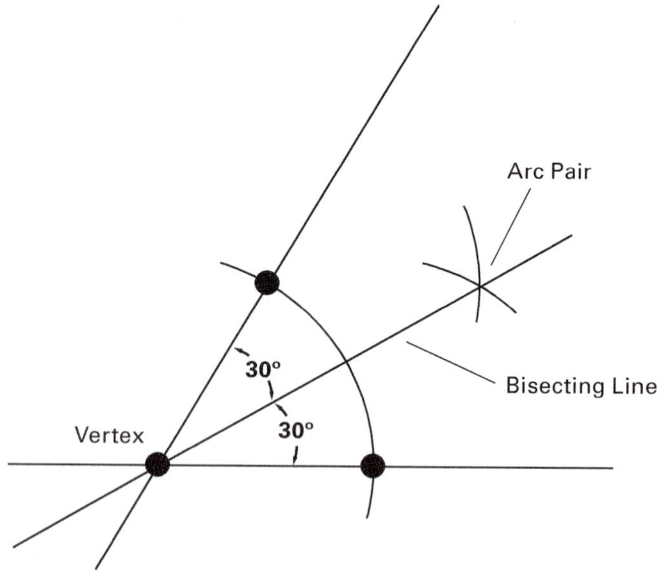

Figure 47 Line scribed from vertex to arc intersection.

Step 1 Using a prick punch, mark each end of the line.

Step 2 Set one point in one of the punch marks.

Step 3 Scribe an arc above the line (*Figure 48*). Any radius works fine.

Step 4 Set one point in the other punch mark.

Step 5 Scribe an arc below the line (*Figure 48*). Use the same radius.

Step 6 Scribe a diagonal line from one endpoint to the circumference of the arc on the same side (*Figure 49*). The line must be **tangent** to the arc (touch it at exactly one point).

Step 7 Repeat the process for the other endpoint and arc (*Figure 49*).

Step 8 Set the dividers to any convenient radius. It should be less than the size of the sections that you are creating ($^1/_5$ of a line, in this case).

Step 9 Set one point in one of the endpoint punch marks.

Step 10 Scribe a short arc through the diagonal line that passes through that endpoint.

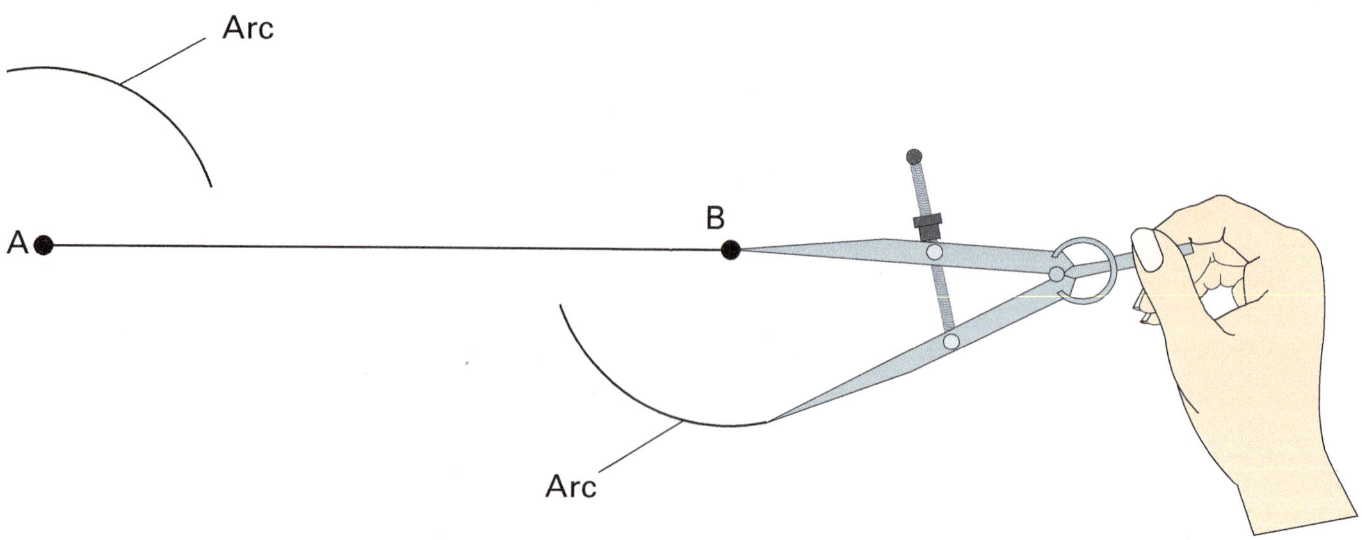

Figure 48 Arcs above and below the line.

Step 11 Punch a mark where the arc crosses the diagonal line.

Step 12 Set one point in the new punch mark.

Step 13 Scribe another short arc through the diagonal line and punch a mark at the intersection.

Step 14 Repeat *Steps 12* and *13* until you have created arcs equal to the number of sections that you are producing (five in this case).

Step 15 Number each arc (1 to 5 in this case).

Step 16 Repeat *Steps 9* through *14* starting from the other endpoint and scribing arcs on the other diagonal line.

Step 17 Number each arc (6 to 10 in this case). *Figure 50* shows the finished result.

Step 18 Scribe lines across the original line between closest opposite points (*Figure 51*). In this case, they are A–10, 1–9, 2–8, 3–7, 4–6, 5–B. These lines split the original line into identical sections.

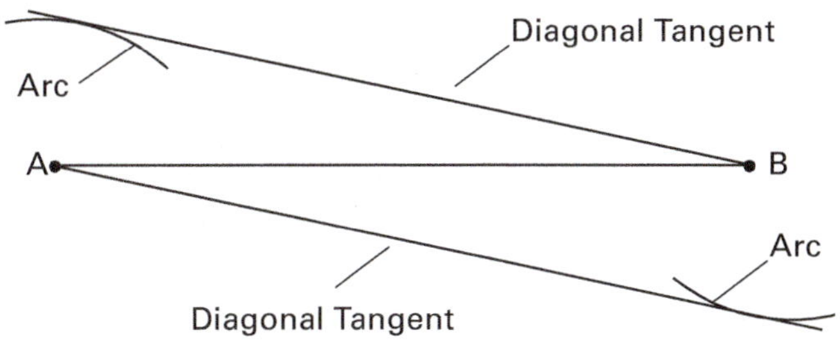

Figure 49 Lines tangent to the arcs.

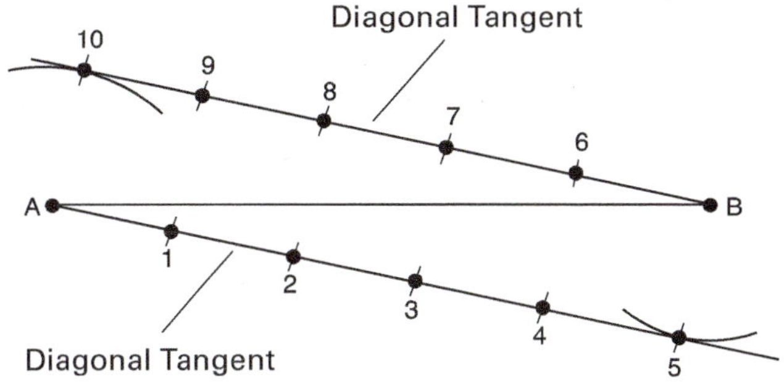

Figure 50 Arcs on the diagonal lines.

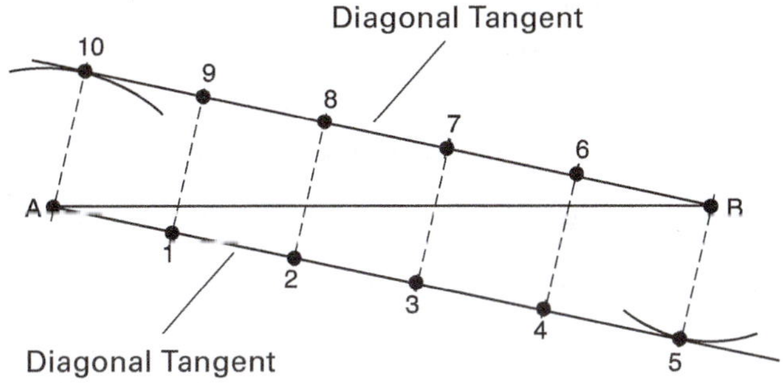

Figure 51 Dividing lines between opposite points.

1.3.10 Dividing Circles into Six Equal Parts

Dividing a circle into an equal number of sectors may seem challenging, but it is easier than dividing a line.

Follow these steps to divide a circle into six equal parts using dividers and a straightedge:

Step 1 Using a prick punch, mark the circle's center point.

Step 2 Set the dividers equal to the circle's radius. This setting must not change.

Step 3 Punch a mark anywhere on the circle's circumference.

Step 4 Set one point in the new mark.

Step 5 Scribe a short arc through the circumference (*Figure 52*).

Step 6 Punch a mark where the arc crosses the circumference.

Step 7 Set one point in the new punch mark.

Step 8 Repeat *Steps 4* through *6* until you arrive back at the starting point (*Figure 53*).

Step 9 Scribe three lines between opposite points through the circle center. The result should be six identical sectors (*Figure 54*).

Step 10 Optionally, you can bisect each 60-degree angle using the technique introduced earlier in this module. The result will be 12 identical sectors.

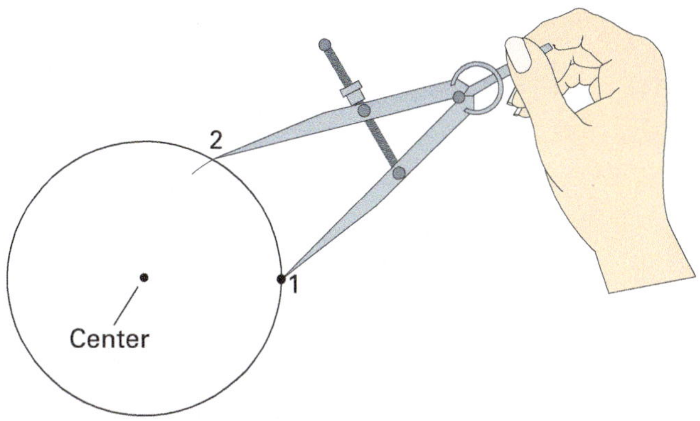

Figure 52 Starting point and first arc.

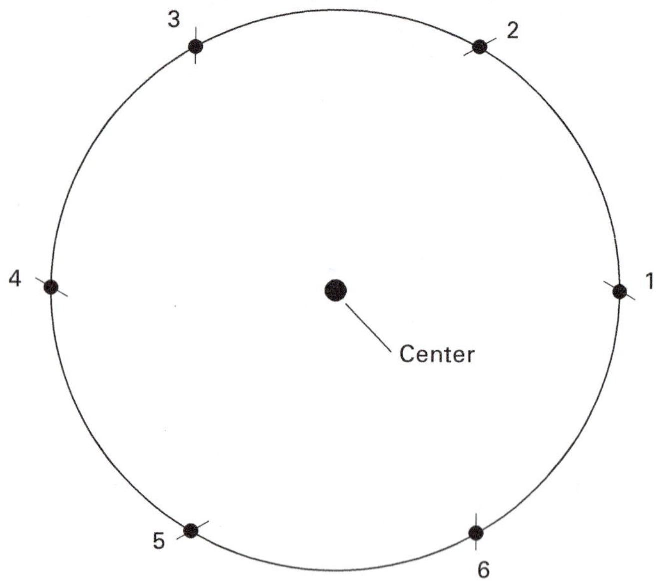

Figure 53 Arcs scribed around circumference.

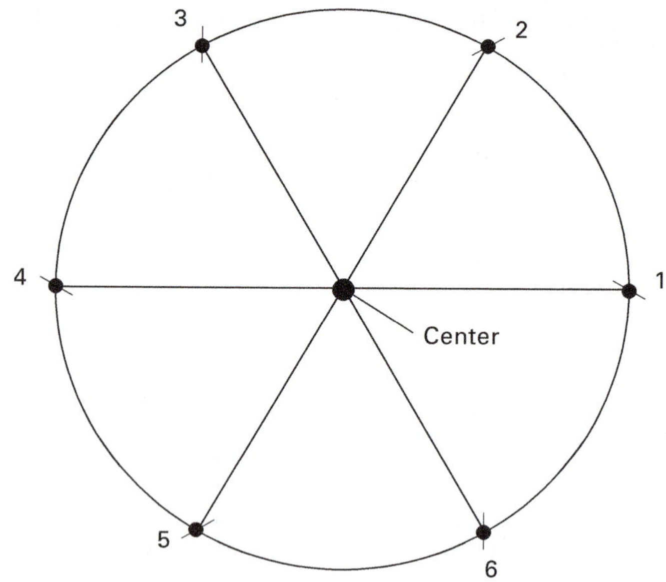

Figure 54 Six equal sectors.

1.0.0 Section Review

1. A combination set includes a blade, square head, center head, and _____.
   a. scribe head
   b. protractor head
   c. framing head
   d. leveling head

2. Millwrights place brass plugs in the floor to mark equipment _____.
   a. centerlines
   b. elevations
   c. baseplate holes
   d. anchors

3. When scribing lines, it is acceptable to make several passes if the line is long.
   a. True
   b. False

Review Questions

1. Which layout tool has the *longest* blade?
   a. Machinist try square
   b. Framing square
   c. Combination set
   d. Dividers

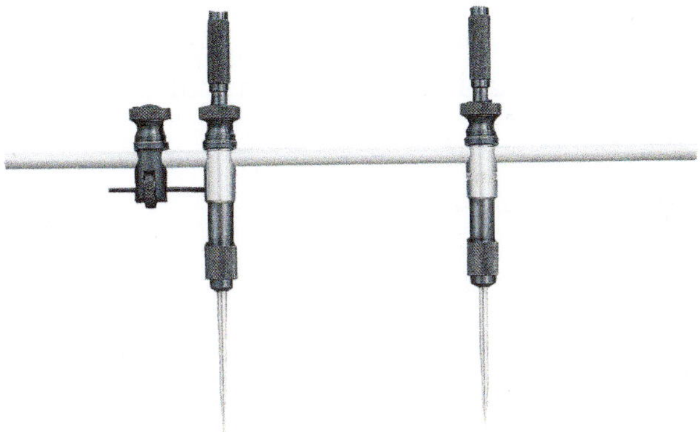

Figure RQ01

2. The layout devices shown in *Figure RQ01* are called _____.
   a. dividers
   b. protractors
   c. trammel points
   d. transfer punches

3. Which tools would you use to prepare a metal plate for drilling holes?
   a. A prick punch and a center punch
   b. A trammel point and a pair of dividers
   c. A center punch and a transfer punch
   d. A scriber and a transfer punch

4. To copy the location of a hole onto another surface, use a _____.
   a. center punch
   b. trammel point
   c. prick punch
   d. transfer punch

5. What does a column identifier look like?
   a. A pair of numbers
   b. An angle and an elevation
   c. A letter and a number
   d. GPS coordinates

6. If a triangle has two sides 15" and 20" long at right angles to each other, how long must the third side be?
   a. 10"
   b. 15"
   c. 25"
   d. 30"

7. What is the *most* accurate way to establish a piece of equipment's elevation?
   a. Use a chalk line.
   b. Use piano wire.
   c. Use a string level.
   d. Use an optical level.

8. To scribe a series of parallel lines 45 degrees to an edge, use a(n) _____.
   a. combination square
   b. framing square
   c. inside caliper
   d. beam compass

9. If you need to scribe a straight line at a 72-degree angle, which tool would be the *best* choice?
   a. Combination square
   b. Protractor
   c. Dividers
   d. Trammel points

10. To bisect an angle, you could use a straightedge and _____.
    a. dividers
    b. a framing square
    c. a chalk line
    d. an optical level

Trade Terms Quiz

Fill in the blank with the correct term learned from your study of this module.

1. The point where two lines or arcs cross is a(n) _______________.
2. To _______________ a line, you cut it into two equal pieces.
3. Two _______________ lines form a 90-degree angle.
4. _______________ are metal markers that identify specific locations.
5. A(n) _______________ is a line that touches a circle's circumference at exactly one point.
6. The 3-4-5 _______________ describes the length relationship between the sides of some right triangles.
7. _______________ are a pair of right-angle reference lines used in layouts.

Trade Terms

Baselines
Benchmarks
Bisect
Intersection
Perpendicular
Ratio
Tangent

Cornerstone of Craftsmanship

Neil Steagall

Senior Rotating Equipment Specialist
Air Liquide

How did you choose a career in the industry?

I've always been a mechanical and hands-on kind of person. So, after high school, I knew I wasn't going to college and needed to do something that I liked and paid well. I tried a couple of different trades before my first millwright opportunity, and after that, I knew that's what I wanted to do.

Who inspired you to enter the industry?

My dad. He urged me to start based on my knowledge and the opportunities in the field.

What types of training have you been through?

I did an apprenticeship for three years at the beginning of my career. Since then, my training has never really stopped. I believe the most notable was attending Siemens' turbine training in Duisburg, Germany, and Sulzer's reciprocating compressor training in Winterthur, Switzerland.

How important is education and training in construction?

It's important to have so you can keep up with the changing work environment and evolving technology.

How important are NCCER credentials to your career?

I was certified by the UBC as a millwright, and some companies would not recognize that certification. I later became NCCER certified to continue working in my field.

How has training/construction impacted your life?

It has provided me with the ability to become a vital employee, remain employed, and eventually progress enough to become a millwright subject matter expert.

What kinds of work have you done in your career?

I was lucky enough to be able to experience many different opportunities. I've worked in nuclear, natural gas, and coal power plants, paper mills, refineries, chemical plants, air separation units, and even a coffee plant.

Tell us about your current job.

I'm currently a senior rotating equipment specialist with Air Liquide. I manage a fleet of 22 integrally geared compressors, 9 reciprocating compressors, 3 centrifugal compressors, and 15 expander turbines across 7 sites. My work includes everything from the overhaul and repair to troubleshooting.

What do you enjoy most about your job?

I enjoy being able to not only rebuild and repair the equipment, but also troubleshoot performance issues based on vibration, temperature, and flows.

What factors have contributed most to your success?

Well, it sounds easy, but being interested, paying attention, and staying involved at every opportunity. Volunteering and working overtime on the hard jobs? That's what helps you grow.

Would you suggest construction as a career to others? Why?

Of course. This industry allows you to make a career and, most of the time, an above-average salary.

What advice would you give to those new to the field?

Ask questions, pay attention, and be valuable. Also, you will not become rich or an SME overnight. Take your time and be patient. The skills you'll acquire will become invaluable in the future.

Interesting career-related fact or accomplishment?

I was nominated twice by my peers! The first nomination was to become a rotating equipment specialist, and now my current role is a senior rotating equipment specialist.

How do you define craftsmanship?

Craftsmanship is dedication, quality, and the ability to communicate effectively and intelligently.

Trade Terms Introduced in This Module

Baselines: Key reference lines around which all layout elements develop.

Benchmarks: Metal markers that precisely identify specific locations.

Bisect: To divide into two equal parts.

Intersection: The point where two lines or arcs cross.

Perpendicular: A line that is at a 90-degree angle to another line.

Ratio: A proportional relationship between two or more numbers.

Tangent: A line that touches a circle's circumference at exactly one point.

Additional Resources

This module presents thorough resources for task training. The following reference material is suggested for further study.

Audel Millwrights and Mechanics Guide. Latest Edition. Thomas B. Davis, Carl A. Nelson. Hoboken, NJ: John Wiley & Sons.

Machine Tool Practices. Latest Edition. Richard R. Kibbe, Roland O. Meyer, Jon Stenerson, Kelly Curran. New York, NY: Pearson.

Figure Credits

iStock@1715d1db_3, Module Opener

The L.S. Starrett Company, Figures 1–6, 8–11, RQ01

Section Review Answer Key

Section 1.0.0

Answer	Section Reference	Objective
1. b	1.1.6	1a
2. a	1.2.4	1b
3. b	1.3.1	1c

This page is intentionally left blank.

NCCER CURRICULA — USER UPDATE

NCCER makes every effort to keep its textbooks up-to-date and free of technical errors. We appreciate your help in this process. If you find an error, a typographical mistake, or an inaccuracy in NCCER's curricula, please fill out this form (or a photocopy), or complete the online form at **www.nccer.org/olf**. Be sure to include the exact module ID number, page number, a detailed description, and your recommended correction. Your input will be brought to the attention of the Authoring Team. Thank you for your assistance.

Instructors – If you have an idea for improving this textbook, or have found that additional materials were necessary to teach this module effectively, please let us know so that we may present your suggestions to the Authoring Team.

NCCER Product Development and Revision

13614 Progress Blvd., Alachua, FL 32615

Email: curriculum@nccer.org
Online: www.nccer.org/olf

❑ Trainee Guide ❑ Lesson Plans ❑ Exam ❑ PowerPoints Other ______________________

Craft / Level: Copyright Date:

Module ID Number / Title:

Section Number(s):

Description:

Recommended Correction:

Your Name:

Address:

Email: Phone:

This page is intentionally left blank.

Reading Mechanical Drawings

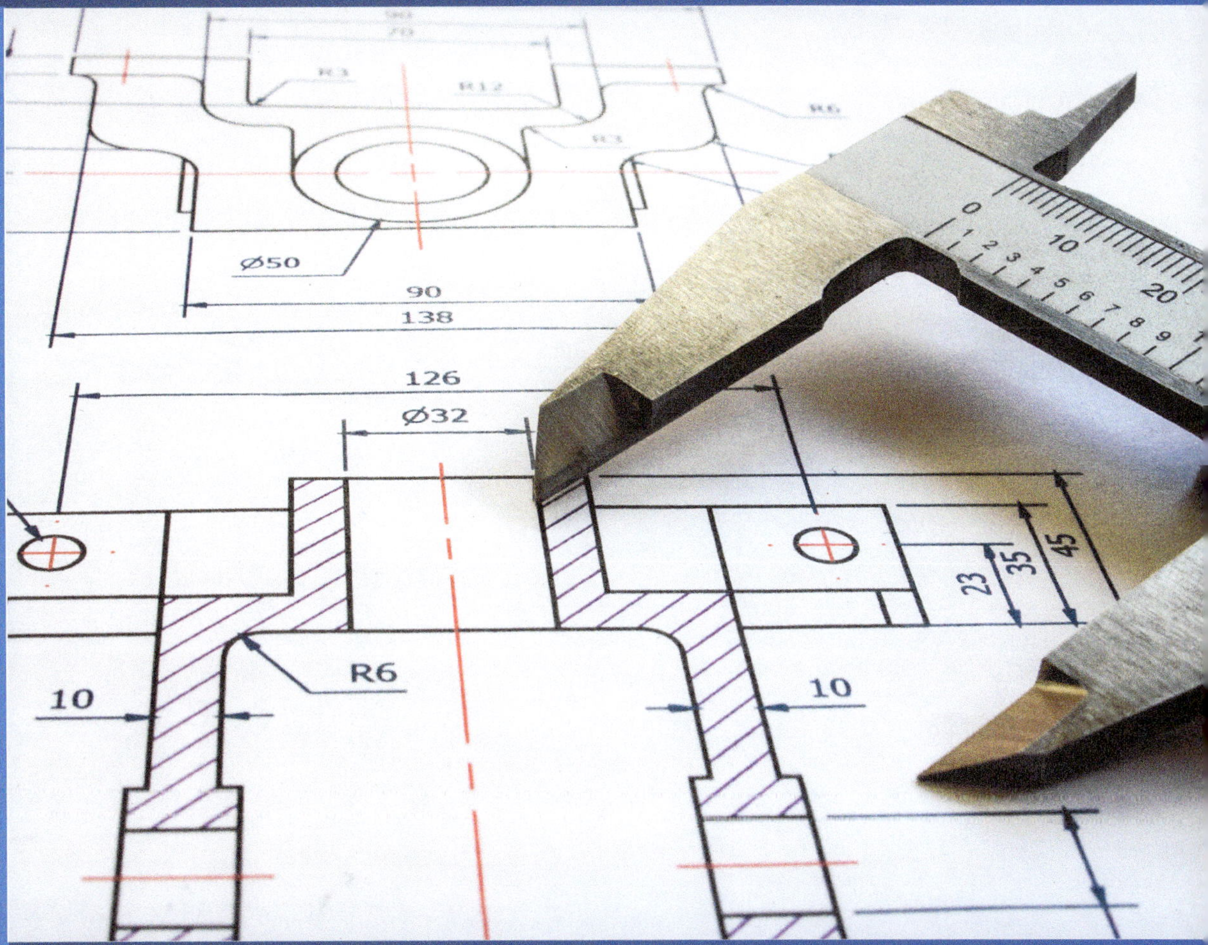

Overview

All craftworkers read mechanical drawings at one time or another. Drawings communicate project details in a standardized way. With proper training, any craftworker can interpret and use them to perform necessary tasks. Some drawings represent mechanical parts in detail with accompanying dimensions. Others show a large system's components and overall layout. And some portray three-dimensional objects almost photo-realistically. This module introduces mechanical drawings and shows you how to understand what they communicate.

Module 15203

Trainees with successful module completions may be eligible for credentialing through the NCCER Registry. To learn more, go to **www.nccer.org** or contact us at 1.888.622.3720. Our website, **www.nccer.org**, has information on the latest product releases and training.

Your feedback is welcome. You may email your comments to **curriculum@nccer.org**, send general comments and inquiries to **info@nccer.org**, or fill in the User Update form at the back of this module.

This information is general in nature and intended for training purposes only. Actual performance of activities described in this manual requires compliance with all applicable operating, service, maintenance, and safety procedures under the direction of qualified personnel. References in this manual to patented or proprietary devices do not constitute a recommendation of their use.

15203 V4.0

15203
READING MECHANICAL DRAWINGS

Objectives

Successful completion of this module prepares you to do the following:

1. Read mechanical drawings, interpreting the information they contain.
   a. Describe and interpret orthographic projections.
   b. Describe and interpret schematic drawings.
   c. Describe and interpret isometric drawings.

Performance Tasks

Under supervision, you should be able to do the following:

1. Interpret an instructor-provided orthographic projection.
2. Interpret instructor-provided hydraulic and pneumatic schematic drawings.
3. Interpret an instructor-provided isometric drawing.

Trade Terms

Bill of materials (BOM)
Centerlines
Dimension line
Extension lines
Hidden lines
Isometric (ISO) drawing
Orthographic projection
Schematics
View

Industry Recognized Credentials

If you are training through an NCCER-accredited sponsor, you may be eligible for credentials from NCCER's Registry. The ID number for this module is 15203. Note that this module may have been used in other NCCER curricula and may apply to other level completions. Contact NCCER's Registry at 1.888.622.3720 or go to **www.nccer.org** for more information.

You can also show off your industry-recognized credentials online with NCCER's digital badges. Transform your knowledge, skills, and achievements into badges that you can share across social media platforms, send to your network, and add to your resume. For more information, visit **www.nccer.org**.

Contents

Figures and Tables

SECTION ONE

1.0.0 MECHANICAL DRAWINGS

Objective

Read mechanical drawings, interpreting the information they contain.

a. Describe and interpret orthographic projections.
b. Describe and interpret schematic drawings.
c. Describe and interpret isometric drawings.

Performance Tasks

1. Interpret an instructor-provided orthographic projection.
2. Interpret instructor-provided hydraulic and pneumatic schematic drawings.
3. Interpret an instructor-provided isometric drawing.

Trade Terms

Bill of materials (BOM): The list of parts and materials required to construct a system or assembly.

Centerlines: Lines made from alternating long and short dashes that identify the middle of parts or features, such as holes.

Dimension line: A size-, location-, or angle-identifying number combined with an arrow or pair of arrows.

Extension lines: Pairs of short lines perpendicular to a drawing element that identify a dimension's limits.

Hidden lines: Lines made from short dashes that reveal features not visible in a particular view.

Isometric (ISO) drawing: A drawing method that represents a three-dimensional object with all three faces equally tilted toward the drawing surface.

Orthographic projection: A drawing method that represents a three-dimensional object by several two-dimensional views.

Schematics: A diagram-type drawing that shows a system's essential components and interconnections.

View: Part of an orthographic projection, a two-dimensional drawing that accurately represents a three-dimensional object from one vantage point.

Mechanical drawings communicate technical information between engineers, architects, craftworkers, and service personnel. This information includes dimensions, layouts, assembly procedures, and overall appearance. Craftworkers like millwrights, industrial mechanics, and instrumentation technicians use this information to assemble and troubleshoot systems. Carpenters, sheet metal craftworkers, and masons use mechanical drawings to build structures or systems. Drawings come in many styles to meet each craft's particular needs.

For example, millwrights and industrial mechanics commonly work with the following drawing types:

- *General assembly drawings* — These drawings communicate how a machine fits together by showing its different parts and their relationships.
- *Fabrication drawings* — These drawings help machinists construct a part or assembly.
- *Shop drawings* — These facility-specific drawings provide information about an internal process, procedure, or equipment installation.

Some people refer to mechanical drawings as *blueprints*. This term goes back to an early copying process invented in the mid-1800s. Blueprints had white lines on a dark blue background. A later process produced blue lines on a white background. You may see these occasionally if you consult older drawings. Modern drawings, commonly called *prints*, have black lines on a white background. Full-color prints show up in certain settings.

Some facilities avoid paper prints altogether, providing drawings electronically through screens, tablets, and smartphones. This method has become very popular on assembly lines. Large screens near the assembler's workstation show drawings and instructions in full color. Realistic three-dimensional models portray the finished assembly, accurate down to the final paint color.

1.1.0 Orthographic Projections

An **orthographic projection** portrays three-dimensional objects through an assembly of two-dimensional drawings. Imagine an object sitting inside a glass box. You walk around the box and use a marker to draw on the glass what you see. Then, you unfold the box and lay it out flat (*Figure 1*).

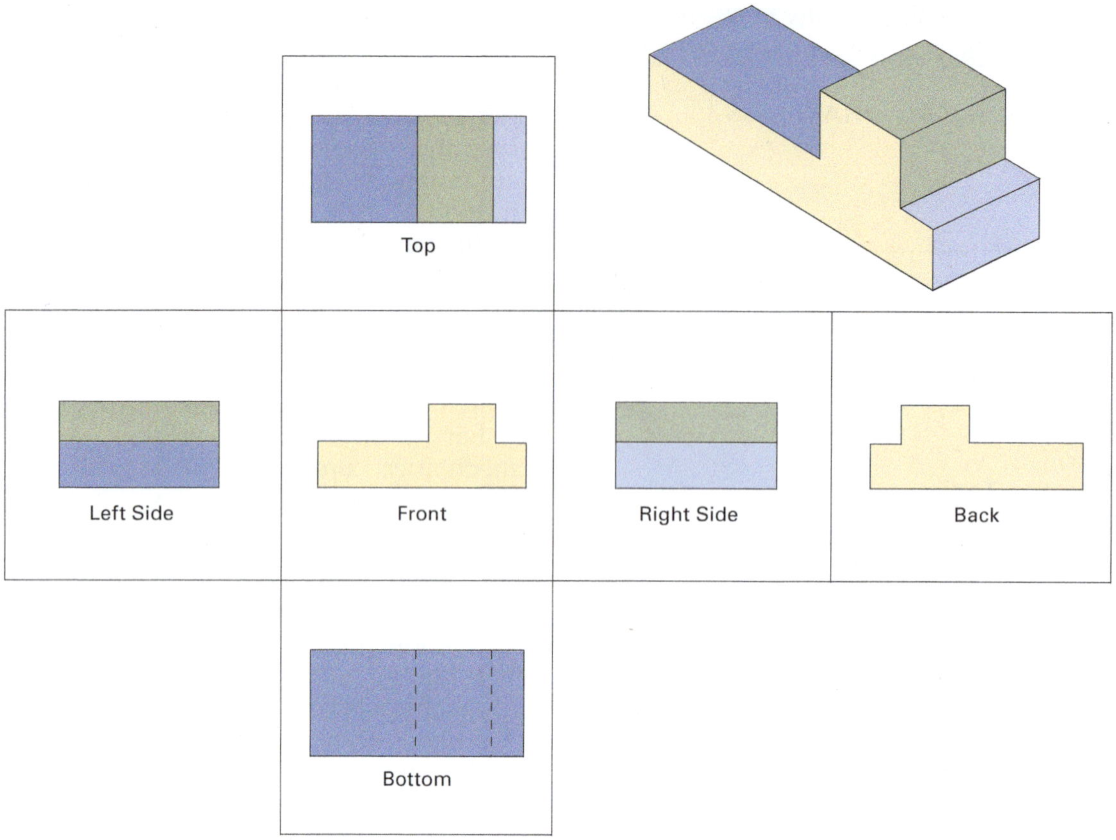

Figure 1 An orthographic projection.

Each individual drawing is an object **view**. Together, the views accurately represent the complete object. As *Figure 1* shows, an orthographic projection may include up to six views. Not all objects will require this many. Notice that each view has a name. View names are relative to the *front view*. Typically, the front view is the most important (*Figure 2*).

1.1.1 Constructing Views

Drafting technicians create orthographic projections by visualizing the object from each view's position. They then "project" perpendicular lines from the object onto the drawing surface (*Figure 2*). These lines help them construct the view. Similar lines also help them create other views from existing ones. After completing the drawing, they erase these lines.

This method produces an accurate and undistorted drawing. If the drawing is *full scale* (1" on the drawing equals 1" on the object), each line will exactly match one on the object. If the drawing is *scaled* (1" on the drawing equals a larger unit on the object), all drawing lines will proportionally match those on the object.

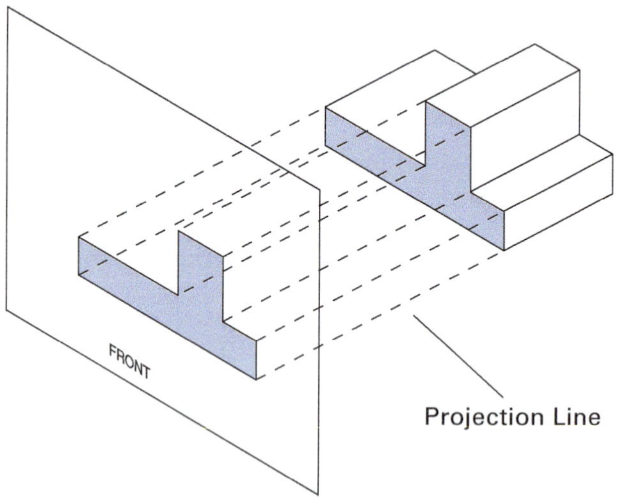

Figure 2 Front view.

In almost all cases, a single view can't represent the object completely. For instance, the front view doesn't reveal the object's depth. *Figure 3* shows two different objects with identical front views.

Adding a top view (*Figure 4*) shows the object's depth accurately. This view also provides extra detail not visible from the front view. Notice that the drafting technician creates the top view by the same method—projecting perpendicular lines from the object onto the drawing surface. The result represents the object more completely than just the one view.

While two views might be enough for some objects, many require more. *Figure 5* shows two different objects with the same front and top views.

Adding a side view (*Figure 6*) completes the orthographic projection. Again, the drafting technician projects perpendicular lines from the object onto the drawing surface. The three views adequately represent the object, so no more are necessary.

The finished drawing resembles an unfolded box with each view appearing around the front view in the proper position (*Figure 7*). For many objects, three views provide enough information. With complex objects, however, they may not show sufficient detail. Adding extra views solves the problem. Referring back to *Figure 1*, you can see that up to six views are possible.

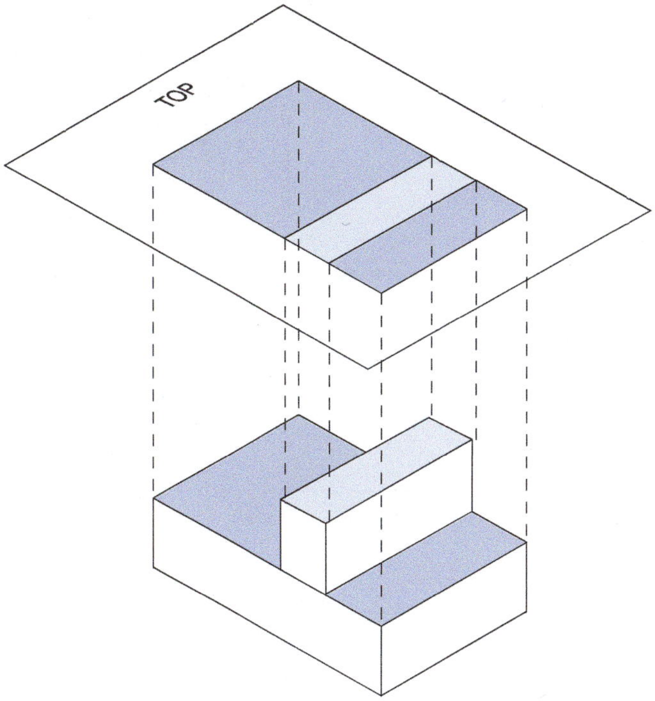

Figure 4 Adding a top view.

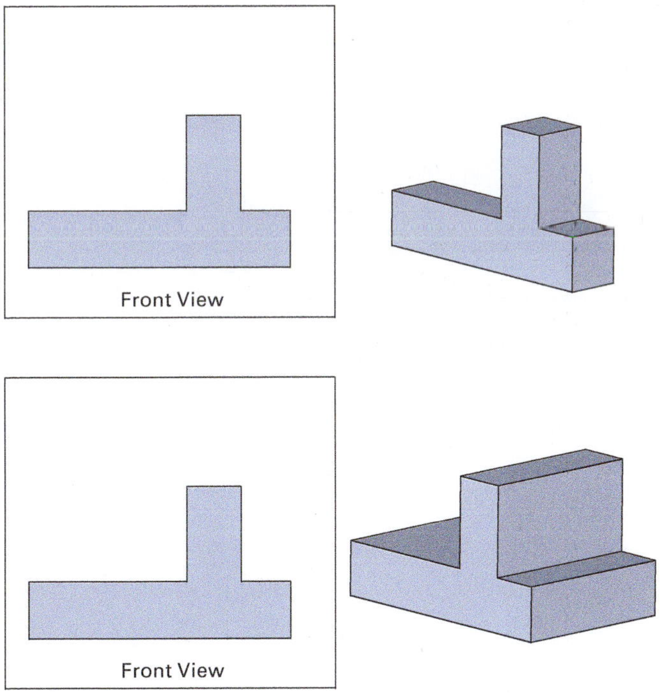

Figure 3 Two objects with the same front view.

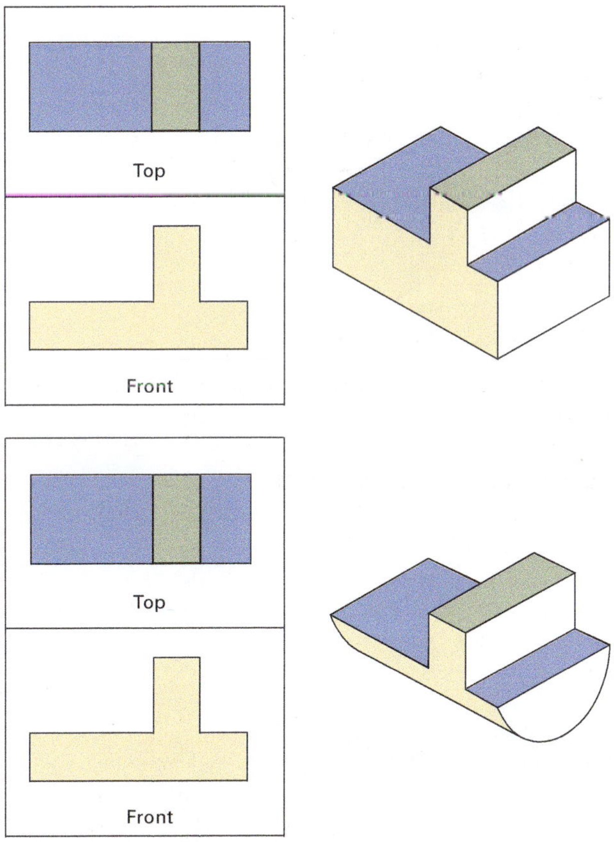

Figure 5 Two objects with the same front and top views.

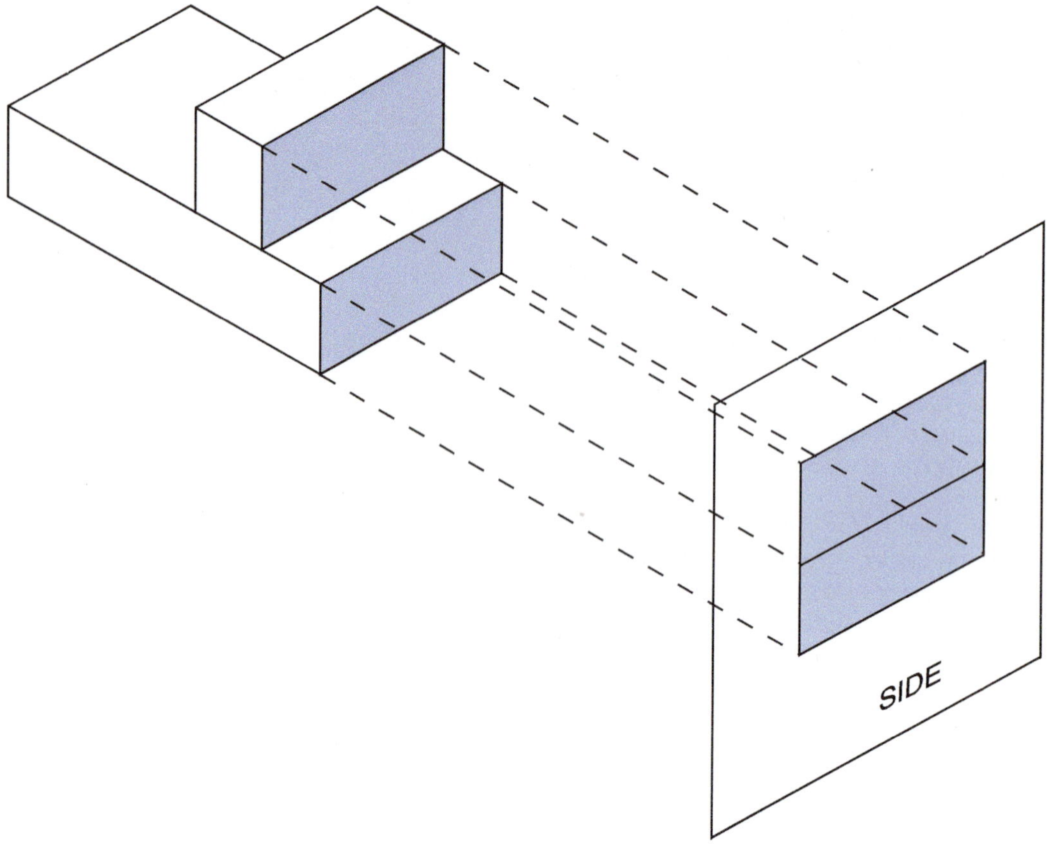

Figure 6 Adding a side view.

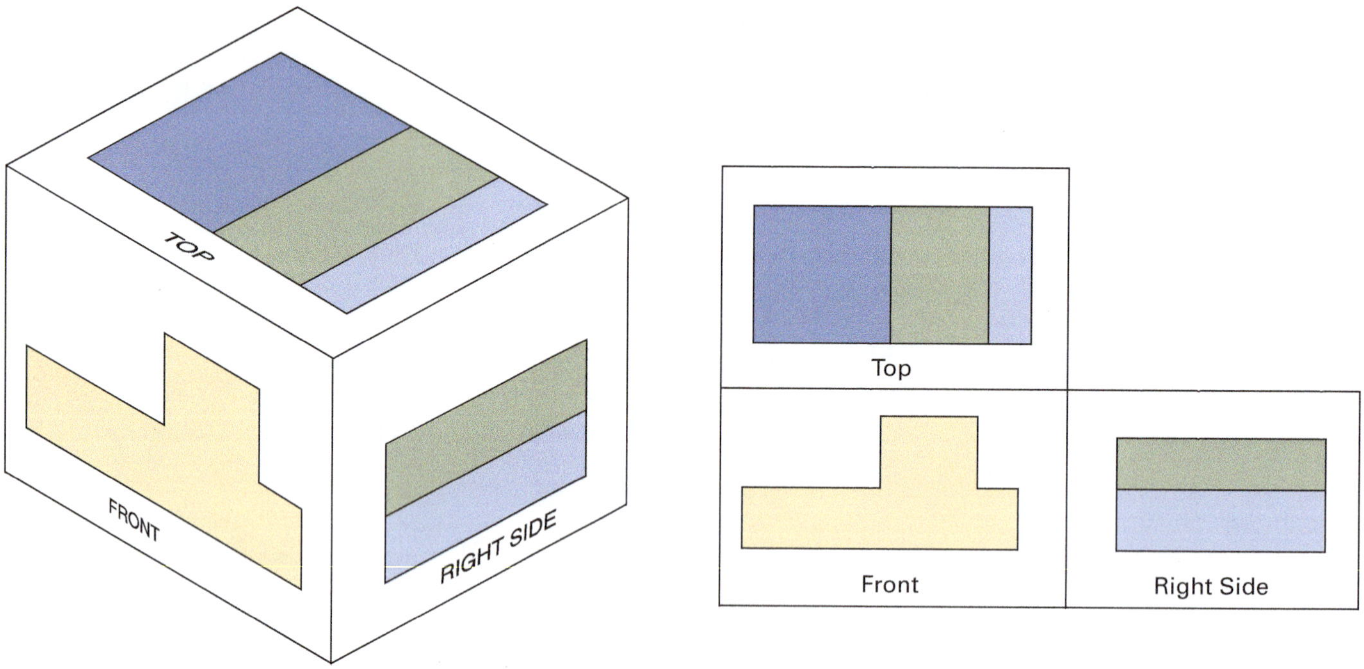

Figure 7 Completed orthographic projection.

Sometimes, internal features, such as blind holes, may not be fully visible on any view. In these cases, the drafting technician adds **hidden lines** to supply the detail. Hidden lines are dashed rather than solid. Going back to *Figure 1*, notice the two hidden lines on the bottom view. These show the position of the middle section above.

In certain situations, a few well-placed hidden lines may eliminate the need for an entire view. But unnecessary hidden lines can cause confusion. For this reason, drafting technicians add them only when required.

1.1.2 Adding Dimensions

After constructing the views, the drafting technician dimensions the drawing. Dimensions provide the details required to fabricate the object, install it, or attach other assemblies to it. *Figure 8* shows a dimensioned three-view orthographic projection.

Notice that almost all dimensions appear between a pair of lines perpendicular to the part. These are **extension lines** and show the dimension's limits. The arrows and the value between them make up the **dimension line** itself. In this case, it indicates the distance between the extension lines. Dimension lines can also specify angles, hole diameters, and an arc or circle's radius.

Centerlines have alternating long and short dashes. They identify hole centers or the centers of other features. Notice that the top view in *Figure 8* includes several dimensions associated with centerlines. These identify hole positions and the center of the structure on the right.

All three views include hidden lines. These show where holes pass through the part. The right view shows two holes that are only partly visible from this perspective. Hidden lines show the rest of the holes.

> **NOTE**
> Some drawings do not include units with dimensions. Examine the drawing's title block to identify the drawing's standard unit (inches, centimeters, etc.)

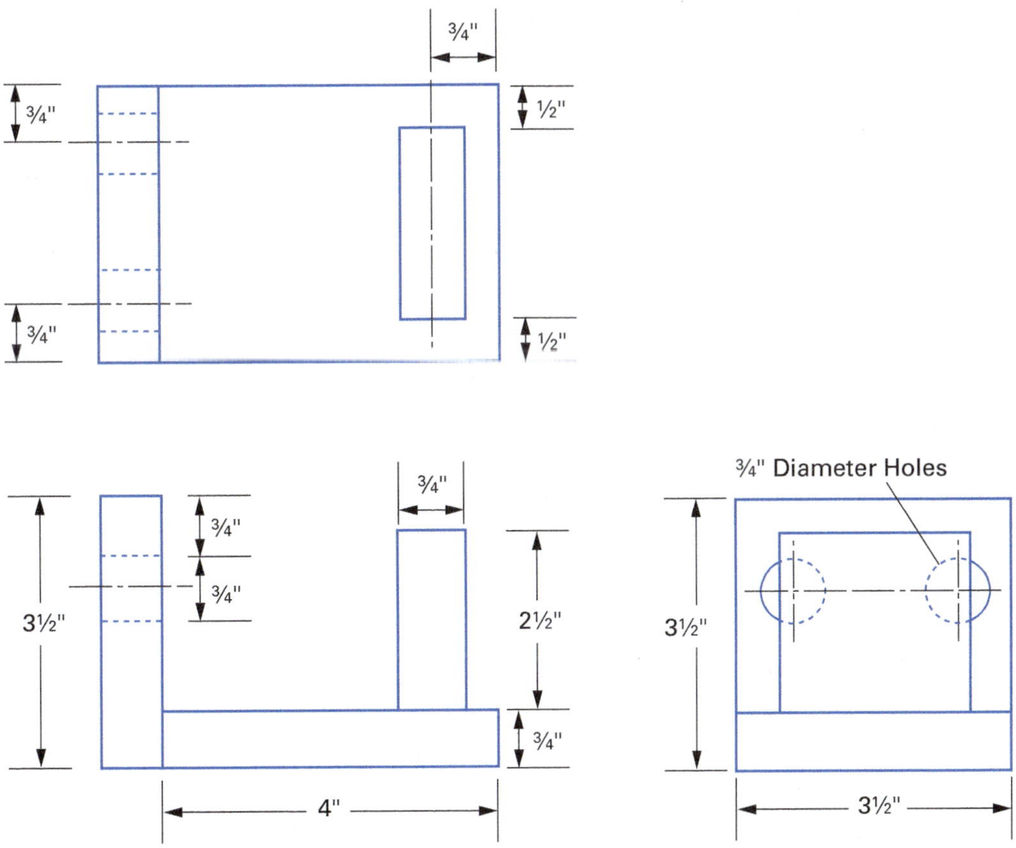

Figure 8 Dimensioned orthographic projection.

1.1.3 Interpreting Orthographic Projections

Now that you understand how orthographic projections work, you should be able to interpret them easily. Always begin by identifying the views. Mentally reconstruct the object, noting the features visible on each view. Use dimensions to determine sizes and locations. You may have to search each view to find a particular dimension. Finally, identify features revealed by hidden lines.

1.2.0 Schematic Drawings

Several different crafts use **schematics** to install, troubleshoot, and repair equipment. When you're tracing a fault, the schematic helps you find the source. Schematics show the system's components and their connections. This information helps you determine where problems might lie and how components affect each other. With a schematic, troubleshooting becomes systematic rather than hit-or-miss. These drawings have the following qualities:

- They make a complicated system easy to understand.
- They tell you how the system should work. Connection lines link components, showing relationships and interactions.
- They represent components with standard symbols. Once you've learned the symbols, you can read almost any schematic.
- Schematics cross language barriers, since anyone who understands the symbols can interpret the schematics.

Many craftworkers rely on schematics, including millwrights, industrial mechanics, instrumentation technicians, electricians, and electronic systems technicians. Schematic type and style depend upon the craft. For example, piping, hydraulic, and pneumatic schematics are common in the millwright, industrial mechanic, and instrumentation worlds.

1.2.1 Piping Schematics

To read a schematic you must know what its symbols represent. Piping schematics use a large symbol library, as *Figure 9A*, *Figure 9B*, *Figure 9C*, and *Figure 9D* illustrate. Study these symbols so you'll recognize them. Engineering firms often provide a symbol key with their schematics. Consult it to identify unfamiliar symbols.

Besides symbols, you must also recognize piping abbreviations. *Table 1A*, *Table 1B*, and *Table 1C* list the most common ones. Study these so you'll recognize them on piping drawings.

Piping schematics show the system's layout. Symbols identify major features, such as valves, unions, flow direction, pipe sizes, and supporting wall arrangements. *Figure 10* shows a building piping system's schematic. With the help of symbol lists, you should be able to identify everything.

Drawings of systems with complex subassemblies manage things differently. Separate pages diagram the subassemblies. On the main drawing, boxes referencing the separate drawings replace the subassemblies. This strategy keeps the main drawing compact and manageable.

Most piping schematics show each joint's connection type. These include the following:

- Flanged
- Screwed
- Butt welded
- Socket welded
- Soldered or brazed

When reading a piping schematic, first identify each component. Then, work out how the components fit together and interact. Be sure to read the Notes section, if present, since it often contains essential details.

Figure 11 shows a different piping schematic style. It's a *piping and instrumentation diagram* (P&ID). These document refineries and chemical plant process equipment. They help you understand the system quickly since they contain only essential information.

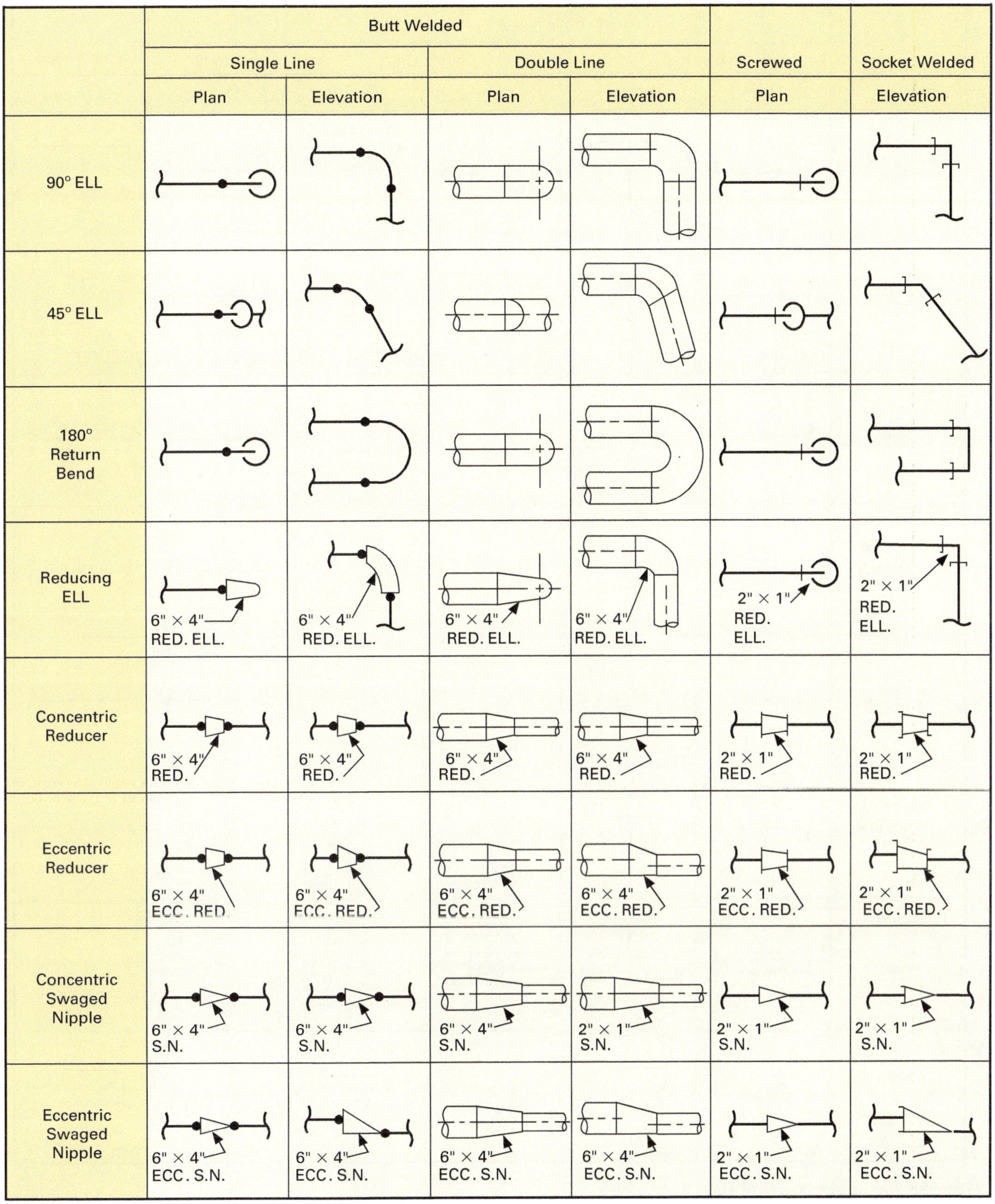

Figure 9A Piping symbols (1 of 4).

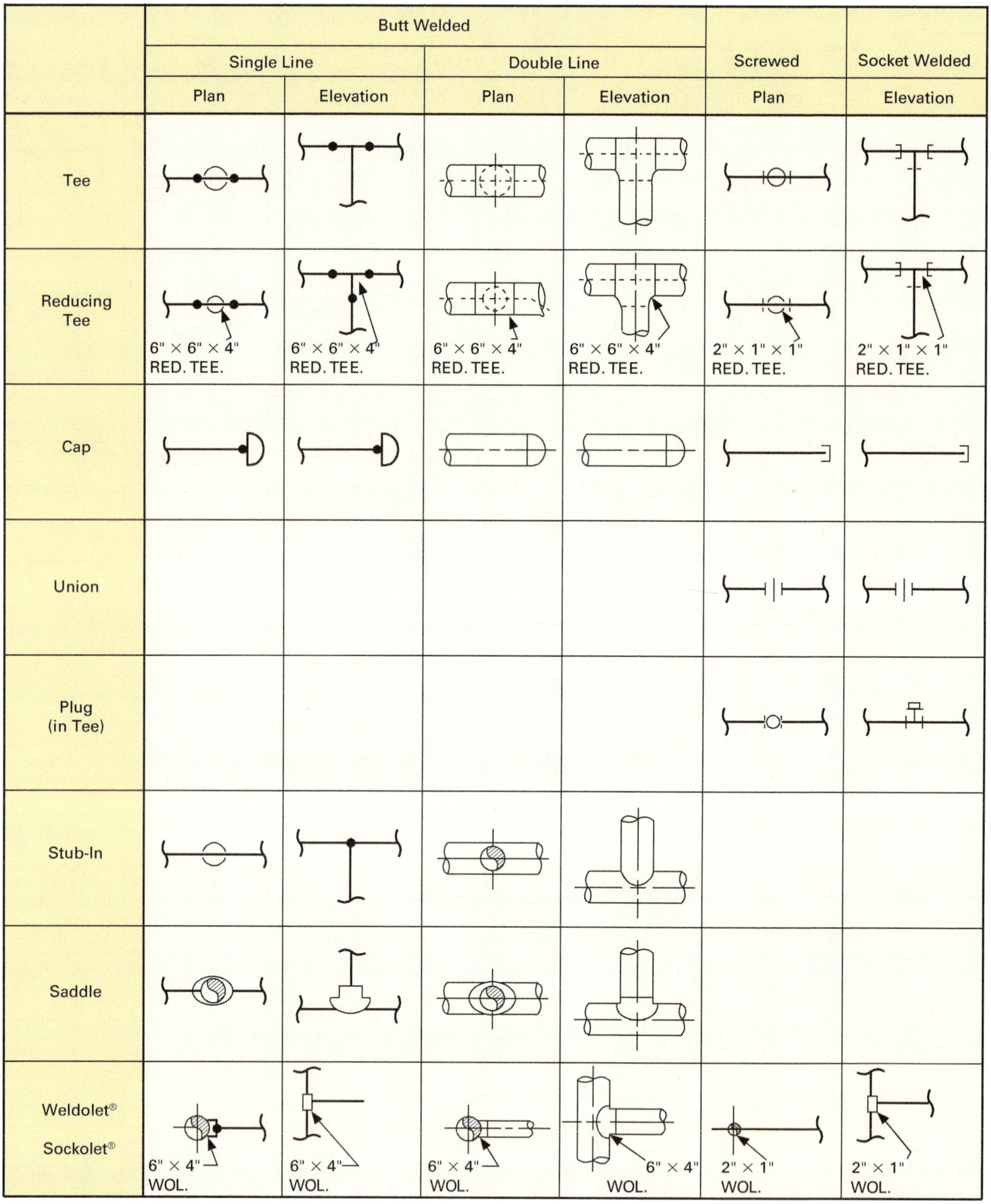

Figure 9B Piping symbols (2 of 4).

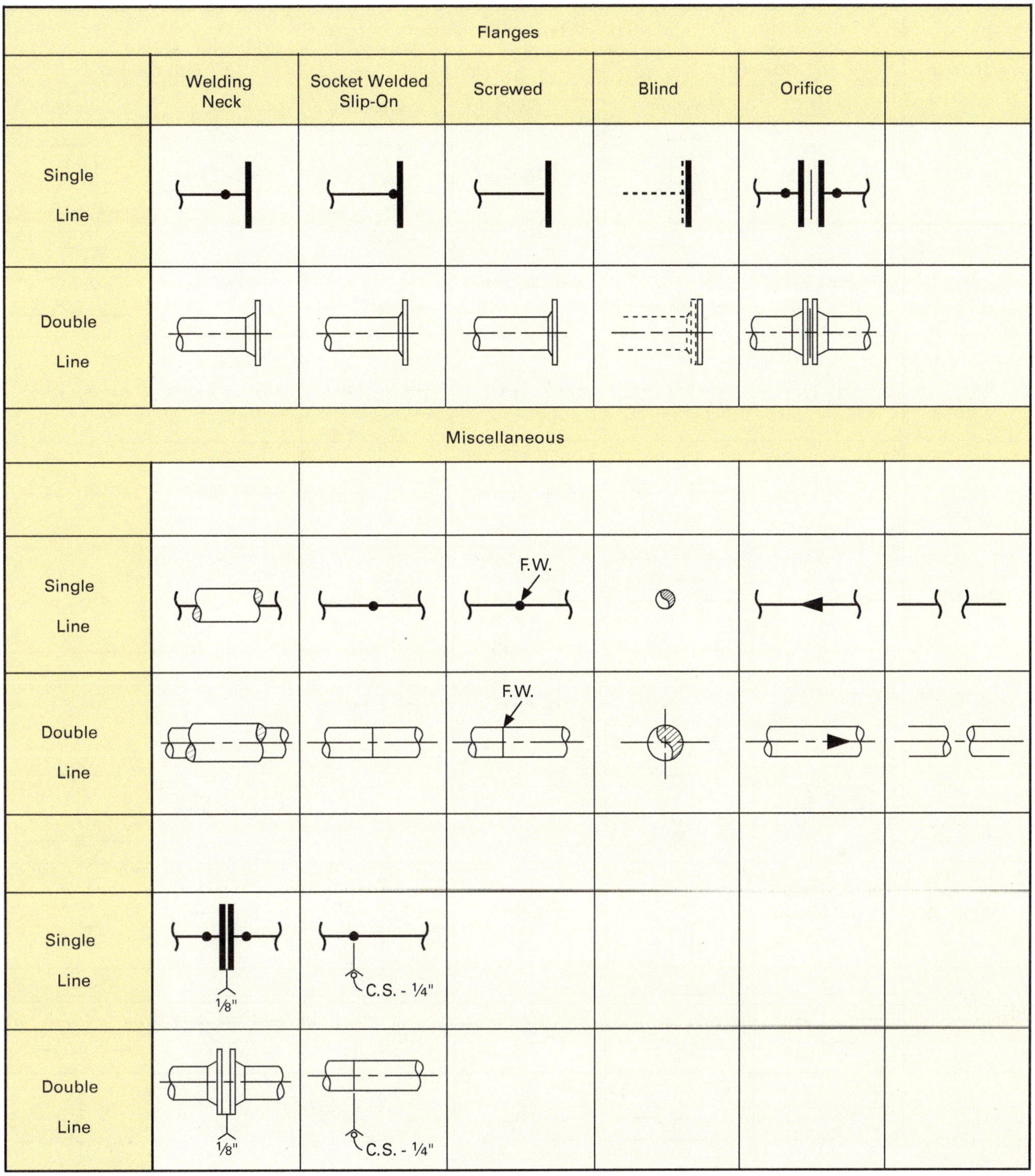

Figure 9C Piping symbols (3 of 4).

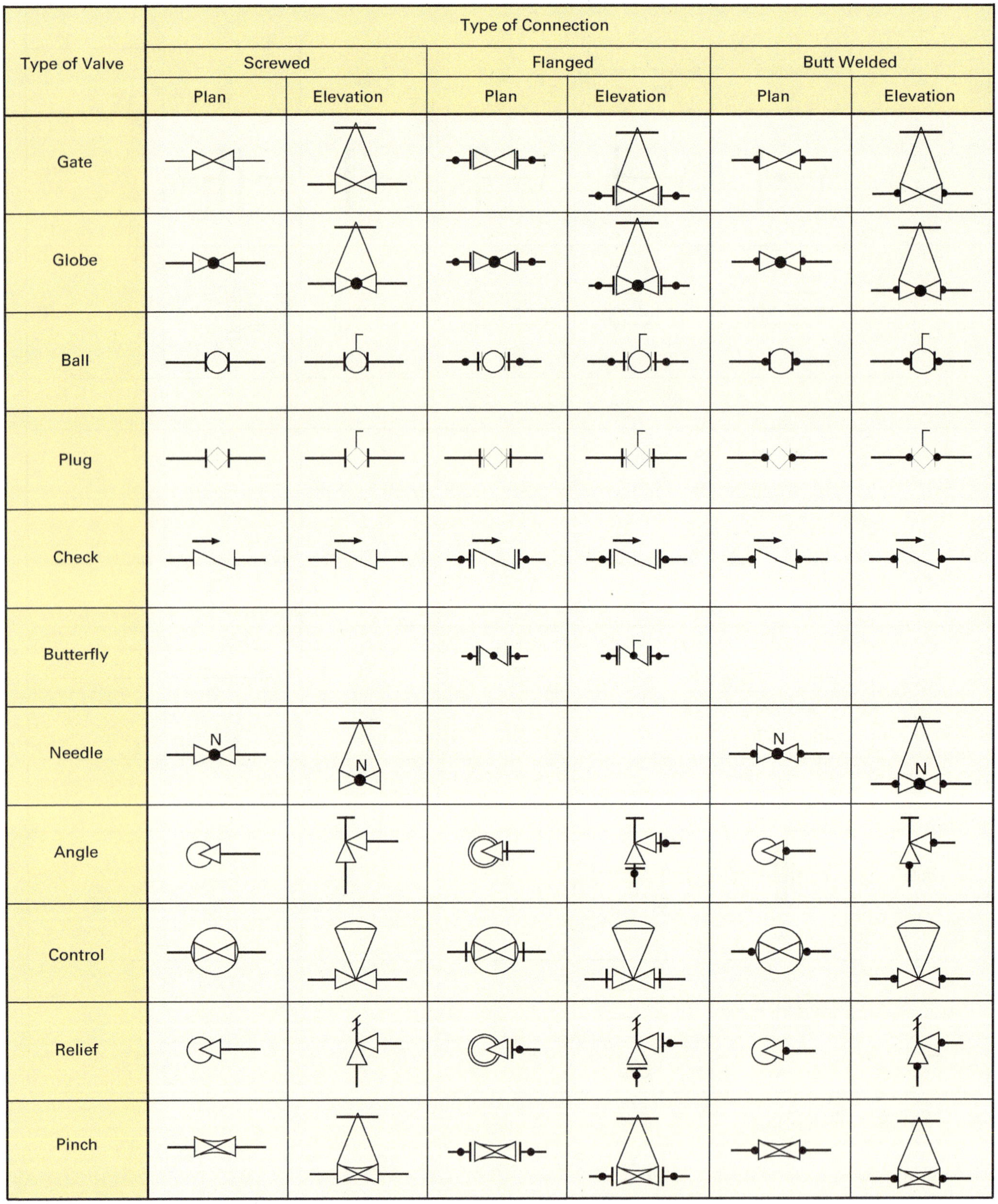

Figure 9D Piping symbols (4 of 4).

Table 1A Common Piping Abbreviations (1 of 3)

Term	Abbreviation	Term	Abbreviation
Adapter	ADPT	Detail	DET
Air preheating	A	Diameter	DIA
American Iron and Steel Institute	AISI	Dimension	DIM
American National Standards Institute	ANSI	Discharge	DISCH
American Petroleum Institute	API	Double extra-strong	XX STRG
American Society for Testing and Materials International	ASTM	Drain	DR
American Society of Mechanical Engineers	ASME	Drain funnel	DF
Ash removal water, sluice or jet	AW	Drawing	DWG
Aspirating air	AA	Drip leg or dummy leg	DL
Auxiliary steam	AS	Ductile iron	DI
Benchmark	BM	Dust collector	DC
Beveled	B	Each	EA
Beveled end	BE	Eccentric	ECC
Beveled, both ends	BBE	Elbolet®	EOL
Beveled, large end	BLE	Elbow	ELB
Beveled, one end	BOE	Electric resistance weld	ERW
Beveled, small end	BSE	Elevation	EL
Bill of materials	BOM	Equipment	EQUIP
Blind flange	BF	Evaporator vapor	EV
Blowoff assembly	BO	Exhaust steam	E
Bottom of pipe	BOP	Expansion	EXP
Butt weld	BW	Expansion joint	EXP JT
Carbon steel or cold spring	CS	Extraction steam	ES
Cast iron	CI	Fabrication (dimension)	FAB
Ceiling	CLG	Face of flange	FOF
Chain operated	CO	Faced and drilled	F&D
Chemical feed	CF	Factory Mutual	FM
Circulating water	CW	Far side	FS
Cold reheat steam	CR	Feed pump balancing line	FB
Compressed air	CA	Feed pump discharge	FD
Concentric or concrete	CONC	Feed pump recirculating	FR
Condensate	C	Feed pump suction	FS
Condenser air removal	AR	Female	F, FM
Continue, continuation	CONT	Female pipe thread	FPT
Coupling	CPLG	Field support / Forged stee	FS

Table 1B Common Piping Abbreviations (2 of 3)

Term	Abbreviation	Term	Abbreviation
Field weld	FW	Lap weld	LW
Figure	FIG	Large male	LM
Fillet weld	W	Length	LG
Finish floor	F/F, FIN FL	Long radius, Long tangent	LR, LT
Finish grade	FIN GR	Long weld neck	LWN
Fitting	FTG	Low pressure	LP
Fitting makeup	FMU	Low-pressure drains	DR
Fitting to fitting	FTF	Low-pressure steam	LPS
Flange	FLG	Lubricating oil	LO
Flat face	FF	Main steam	MS
Flat on bottom	FOB	Main system blowouts	BL
Flat on top	FOT	Makeup water	MU
Floor drain	FD	Male	M
Foundation	FDN	Malleable iron	MI
Fuel gas	FG	Manufacturer	MFR
Fuel oil	FO	Manufacturer's Standard Society	MSS
Gage	GA	National Pipe Thread	NPT
Galvanized	GALV	Nipolet®	NOL
Gasket	GSKT	Nipple	NIP
Grating	GRTG	Nominal	NOM
Hanger	HGR	Not to scale	NTS
Hanger rod	HR	Nozzle	NOZ
Hardware	HDW	Outside battery limits	OSBL
Header	HDR	Outside diameter	OD
Heat traced, heat tracing	HT	Outside screw and yoke	OS&Y
Heater drains	HD	Overflow	OF
Heating system	HS	Pipe support	PS
Heating, ventilating, and air conditioning	HVAC	Pipe tap	PT
Hexagon	HEX	Piping and instrumentation diagram	P&ID
Hexagon head	HEX HD	Plain	P
High point	HPT	Plain end	PE
High pressure	HP	Plain, both ends	PBE
Horizontal	HORIZ	Plain, one end	POE
Hot reheat steam	HR	Point of intersection	PI
Hydraulic	HYDR	Point of tangent	PT
Increaser	INCR	Pounds per square inch	PSI
Input/output	I/O	Process flow diagram	PFD
Inside diameter	ID	Purchase order	PO
Insulation	INS	Radius	RAD
Invert elevation	IE	Raised face	RF
Iron pipe size	IPS	Raised face slip-on	RFSO
Isometric	ISO		
Issued for construction	IFC		

Table 1C Common Piping Abbreviations (3 of 3)

Term	Abbreviation	Term	Abbreviation
Raised face smooth finish	RFSF	Temperature or temporary	TEMP
Raised face weld neck	RFWN	That is	I.E.
Raw water	RW	Thick	THK
Reducer, reducing	RED	Thousand	M
Relief valve	PRV-PSV	Thread, threaded	THRD, T
Ring-type joint	RTJ		
Rod hanger or right hand	RH	Threaded end	TE
Safety valve vents	SV	Threaded, both ends	TBE
Sanitary	SAN	Threaded, large end	TLE
Saturated steam	SS	Threaded, one end	TOE
Schedule	SCH	Threaded, small end	TSE
Screwed	SCRD	Threadolet®	TOL
Seamless	SMLS	Top of pipe or top of platform	TOP
Section	SECT	Top of steel or top of support	TOS
Service and cooling water	SW	Treated water	TW
Sheet	SH	Turbine	TURB
Short radius	SR	Typical	TYP
Slip-on	SO	Underwriters Laboratories	UL
Socket weld	SW	Vacuum	VAC
Sockolet®	SOL	Vacuum cleaning	VC
Stainless steel	SS	Vents	V
Standard weight	STD WT	Vitrified tile	VT
Steel	STL	Wall thickness or weight	WT
Suction	SUCT	Weld neck	WN
Superheater drains	SD	Weldolet®	WOL
Swage	SWG	Well water	WW
Swaged nipple	SN	Wide flange	WF

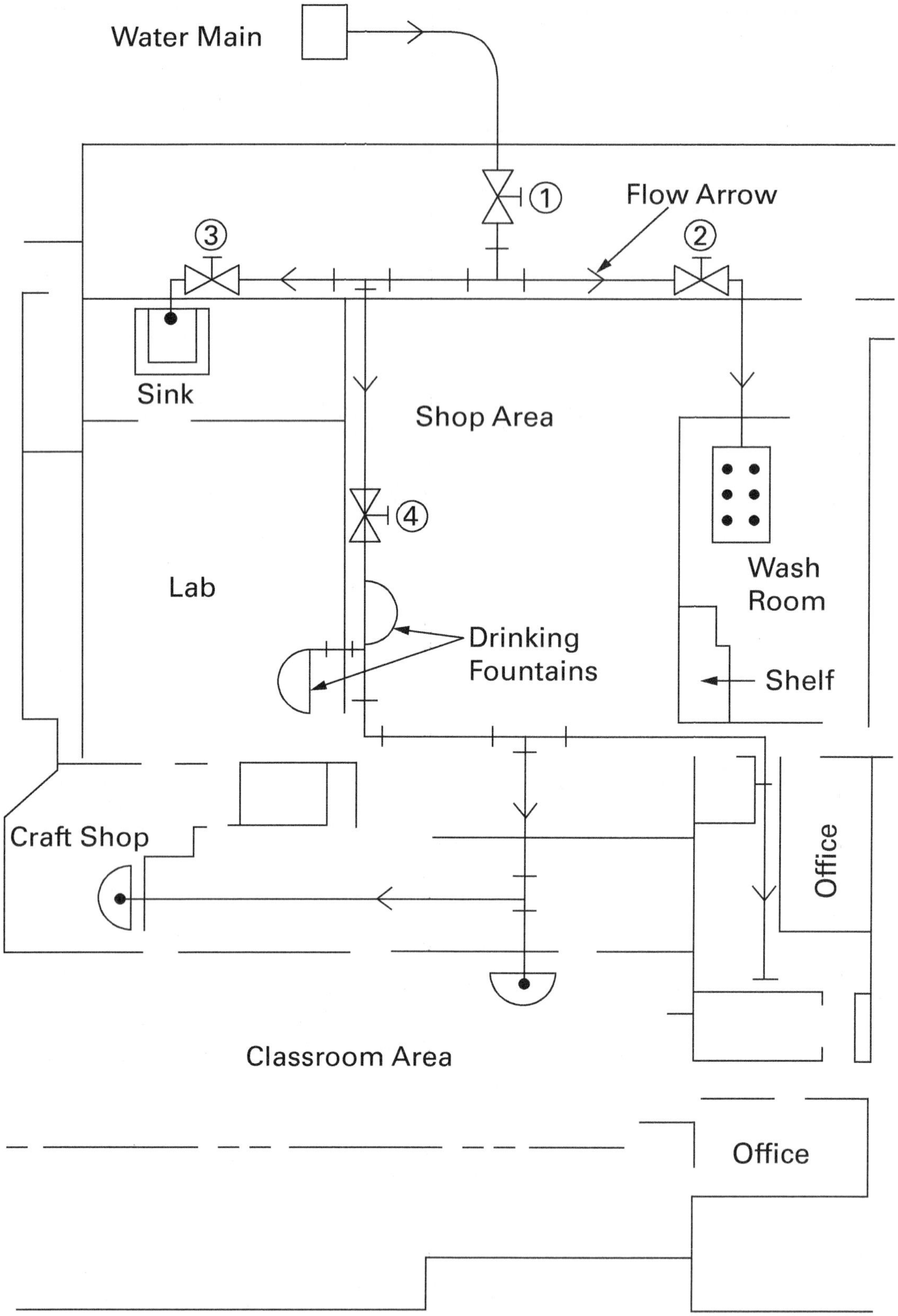

Figure 10 Piping schematic.

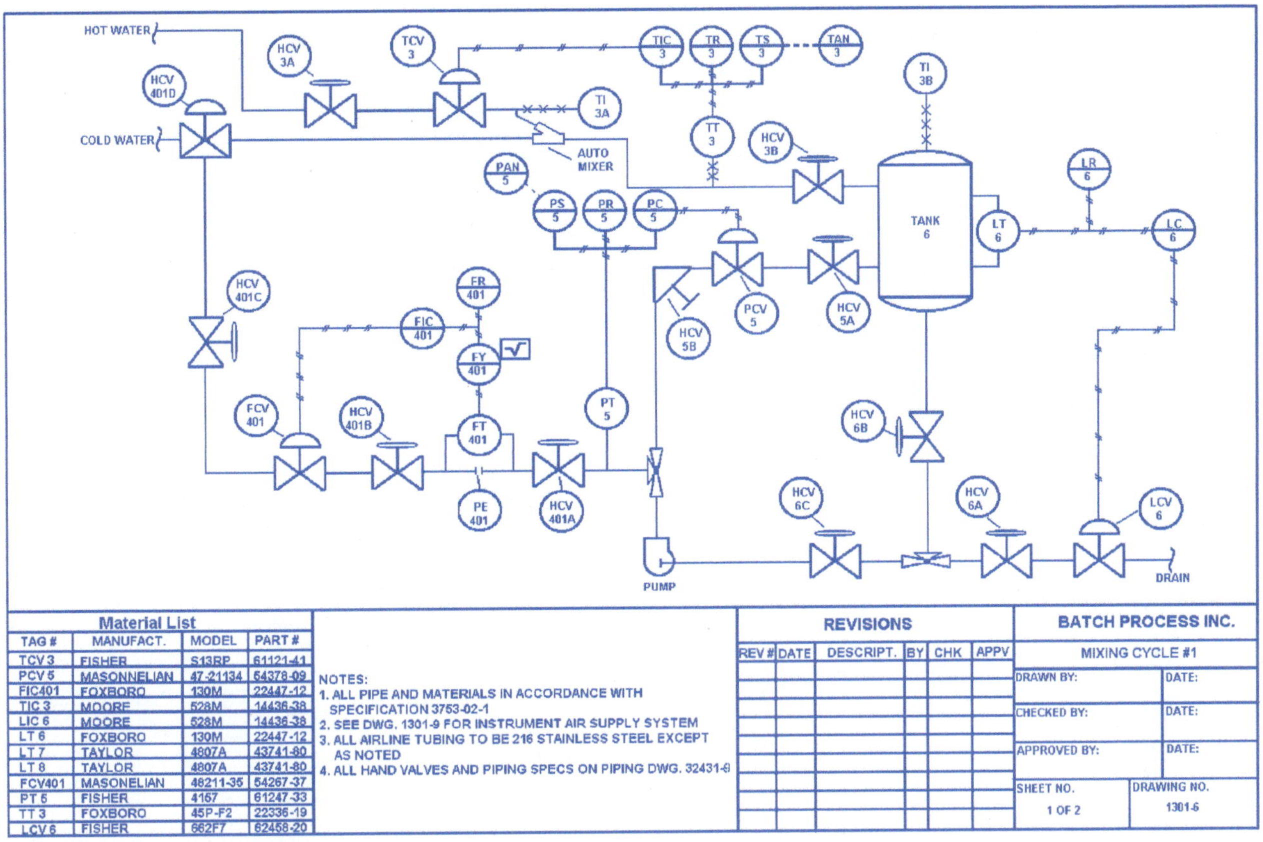

Figure 11 Another piping schematic (a P&ID).

1.2.2 Hydraulic and Pneumatic Schematics

Many industrial installations rely on hydraulic or pneumatic systems to control and power machinery. In many ways, they're similar since both use a fluid (oil or air) to transmit power. Pipes, tubes, and hoses carry the fluid, while valves control it.

Like piping schematics, hydraulic and pneumatic schematics use standardized symbols and abbreviations. Hydraulic and pneumatic symbols are almost identical. Consult the drawing's title block and Notes section to identify the system type. *Figure 12A* and *Figure 12B* show common schematic symbols. *Table 2* lists common abbreviations. Study these so you'll recognize them. By identifying the symbols and the connections between them, you'll understand how the system works.

Valves and Basic Symbols	
Valve, Check	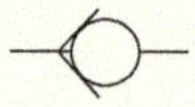
Valve, Manual Shutoff	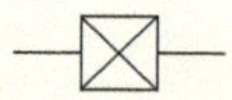
Valve, Maximum Pressure (Relief)	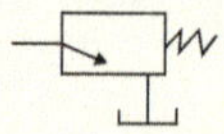
Valve, Basic Symbol Single Flow Path is Modified	
Valve, Basic Symbol Multiple Flow Paths are Changed	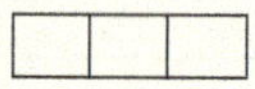
Valve, Single Flow Path, Normally Open	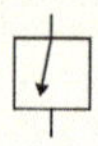
Valve, Single Flow Path, Normally Closed	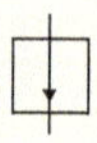
Valve, Blocked	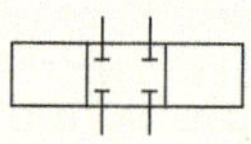
Valve, Multiple Flow Paths, Open (Arrows Denote Difection of Flow)	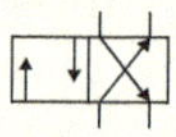
Valve, Special (Identify and Connect All Lines)	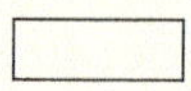
Valve Examples	
Valve, Relief Remotely Operated (Unloading Valve)	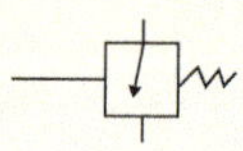

Valve Examples (Cont'd)	
Valve, Deceleration Normally Open	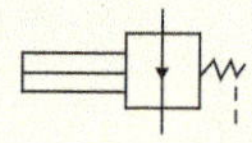
Valve, Sequence, Directly Operated	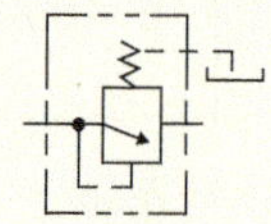
Valve, Pressure Reducing	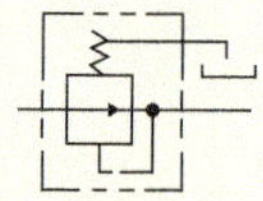
Valve, Counterbalance With Integral Check	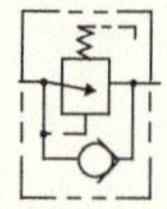
Valve, Flow Rate Control, Variable, Pressure Compensated	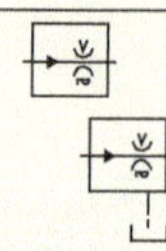
Valve, Directional, 2 Position 3 Connection	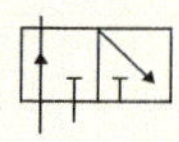
Valve, Directional, 3 Position 4 Connection Open Center	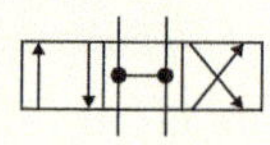
Valve, Directional, 3 Position 4 Connection Closed Center	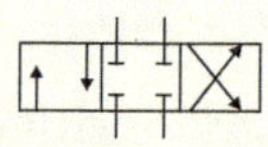
Methods of Control	
Spring	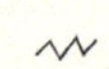
Pilot Operated	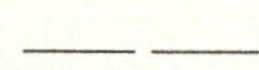
Pilot Operated, Differential Area	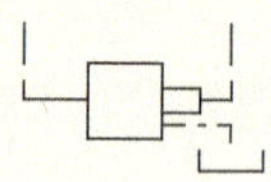

Figure 12A Hydraulic and pneumatic symbols (1 of 2).

Lines and Line Functions	
Line, Working	
Line, Pilot (L 20W)	
Line, Drain (L 5W)	
Connector (Dot to Be 5 × Width of Lines)	
Line, Flexible	
Line, Joining	
Line, Passing	
Direction of Flow	
Line to Reservoir Above Fluid Level Below Fluid Level	
Line to Vented Manifold	
Plug or Plugged Connection	
Testing Station (Gauge Connection)	
Power Takeoff (Hyd.)	
Restriction, Fixed	
Restriction, Variable	

Pumps	
Hydraulic Pump Fixed Displacement	
Hydraulic Pump Variable Displacement	

Motors and Cylinders	
Motor, Rotary, Fixed Displacement	
Motor, Rotary, Variable Displacement	
Cylinder, Single Acting	
Cylinder, Double Acting Single End Rod Double End Rod	

Miscellaneous Units	
Rotating Shaft (Arrow in Front of Shaft)	
Component Enclosure	
Reservoir	
Pressure Gauge (Manometer)	
Accumulator	
Receiver	
Lubricator	
Filter	
Cooler	
Electric Motor	

Figure 12B Hydraulic and pneumatic symbols (2 of 2).

Table 2 Common Hydraulic and Pneumatic Abbreviations

Term	Abbreviation
Accumulator	ACC
Centrifugal	CENT
Compensator	COMP
Cylinder	CYL
Detent	DET
Electric motor	ELEC MOT
Engine	ENG
Filter	FLT
Flowmeter	FM
Heat exchanger	HE
Hydraulic motor	HYD MOT
Intensifier	INT
Manual	MAN
Mechanical	MECH
Pressure switch	PS
Servo	SERV
Solenoid	SOL
Solenoid controlled, pilot operated	SOL PLT
Strainer	STR
Tachometer	TACH
Thermal	THRM

NOTE

The symbols in *Figure 12A* and *Figure 12B* are typical of those that you'll encounter. Be aware, however, that symbols can vary in appearance. Similarly, symbols used in other countries, while similar, may look slightly different. Be sure that you understand all symbols before working with a system.

1.2.3 Pneumatic Systems

Pneumatic systems are *open*. Their working fluid (air) comes straight from the atmosphere. A compressor pumps it into a tank. After it does its work, it simply escapes back to the atmosphere. Pneumatic schematics contain seven key elements:

- Compressor
- Receiver
- Relief valves
- Conditioning devices
- Lines
- Valves
- Actuators

The compressor pumps air into the receiver under pressure. Relief valves protect the system from excess pressure. Conditioning devices process and clean the air. Lines deliver it to the destination. Valves control the flow. Actuators turn the fluid power into motion. *Figure 13* shows a simple pneumatic system schematic.

To understand the system, work through the schematic carefully. Identify each symbol. Examine the links between the symbols to determine relationships. For the system in *Figure 13*, the first component is the compressor (A). It feeds air into the receiver equipped with a relief valve (B). Next, comes a filter (C) that removes particles from the air.

The pressure-reducing valve (D) is adjustable. It lowers the air pressure from the receiver to whatever the system requires. A lubricator (E) adds an oil mist to the air, something many pneumatic tools require. Finally, the pressure gauge (F) indicates the output air pressure. Since this is a pneumatic system, there are no return lines coming back to the receiver.

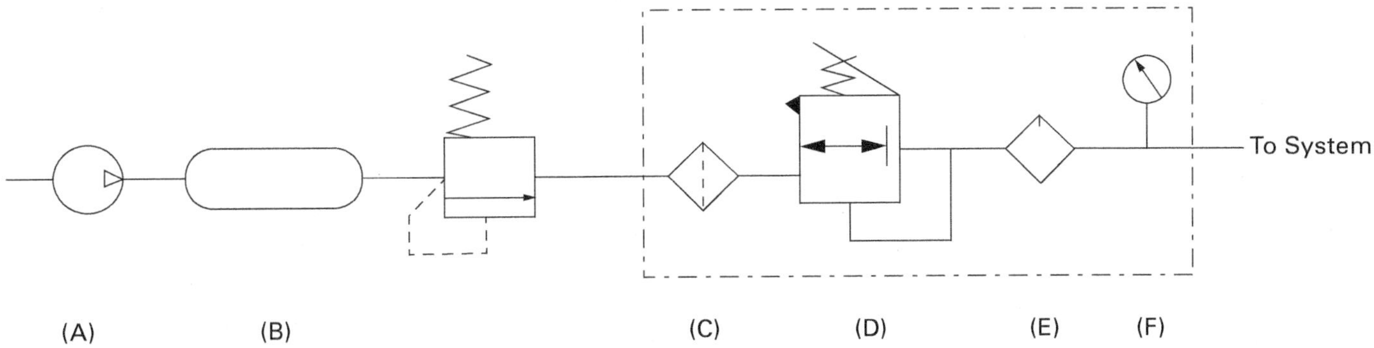

Figure 13 Pneumatic system schematic.

1.2.4 Hydraulic Systems

Hydraulic systems are *closed*. The working fluid (oil) comes from a reservoir. A pump pressurizes the oil. After it does its work, the oil returns to the reservoir, where the cycle begins again. Hydraulic schematics contain seven key elements:

- Reservoir
- Pump
- Accumulator
- Relief valves
- Lines
- Valves
- Actuators

The reservoir holds the hydraulic oil. The pump transfers it to the accumulator, which maintains it under pressure. Relief valves protect the system. Lines deliver it to the destination. Valves control the flow. Actuators turn the fluid power into motion. Used oil returns to the reservoir. *Figure 14* shows a hydraulic system schematic.

To understand the system, work through the schematic carefully. Identify each symbol. Examine the links between the symbols to determine relationships. The system in *Figure 14* drives a hydraulic cylinder, which can move in two directions. The operator selects the direction with a control valve. Since this system has many components, dotted lines divide it into numbered sections called *local areas*. Each performs a specific function:

- *Area 1* — Contains the reservoir and a strainer to clean the oil before it enters the next stage.
- *Area 2* — Contains an electric motor that drives the hydraulic pump. The accumulator maintains the oil under pressure.
- *Area 3* — Contains a pressure-relief valve that protects the system. A pressure gauge reports the oil's pressure at this point.
- *Area 4* — Contains the control valve. It directs the pressurized oil either to the top or bottom of the cylinder. It also allows the oil from the opposite end to flow back into the reservoir.
- *Area 5* — Contains the actuator, in this case a hydraulic cylinder. If pressurized oil enters the bottom it moves up. If oil enters the top, it moves down.
- *Area 6* — Contains another pressure-relief valve and a pressure gauge. They help prevent a pressure buildup in the actuator's lower section.

Since this is a hydraulic system, you may wonder why no return line goes back to the reservoir in Area 1. Notice that there are several small reservoir symbols in other areas. These are not separate reservoirs. Instead, they indicate that oil can flow back to the reservoir in Area 1 from several different places. Each pressure-relief valve can return oil to the reservoir if the valve opens. The main oil return comes from the control valve in Area 4. It sends the oil back after it leaves the actuator.

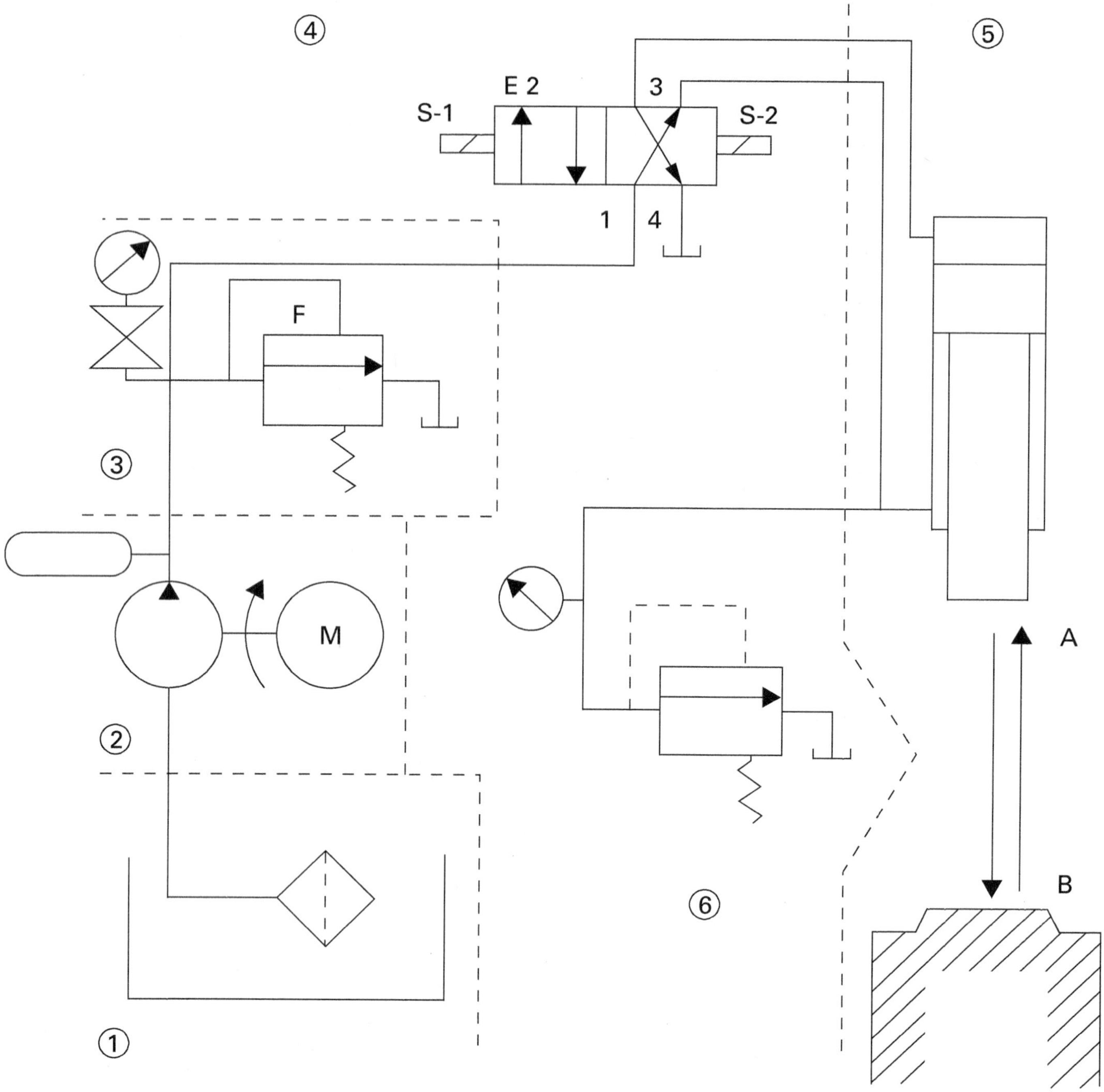

Figure 14 Hydraulic system schematic.

1.3.0 Isometric Drawings

An **isometric (ISO) drawing** portrays a three-dimensional object somewhat realistically. As *Figure 15* shows, an ISO is a single view, typically from one corner. Sometimes, isometric drawings include several orthographic-style views. This information simplifies fabricating the object. Realistic 3D models are common today thanks to solid-modeling software.

Isometric drawings usually include dimensions, like those in *Figure 15*. If units do not appear with dimensions, consult the drawing title block to determine the standard unit. Also remember that the Ø symbol indicates a diameter value, while an R indicates a radius value.

Isometric-style schematics can illustrate piping systems (*Figure 16*). The ISO shows the details required to fabricate and install the system. They use the same symbols as other piping schematics.

To read a piping ISO, begin at the end nearest the source and work in the flow direction. Usually, arrows provide this information. Trace the line from the first symbol, identifying each symbol and abbreviation as you progress. Note component dimensions. You'll need this information to calculate take-out for fittings and flanges. Pay attention to pipeline elevations too.

ISOs usually includes a **bill of materials (BOM)**, which lists the system's components and materials (*Figure 17*). Examine the BOM since it provides extra details that may help you interpret the drawing.

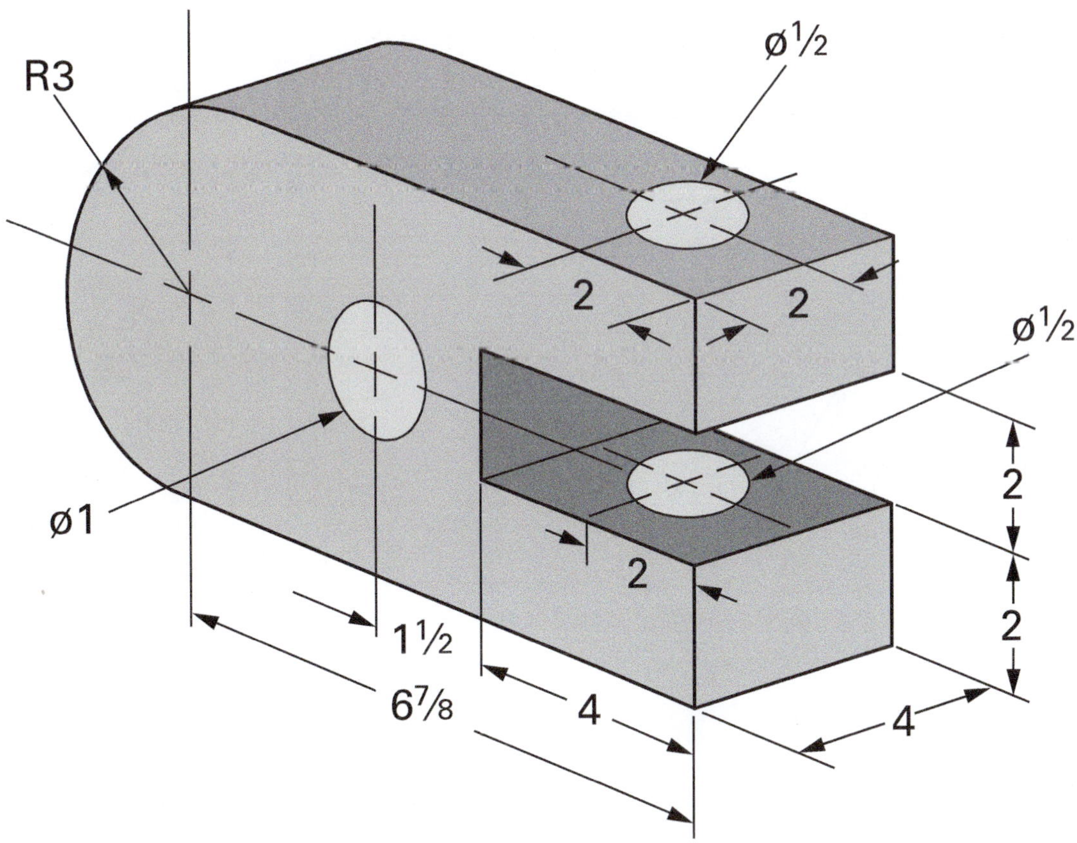

Figure 15 Isometric drawing.

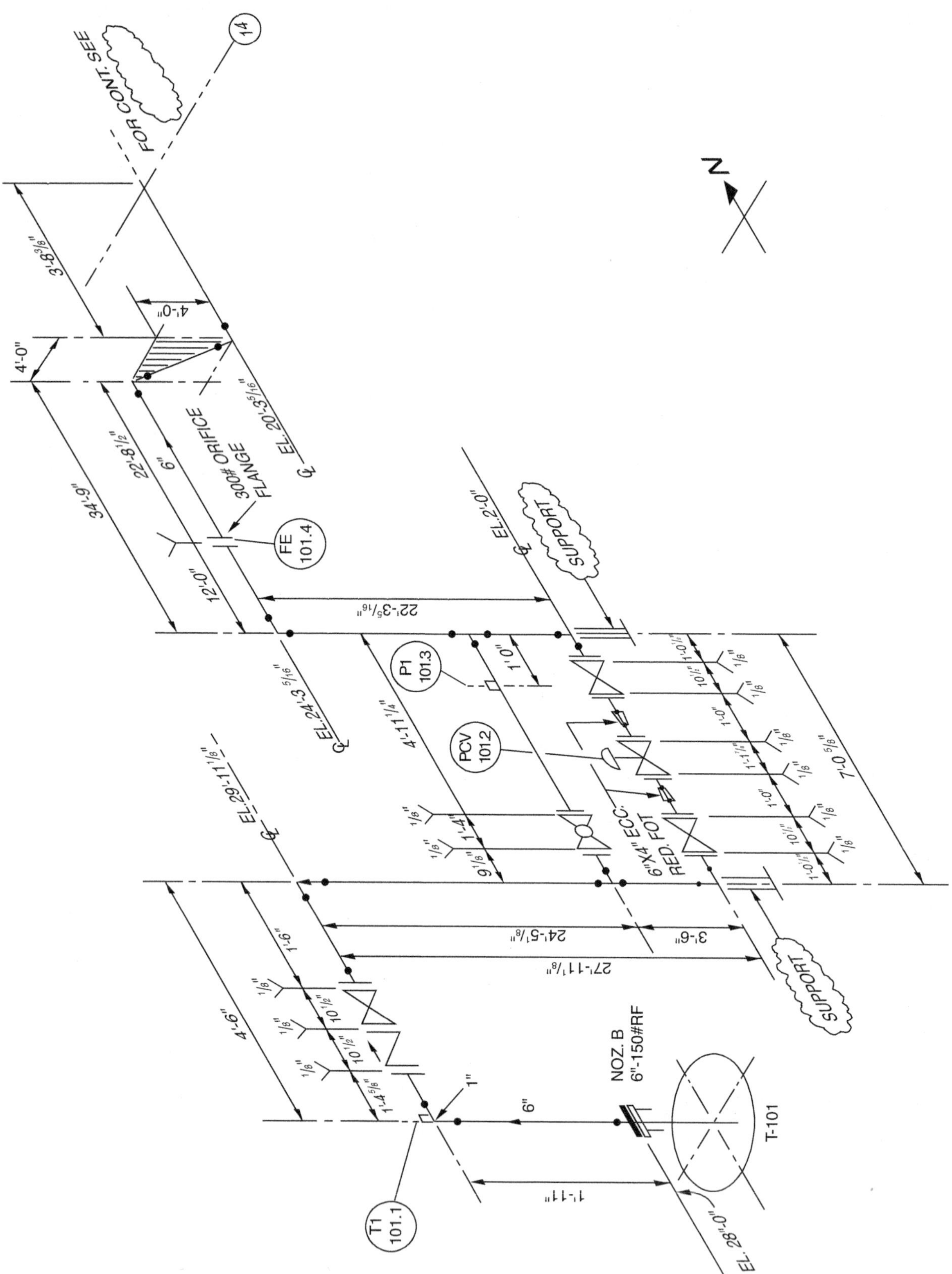

Figure 16 Piping isometric.

BILL OF MATERIAL

		SIZE	6"		4"		6×4
	SCHEDULE	ASTM	40		40		40
PIPE	Seamless	A53	100				
WELD FITTINGS	90 ELL	A234	7				
	TEE	A234	2				
	ECC. RED.						2
FLANGES	150# R.F.W.N.	A181	9				
	300# R.F.W.N. ORIFICE	A181	2				
	150# R.F.W.N.	A181			2		

BOLTS	NO.	SIZE
	80	3⁄4" × 4"
	12	3⁄4" × 5"
	16	5⁄8" × 3 1⁄2"
		Stud
		Bolts

RINGS & GASKETS	NO.	SIZE
	10	6" – 150#
	2	6" – 300#
	2	4" – 150#
		Flexitalic

	NO.	SIZE	DESCRIPTION
VALVES	1	6"	Check – Crane #147-x
	1	6"	Globe – Crane #143-x
	3	6"	Gate – Crane #47-x
90° ELLS			
45°			
CAPS			

	NO.	SIZE	DESCRIPTION
TEES			
UNIONS			
PLUGS			
MISCELLANEOUS			

	NO.	SIZE	DESCRIPTION
COUPLINGS	2	1"	3000# F.S. SCR'D
REDUCERS/SWAGE NIPPLES			
WELDOLETS			

REVISIONS

NO.	DRAWN	CHK'D	APP'D	DATE

PROJ. NO.	MR-237A
DR.	M.L.
CK.	M.T.
APP.	J.J.
DATE	2-14-07

DRAWING 2-4

PROJECT	Metals Recovery Inc.	
	Lowman	Idaho
LINE NUMBER	PG-0347-6"-A	
AREA PLAN DWG. NO.	03-P201	DETAIL SHEET NO. 486

Figure 17 Bill of materials.

1.0.0 Section Review

1. The *most* important view in an orthographic projection is the _____.
   a. perpendicular view
   b. isomorphic view
   c. auxiliary view
   d. front view

2. What do schematic drawings use to represent components?
   a. Isometric drawings
   b. Specialized symbols
   c. Connection lines
   d. Orthographic views

3. What kind of drawing provides a somewhat realistic 3D representation?
   a. Isometric drawing
   b. Schematic drawing
   c. Symbolic drawing
   d. Front-view drawing

Review Questions

1. The two-dimensional drawings making up an orthographic projection are _____.
   a. ports
   b. angles
   c. vantages
   d. views

2. What is the *maximum* possible number of two-dimensional drawings in an orthographic projection?
   a. 1
   b. 3
   c. 6
   d. 12

3. Which drawing element reveals features *not* visible from a particular perspective?
   a. Centerlines
   b. Hidden lines
   c. Isometric projections
   d. Perpendiculars

4. Dimension lines indicating size usually fit between a pair of _____.
   a. extension lines
   b. hidden lines
   c. centerlines
   d. projection lines

5. If dimensions *don't* include units, consult the drawing's _____.
   a. symbol key
   b. subassembly sheets
   c. title block
   d. auxiliary views

6. In schematic drawings, what do symbol connection lines communicate?
   a. Dimension boundaries
   b. Component relationships
   c. Physical limits
   d. Wall positions

7. Schematics of large systems usually show complex subassemblies _____.
   a. in the drawing's Notes section
   b. with a general-purpose symbol
   c. on separate pages
   d. inside of dotted lines

8. Pneumatic systems store pressurized air in a(n) _____.
   a. accumulator
   b. receiver
   c. reservoir
   d. sump

9. What do *all* hydraulic systems have that pneumatic systems don't need?
   a. A return line
   b. Pressure gauges
   c. Relief valves
   d. Control valves

10. What may appear alongside an isometric projection to make component fabrication easier?
    a. Abbreviation tables
    b. Pneumatic specifications
    c. Orthographic views
    d. Pipeline symbols

Trade Terms Quiz

Fill in the blank with the correct term learned from your study of this module.

1. _______________ show a dimension's limits.
2. A(n) _______________ is a single two-dimensional perspective of a three-dimensional object.
3. The _______________ lists all components and materials used in a system or assembly.
4. To determine a distance, angle, or position on a drawing, consult the _______________.
5. _______________ are drawings that show a system as a collection of symbols linked by lines.
6. Drawing lines composed of short and long dashes are _______________.
7. To show obscured drawing features in a particular view, use _______________.
8. A(n) _______________ depicts a three-dimensional object with several two-dimensional drawings.
9. To represent an object or schematic realistically with a single drawing, create a(n) _______________.

Trade Terms

Bill of materials (BOM)
Centerlines
Dimension line
Extension lines
Hidden lines
Isometric (ISO) drawing
Orthographic projection
Schematics
View

Cornerstone of Craftsmanship

Dave Roberts

Technical Services Manager
FLSmidth, Inc.

How did you choose a career in the industry?

I've always been interested in mechanical items. As a kid, I was constantly taking things apart to see how they worked (I didn't always get them put back together). I also found troubleshooting and fixing things very satisfying. So, I studied electrical engineering and computer science in college and then started working at IBM. Then, one Christmas, my brother-in-law introduced me to FLSmidth.

Who inspired you to enter the industry?

Definitely my brother in-law. He was an ironworker, and one Christmas I mentioned that I enjoyed hands-on work and travel. He suggested I check out the company he worked for, and I did. Now I've been there for over 30 years.

What types of training have you been through?

I went to college for four years, and I also have training in optical tooling, shaft alignments, SFK bearings, crushers, mills and more. I'm also a Certified Welding Inspector and Certified Radiographer.

How important is education and training in construction?

To truly be successful in construction and maintenance careers, education and training never stops. The quality and efficiency that results from good training not only gives personal satisfaction, but it drives entire companies to succeed.

How important are NCCER credentials to your career?

In this industry it's rare for workers to stay with the same company for their entire career. A portable credential that is widely recognized is one of the most important assets you can hold.

How has training/construction impacted your life?

As you gain knowledge and expertise, doors open that would simply not be available otherwise. The desire to seek out and complete training has been my most valuable asset over the years. It is also very fulfilling to be able to teach others and help them be successful.

What kinds of work have you done in your career?

I've done plant maintenance, oil exploration, computer science, crane operations, welding inspection, training, quality assurance, safety, management, and more.

Tell us about your current job.

Currently, I'm the Technical Services Manager at FLSmidth Inc (formerly Phillips Kiln Services). I've been with the company for over 30 years, and I've held many jobs from Field Engineer to Director of Repair Services. I'm nearing retirement and currently focus on developing training and providing technical support.

What do you enjoy most about your job?

FLSmidth has provided me with the opportunity to travel the world and solve complex mechanical problems for a wide variety of customers. Every day is a new challenge, and I literally always learn something new. It has been rewarding and is certainly not boring!

What factors have contributed most to your success?

Always wanting to learn and get better. Being honest, asking for help when needed, and dedicating myself to always doing my best.

Would you suggest construction as a career to others? Why?

Working in the mechanical service industry has been very rewarding to me. If you don't mind getting dirty occasionally, like to solve problems, and get satisfaction from repairing and building things, I'd highly recommend it.

What advice would you give to those new to the field?

Show up on time, listen more than you talk, spend extra time seeking and completing training, and take pride in your work.

Interesting career-related fact or accomplishment?

I've worked in North America, South America, Africa, Asia, Europe, Australia, and India. The only place left is Antarctica!

How do you define craftsmanship?

When you do a job, have proper training and a plan before you start. If you have questions, ask. Then do the work to the best of your ability and have pride in the result. If you have problems, document them, find out what went wrong, and figure out how to prevent it the next time. Always share your knowledge with others.

Trade Terms Introduced in This Module

Bill of materials (BOM): The list of parts and materials required to construct a system or assembly.

Centerlines: Lines made from alternating long and short dashes that identify the middle of parts or features, such as holes.

Dimension line: A size-, location-, or angle-identifying number combined with an arrow or pair of arrows.

Extension lines: Pairs of short lines perpendicular to a drawing element that identify a dimension's limits.

Hidden lines: Lines made from short dashes that reveal features not visible in a particular view.

Isometric (ISO) drawing: A drawing method that represents a three-dimensional object with all three faces equally tilted toward the drawing surface.

Orthographic projection: A drawing method that represents a three-dimensional object by several two-dimensional views.

Schematics: A diagram-type drawing that shows a system's essential components and interconnections.

View: Part of an orthographic projection, a two-dimensional drawing that accurately represents a three-dimensional object from one vantage point.

Additional Resources

This module presents thorough resources for task training. The following reference material is suggested for further study.

Architectural Drawing and Light Construction, 8th Edition. Philip A. Grau III, Edward J. Muller, James G. Fausett. Upper Saddle River, NJ: Pearson.

Blueprint Reading for Machine Trades, 7th Edition. Russ Schultz and Larry Smith. Upper Saddle River, NJ: Pearson.

Figure Credits

iStock@Rutmer Visser, Module Opener

Section Review Answer Key

Section 1.0.0

Answer	Section Reference	Objective
1. d	1.1.0	1a
2. b	1.2.1; 1.2.2	1b
3. a	1.3.0	1c

This page is intentionally left blank.

NCCER CURRICULA — USER UPDATE

NCCER makes every effort to keep its textbooks up-to-date and free of technical errors. We appreciate your help in this process. If you find an error, a typographical mistake, or an inaccuracy in NCCER's curricula, please fill out this form (or a photocopy), or complete the online form at **www.nccer.org/olf**. Be sure to include the exact module ID number, page number, a detailed description, and your recommended correction. Your input will be brought to the attention of the Authoring Team. Thank you for your assistance.

Instructors – If you have an idea for improving this textbook, or have found that additional materials were necessary to teach this module effectively, please let us know so that we may present your suggestions to the Authoring Team.

NCCER Product Development and Revision
13614 Progress Blvd., Alachua, FL 32615
Email: curriculum@nccer.org
Online: www.nccer.org/olf

❑ Trainee Guide ❑ Lesson Plans ❑ Exam ❑ PowerPoints Other ____________________

Craft / Level: Copyright Date:

Module ID Number / Title:

Section Number(s):

Description:

Recommended Correction:

Your Name:

Address:

Email: Phone:

Field Sketching

Overview

Sometimes a millwright in the field must help an engineer or machinist understand a part or assembly's shape and dimensions. A millwright who knows how to sketch can provide this information. Field sketching does not require great artistic talent. Understanding the part or assembly, plus a few simple techniques, goes a long way toward producing a useful sketch. This module introduces sketching by breaking it down into a series of basic steps.

Module 15202

Trainees with successful module completions may be eligible for credentialing through the NCCER Registry. To learn more, go to **www.nccer.org** or contact us at 1.888.622.3720. Our website, **www.nccer.org**, has information on the latest product releases and training.

Your feedback is welcome. You may email your comments to **curriculum@nccer.org**, send general comments and inquiries to **info@nccer.org**, or fill in the User Update form at the back of this module.

This information is general in nature and intended for training purposes only. Actual performance of activities described in this manual requires compliance with all applicable operating, service, maintenance, and safety procedures under the direction of qualified personnel. References in this manual to patented or proprietary devices do not constitute a recommendation of their use.

15202 V4.0

15202

Field Sketching

Objectives

Successful completion of this module prepares you to do the following:

1. Summarize sketching techniques and object views.
   a. Describe how to sketch straight lines and angles.
   b. Describe how to sketch arcs, circles, and ellipses.
   c. Explain how to dimension sketches.
   d. Contrast the methods used for producing three-dimensional sketches.

Performance Tasks

Under supervision, you should be able to do the following:

1. Sketch straight lines and angles.
2. Sketch arcs, circles, and ellipses.
3. Dimension a sketch.
4. Produce an orthographic sketch of an object.
5. Produce an oblique sketch of an object.
6. Produce an isometric sketch of an object.

Trade Terms

Ellipses
Isometric
Oblique
Orthographic
Pictorial
Sketch

Industry Recognized Credentials

If you are training through an NCCER-accredited sponsor, you may be eligible for credentials from NCCER's Registry. The ID number for this module is 15202. Note that this module may have been used in other NCCER curricula and may apply to other level completions. Contact NCCER's Registry at 1.888.622.3720 or go to **www.nccer.org** for more information.

You can also show off your industry-recognized credentials online with NCCER's digital badges. Transform your knowledge, skills, and achievements into badges that you can share across social media platforms, send to your network, and add to your resume. For more information, visit **www.nccer.org**.

NOTE

This is an elective module and is not required for successful completion of *Millwright Level One*, nor for the successful completion of the *Millwright* craft credential.

NOTE

This module uses US standard and metric units in up to three different ways. This note explains how to interpret them.

Exact Conversions

Exact metric equivalents of US standard units appear in parentheses after the US standard unit. For example: "Measure 18" (45.7 cm) from the end and make a mark."

Approximate Conversions

In some cases, exact metric conversions would be inappropriate or even absurd. In these situations, an approximate metric value appears in parentheses with the ~ symbol in front of the number. For example: "Grip the tool about 3" (~8 cm) from the end."

Parallel but not Equal Values

Certain scenarios include US standard and metric values that are parallel but not equal. In these situations, a slash (/) surrounded by spaces separates the US standard and metric values. For example: "Place the point on the steel rule's 1" / 1 cm mark."

Contents

Figures

Section One

1.0.0 Sketching Basics

Objective

Summarize sketching techniques and object views.

a. Describe how to sketch straight lines and angles.
b. Describe how to sketch arcs, circles, and ellipses.
c. Explain how to dimension sketches.
d. Contrast the methods used for producing three-dimensional sketches.

Performance Tasks

1. Sketch straight lines and angles.
2. Sketch arcs, circles, and ellipses.
3. Dimension a sketch.
4. Produce an orthographic sketch of an object.
5. Produce an oblique sketch of an object.
6. Produce an isometric sketch of an object.

Trade Terms

Ellipses: Stretched circles, similar to ovals.

Isometric: A drawing method that represents a three-dimensional object with all three faces equally tilted toward the paper surface.

Oblique: A drawing method that represents a three-dimensional object with one face parallel to the paper surface.

Orthographic: A drawing method that represents a three-dimensional object by several two-dimensional views.

Pictorial: Any drawing method that represents three-dimensional objects somewhat realistically.

Sketch: A rough drawing that represents an object's key features.

Millwrights read and interpret drawings, but they often create them as well. Freehand sketches help the millwright communicate key ideas to others. Sketches do not have to be perfect. But they must communicate basic ideas accurately and with adequate detail.

Learning to sketch takes some time and practice, but it's a skill worth acquiring. Think of a **sketch** as a collection of simple drawing elements like lines, angles, circles, arcs, and **ellipses**. Together, these create the complete picture. It must show correct proportions and may require added dimensions to communicate sizes accurately.

You can sketch with just a pencil and paper. Or you might prefer to use a tablet device with a stylus and sketching app (*Figure 1*). Some millwrights take digital photographs and add dimensions or other sketched elements. Sending electronic sketches to others by email or a text message is quick and convenient.

> **NOTE**
> Even if you prefer sketching with a tablet device, pencil-and-paper techniques can be a useful backup. Some companies and jobsites do not allow photography or using electronic devices without permission. Always ask first.

1.1.0 Sketching Lines

Every sketch contains lines, so learning to draw them properly is crucial. Lines can be horizontal, vertical, or angled. The techniques introduced here will help you produce credible lines of all types.

1.1.1 Drawing Straight Lines

Drawing a straight line freehand takes some practice. For best results, hold the pencil lightly about 1" (~25 mm) from the tip. Make the line using arm motion rather than hand and finger movement. The result will be smoother and straighter.

Figure 1 Sketching with a tablet and stylus.

Follow these steps to sketch straight lines:

Step 1 Identify the line's endpoints and mark them with small dots. For horizontal and vertical lines, use the edges of the paper to help align them. Place the endpoints of angled lines by estimating their positions visually.

Step 2 If you're drawing a horizontal line, place the pencil on the left endpoint (left-handed people should start on the right endpoint). For vertical or angled lines, place the pencil on the top endpoint.

Step 3 Keeping the wrist straight and using your whole arm, draw the line in one smooth sweep (*Figure 2*). For horizontal and vertical lines, use the parallel edge of the page to help guide you.

NOTE

Keep your eye on the second endpoint as you draw the line. For very long lines, mark extra dots between the endpoints and use them as guides to help you keep the line straight.

1.1.2 45-Degree Angles

Sketching specific angles is harder than drawing an ordinary angled line. Several common angles, however, are easy to draw. A 45-degree angle, for example, is just half a 90-degree angle.

Follow these steps to sketch a 45-degree angle from a vertical line:

Step 1 Sketch a dotted line perpendicular to the vertical line (*Figure 3*). Mark a dot at the *vertex* (the place where the lines come together).

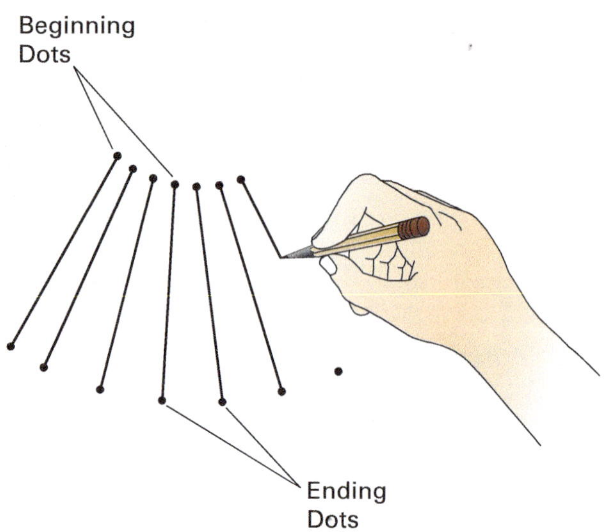

Figure 2 Sketching straight lines.

Step 2 Visually identify a location halfway between the two lines and mark a dot (*Figure 3*).

Step 3 Sketch an angled line between the two dots to form the 45-degree angle (*Figure 3*).

NOTE

Since a 45-degree angle bisects a 90-degree angle, the line also forms a 45-degree angle with the horizontal line.

1.1.3 30-Degree Angles

Sketching a 30-degree angle employs a similar technique. A 30-degree angle is one-third of a 90-degree angle.

Follow these steps to sketch a 30-degree angle from a vertical line:

Step 1 Sketch a dotted line perpendicular to the vertical line (*Figure 4*). Mark a dot at the vertex.

Step 2 Visually identify two locations that divide the space into three equal parts. Mark dots at each place (*Figure 4*).

Step 3 Sketch an angled line between the vertex and the dot closest to the vertical line to form the 30-degree angle (*Figure 4*).

NOTE

To sketch a 30-degree angle from the horizontal line, draw a line between the vertex and the dot closest to the horizontal line.

1.1.4 15-Degree Angles

To sketch a 15-degree angle, simply draw a 30-degree angle and cut it in half.

Follow these steps to sketch a 15-degree angle from a vertical line:

Step 1 Sketch a 30-degree angle by following the previous procedure.

Step 2 Visually identify a location halfway between the vertical line and the 30-degree line. Mark a dot at this place.

Step 3 Sketch an angled line between the new dot and the vertex to form the 15-degree angle (*Figure 5*).

NOTE

To sketch a 15-degree angle from the horizontal line, create a 30-degree angle from the horizontal line and divide it in half.

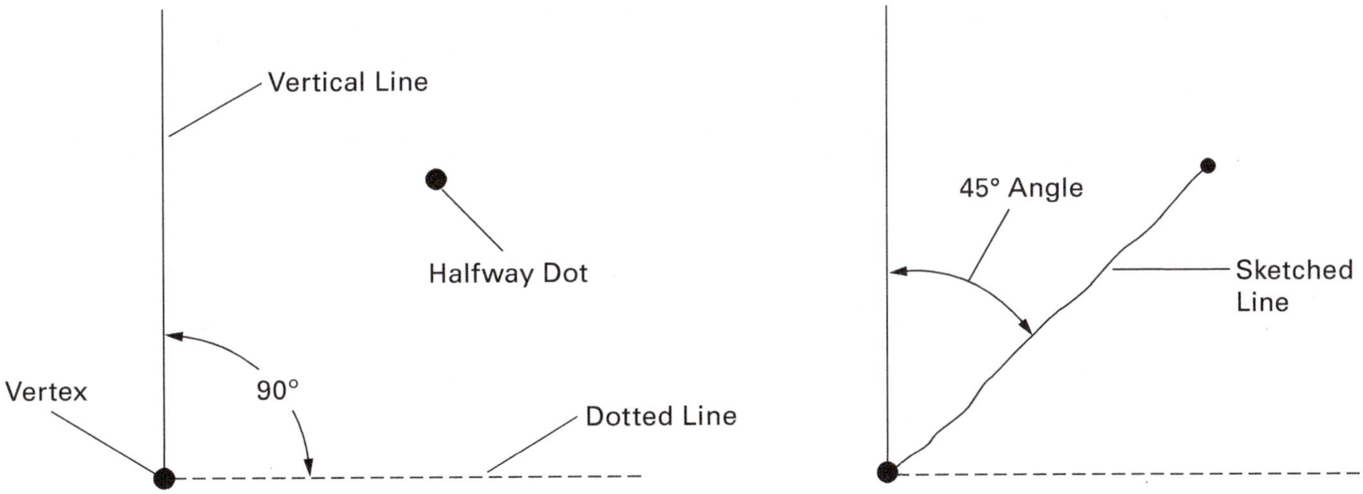

Figure 3 Sketching a 45-degree angle.

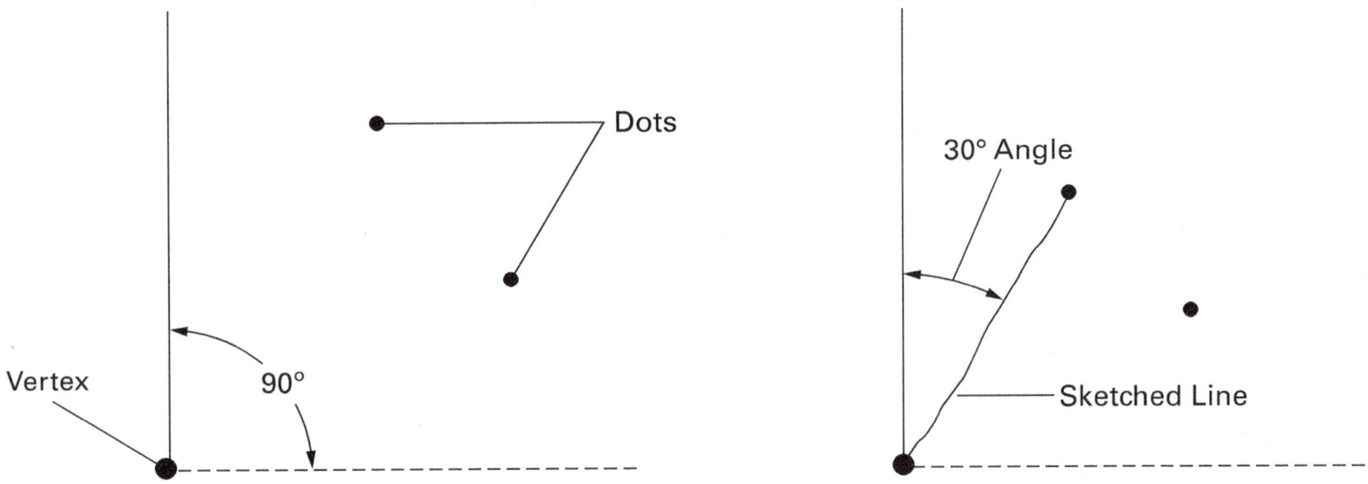

Figure 4 Sketching a 30-degree angle.

1.1.5 Sketching Lines with a Tablet

All sketching apps support freehand drawing, so the pencil-and-paper techniques work on tablets too. But sketching apps almost always include line-drawing tools that create straight lines at specific angles. Whenever possible, use these tools to get better results.

1.2.0 Sketching Circles, Arcs, and Ellipses

Sketching realistic curves like circles and arcs may seem harder than producing straight lines. Some people can draw them freehand with little effort. If you can do this, then draw them in whatever way works best for you. But if you struggle, the following techniques can help you produce better results.

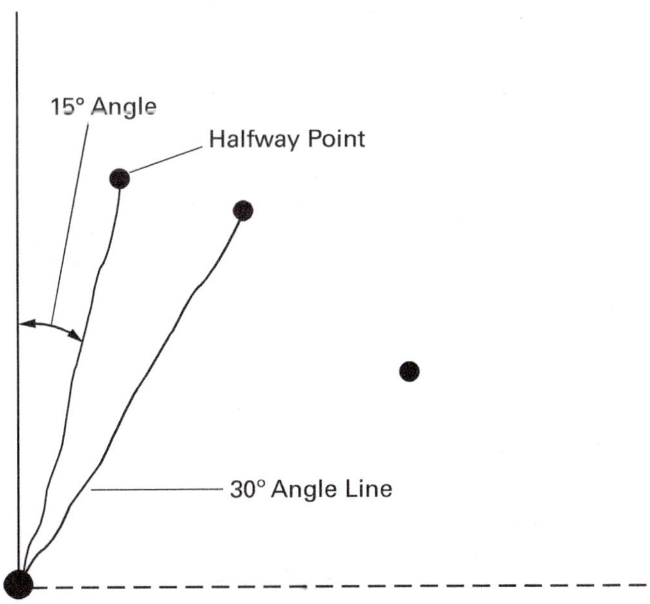

Figure 5 Sketching a 15-degree angle.

1.2.1 Arcs

Arcs are open curves with two endpoints. They commonly form a rounded corner (*fillet*) between two lines. The technique provided produces fillets, as well as arcs for any general-purpose sketching situation.

Follow these steps to sketch an arc from a pair of perpendicular lines:

Step 1 Create a pair of perpendicular lines. If necessary, extend existing lines so they come together. Draw two small lines that cross where the arc's endpoints will be (*Figure 6*).

Step 2 Sketch a straight line between the endpoints to form a triangle.

Step 3 Draw a dot in the triangle's center (*Figure 7*).

Step 4 Sketch a smooth curve that connects the endpoints while passing through the dot (*Figure 8*).

Step 5 Erase all unneeded construction elements, leaving behind the completed arc (*Figure 9*).

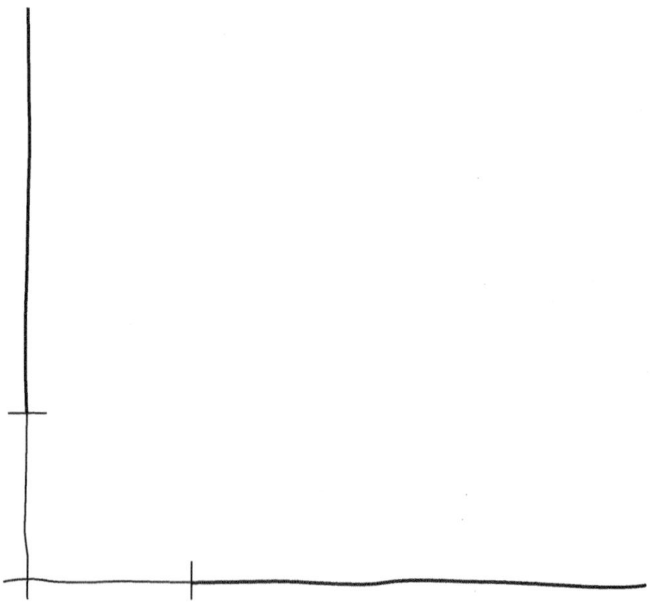

Figure 6 Perpendicular lines with arc endpoints.

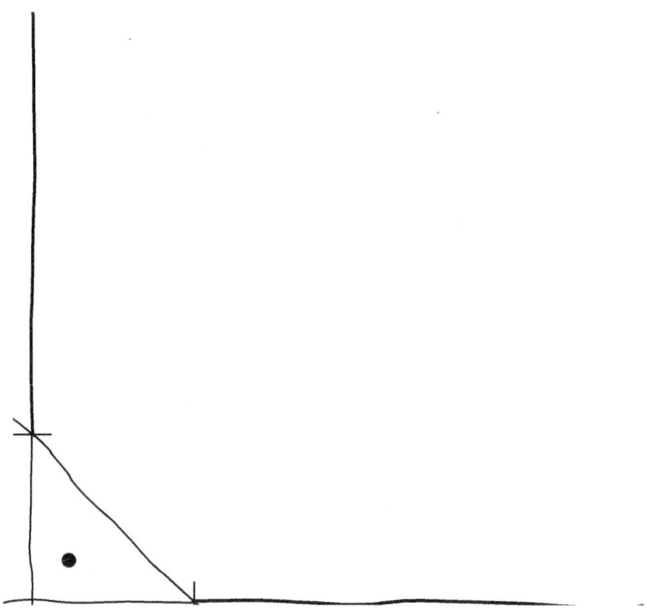

Figure 7 Triangle and dot.

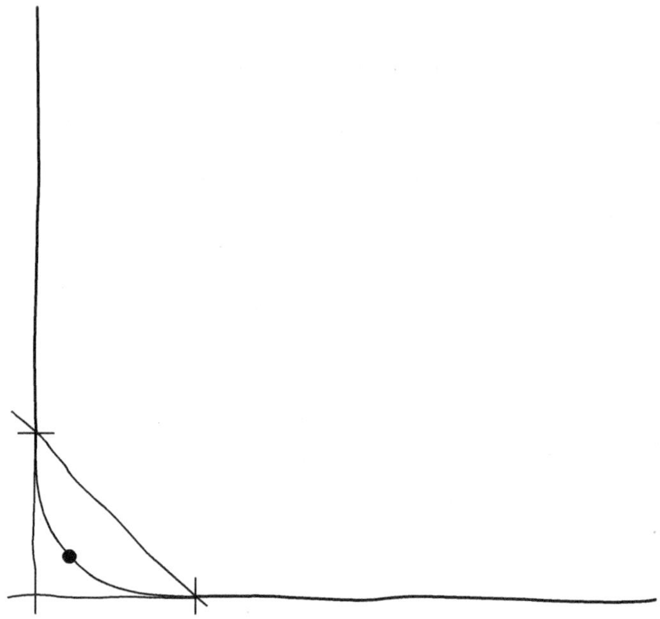

Figure 8 Sketched arc.

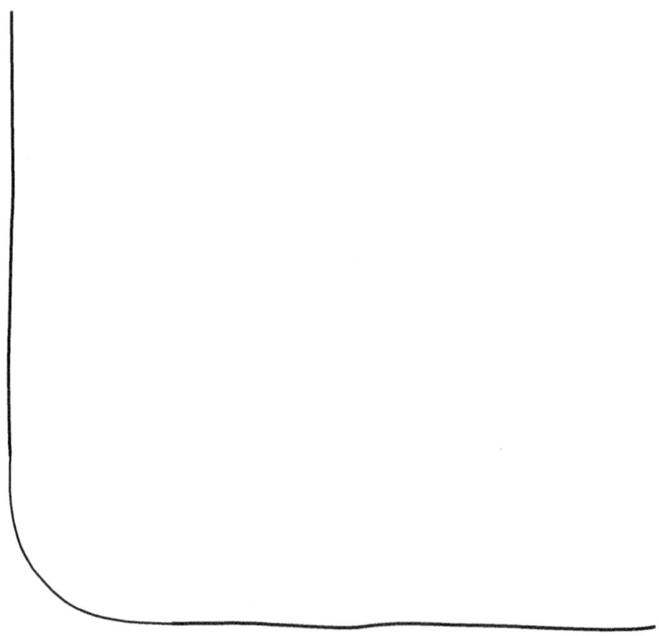

Figure 9 Finished arc.

1.2.2 Circles

Freehand sketching a reasonably round circle may seem difficult. But a circle is just four arcs, so you can sketch one with the technique that you have just learned.

Follow these steps to sketch a circle:

Step 1 Sketch two perpendicular lines that cross at the circle's center point. The lines should equal the planned circle's diameter (twice its radius).

Step 2 Sketch a square around these lines (*Figure 10*).

Step 3 Connect the perpendicular lines' endpoints to form a diamond inside the square (*Figure 11*).

Step 4 Draw a dot in the center of each of the outside triangles.

Step 5 Sketch a smooth curve connecting the endpoints while passing through the dot (*Figure 12*).

Step 6 Erase all unneeded construction elements, leaving behind the completed circle.

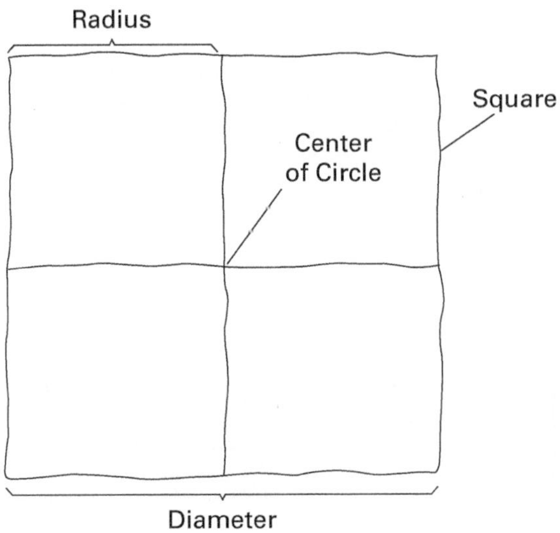

Figure 10 Perpendicular lines and square.

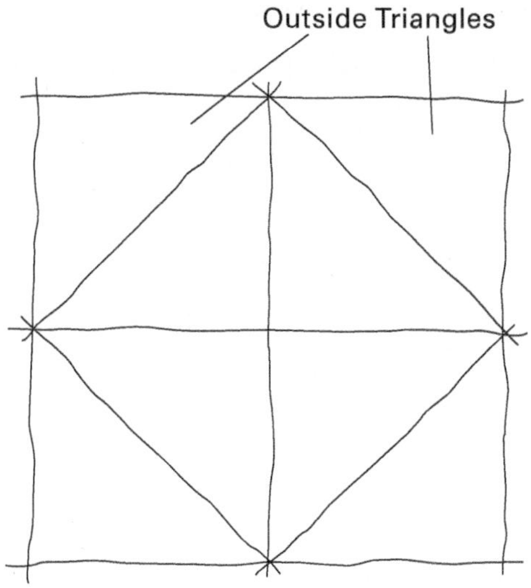

Figure 11 Diamond inside the square.

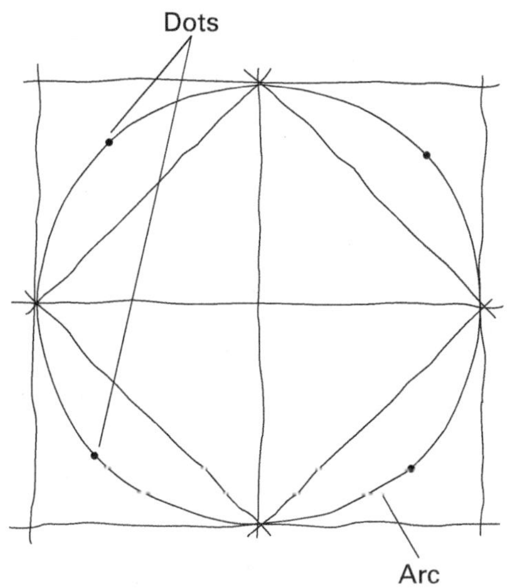

Figure 12 Sketching the circle.

Engineering Drawings

From the beginning, millwrights have had to create effective drawings and sketches. Over time, they developed techniques that today go by the names *engineering drawing, mechanical drawing,* or *drafting.* While sketching takes place freehand (without drawing tools), it uses many of the same principles. Improving your drawing skills will make you a better millwright.

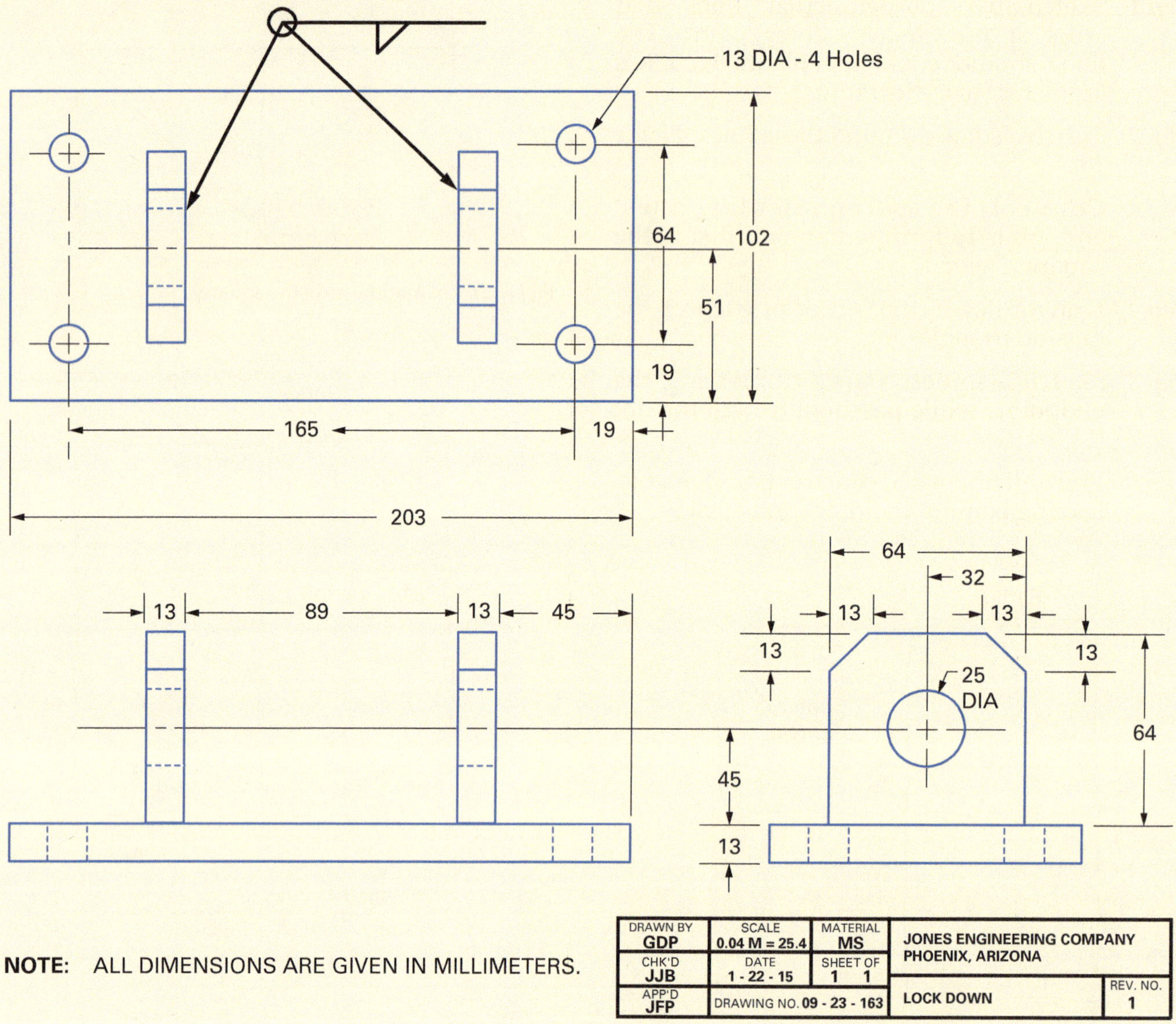

METRIC MEASUREMENTS

1.2.3 Ellipses

An ellipse is a stretched circle. Round objects often look elliptical when viewed at an angle (*Figure 13*), so sketching ellipses can be very useful.

Follow these steps to sketch an ellipse:

Step 1 Determine the ellipse's width and length.

Step 2 Sketch two perpendicular lines that cross at the ellipse's center. One line should equal the ellipse's width, while the other should equal its length (*Figure 14*).

Step 3 Sketch a rectangle around the lines (*Figure 14*).

Step 4 Sketch curved lines to join the perpendicular lines (*Figure 15*). Getting these lines to look smooth is harder than sketching a circle. If necessary, erase and try again.

Step 5 Erase all unneeded construction elements, leaving behind the completed ellipse.

1.2.4 Sketching Curves with a Tablet

A drawing app saves you a lot of time when you must draw curves. Almost all include tools for drawing circles, ellipses, and arcs. In most cases, you can enter a specific radius and then position the curve. But it's equally easy to tweak the shape directly by dragging the control points that surround it.

1.3.0 Dimensions

Sketches usually communicate two pieces of information. They visually describe what objects look like or how they fit together. And they provide detailed size and position information—*dimensions*. Adding a *dimension line* requires sketching specific elements and then adding numbers. Common dimension types include the following:

- Size
- Location
- Angle
- Arc and circle

Always draw dimension lines carefully to avoid confusion. Each dimension type requires a specific style that includes certain dimension elements. These include extension lines, arrowheads, and centerlines.

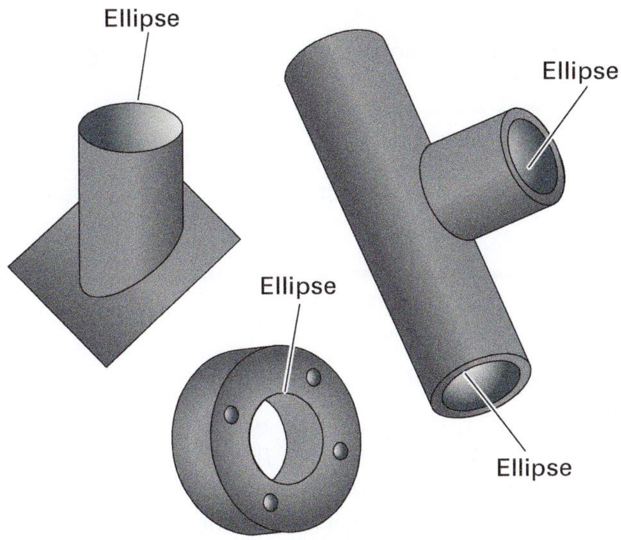

Figure 13 Representing circular objects with ellipses.

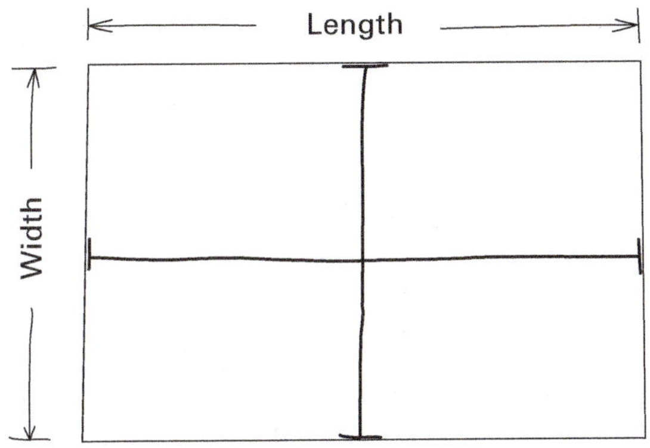

Figure 14 Ellipse width and length.

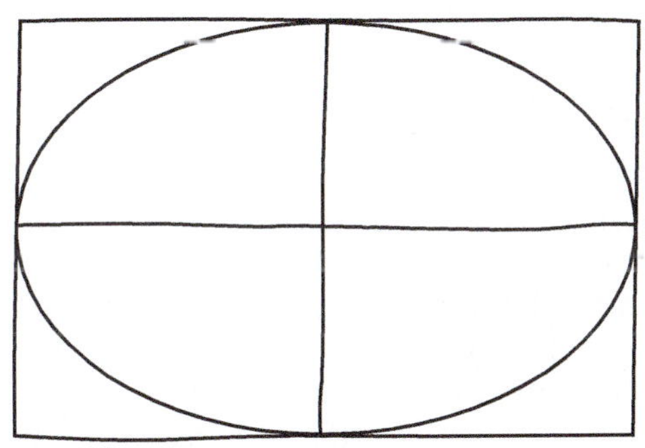

Figure 15 Completing the ellipse.

1.3.1 Extension Lines

Extension lines designate dimension boundaries on the part. In most cases, draw extension lines 90 degrees to the part. Leave a small gap between the extension lines and the part to keep them separate (*Figure 16*).

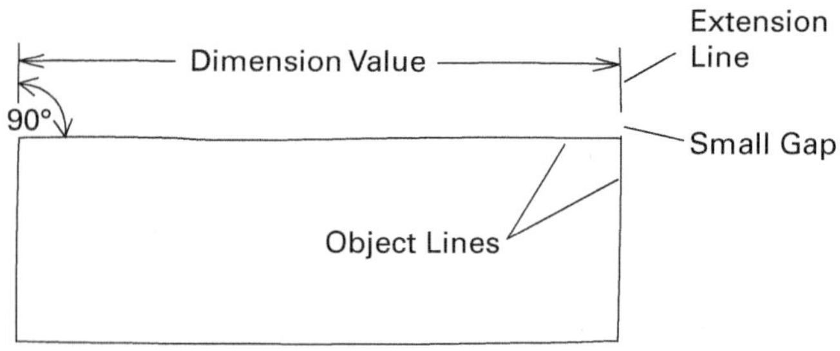

Figure 16 Extension lines.

1.3.2 Arrowheads

Place arrowheads on the ends of the dimension line, just touching the extension lines. Arrowheads should be clear and distinct. Choose a size and style appropriate to the part's size and the available space. *Figure 17* shows correct and incorrect arrowheads.

1.3.3 Centerlines

Centerlines identify the middle of objects or other features. For example, a sketch of a machine base may include a pair of centerlines. A pair of perpendicular centerlines specifies the middle of a hole. A single centerline identifies the axis that runs down the middle of a cylinder. Draw centerlines as a series of alternating long and short dashes (*Figure 18*). In some drawings, the ℄ symbol identifies centerlines.

1.3.4 Size Dimensions

Size dimensions communicate a part's physical size and sometimes its shape. Draw them outside the part as much as possible. Putting size dimensions inside the part creates confusion and makes the drawing harder to understand. *Figure 19* shows correct and incorrect size dimension lines.

The actual numbers can appear above, below, or within the dimension line. If you sketch the dimension within the line, leave a space in the center for the number. This space must be large enough to fit the number but not too large. *Figure 20* shows correct and incorrect dimension spaces.

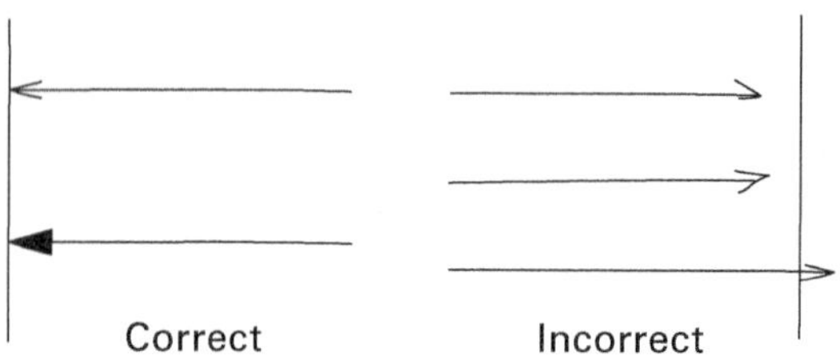

Figure 17 Correct and incorrect arrowheads.

1.3.5 Location Dimensions

Location dimensions show hole positions, radius centers, or specialized features. Sketch these elements carefully to avoid creating problems during fabrication. *Figure 21* shows typical location dimensions.

In some cramped situations, the numbers won't fit between the extension lines. In these cases, sketch the dimension lines outside the extension lines (*Figure 22*).

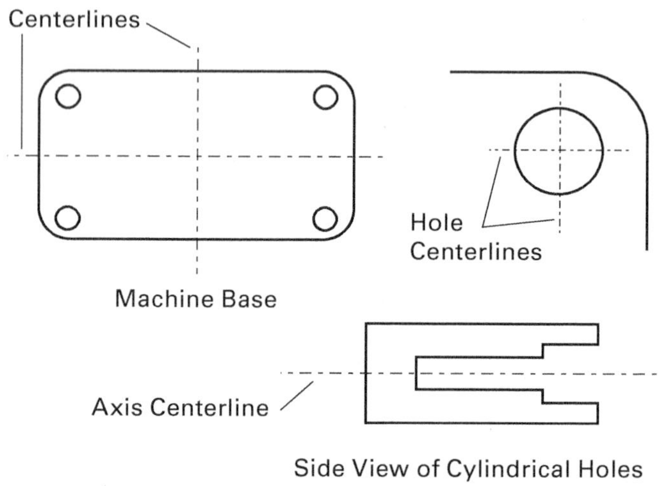

Figure 18 Typical centerlines.

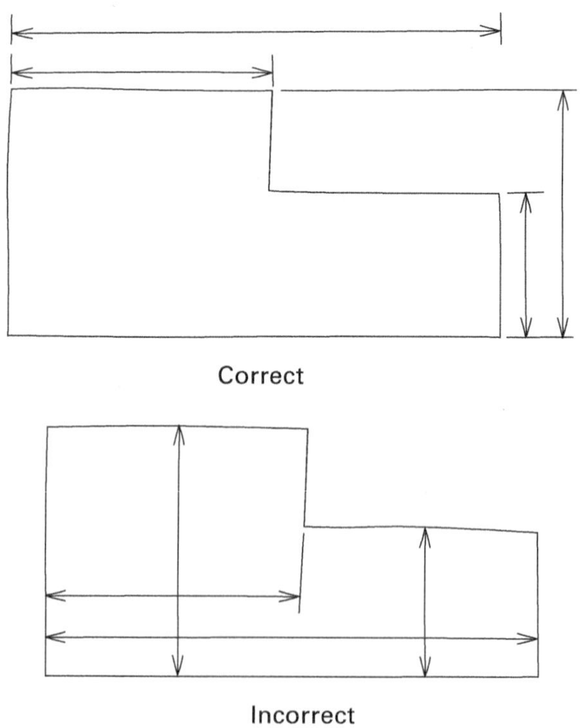

Figure 19 Correct and incorrect size dimension lines.

All About Ellipses

An ellipse is a special kind of closed curve created by cutting a cone at a particular angle. Ellipses usually are wider in one direction than the other. The wider dimension is the *major axis,* while the narrower one is the *minor axis.* Some ellipses have identical major and minor axes. When they do, they have a special name—*circle.* Yes, circles are actually ellipses!

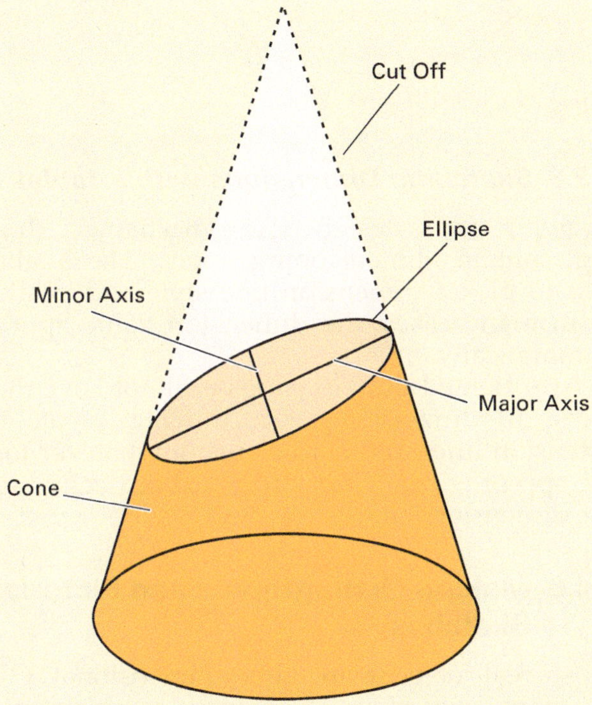

Some people think that ellipses are ovals, but this is not true. While they look quite similar, an ellipse is symmetrical, while an oval is not. Ovals are egg-shaped—bigger at one end than the other.

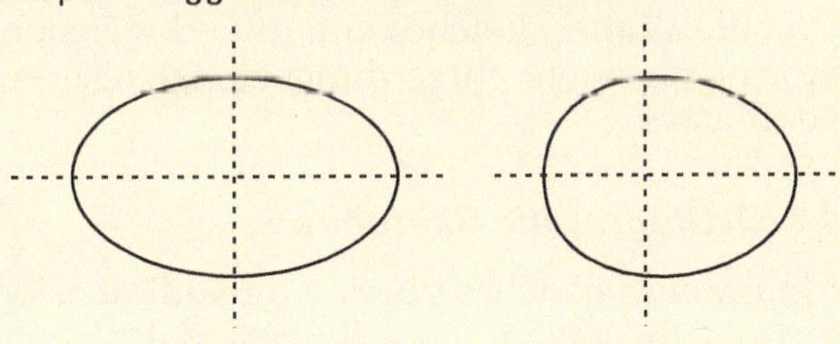

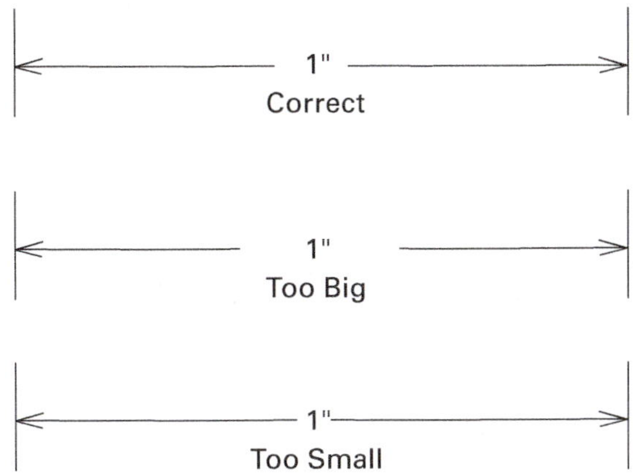

Figure 20 Correct and incorrect dimension spaces.

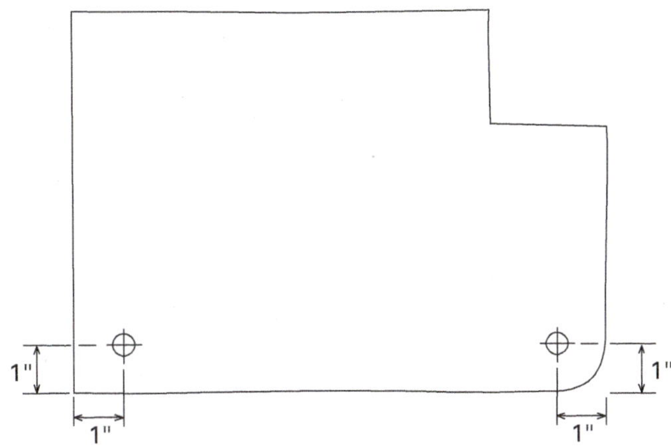

Figure 21 Location dimensions.

1.3.6 Angle Dimensions

Angle dimension lines show the angle between a part's sections. These dimensions usually are curved with their arrowheads touching the extension lines. The angle value fits between the extension lines (*Figure 23*).

1.3.7 Arc and Circle Dimensions

Arc and circle dimensions do not use extension lines. Instead, dimension an arc by showing its center with a pair of centerlines. Specify its radius with an arrow pointing from the center to the arc itself. Place the radius value near the arrow with the letter R in front of it (*Figure 24*).

Dimension a circle by showing its center with a pair of centerlines. Specify its diameter with a two-headed arrow across its diameter. Place the diameter value near the arrow with a diameter symbol (Ø) in front of it (*Figure 25*).

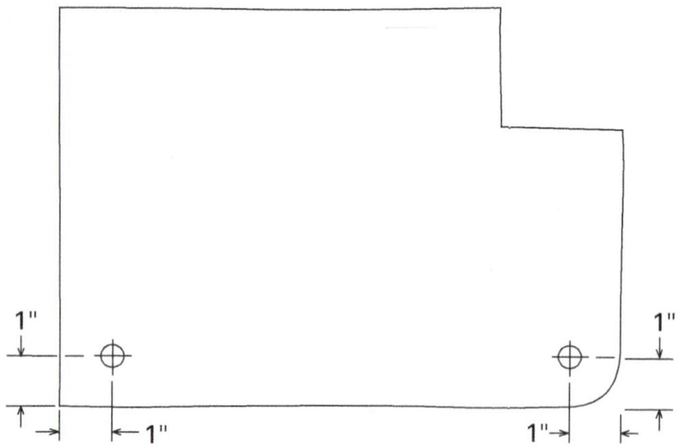

Figure 22 Outside dimension lines.

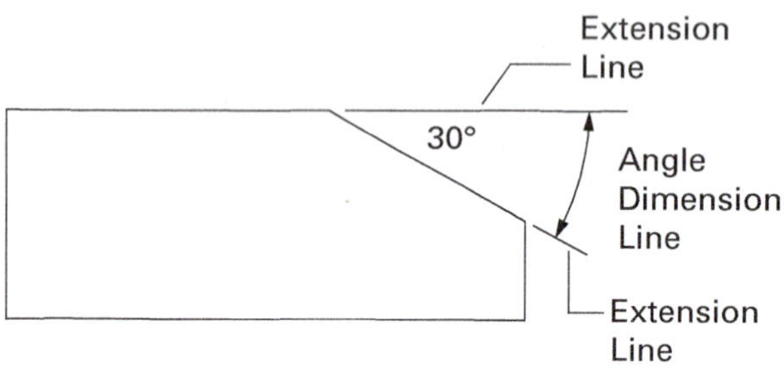

Figure 23 Angle dimension line.

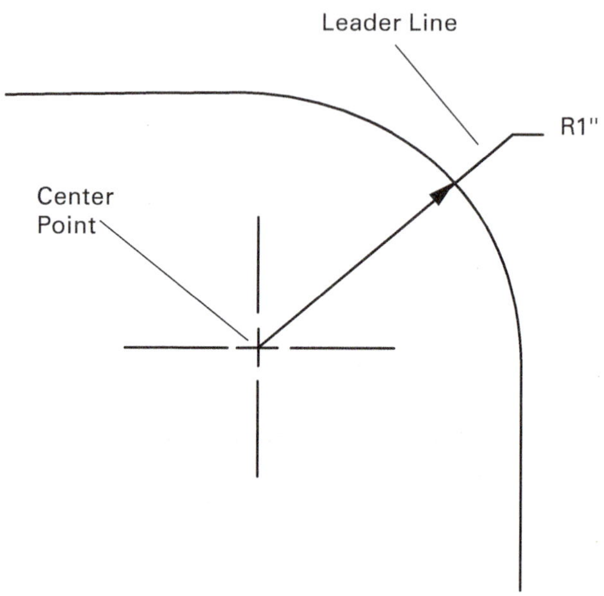

Figure 24 Arc dimension lines.

If an arc or circle is too small to hold the radius or diameter value, you can place it nearby. Use a *leader line,* which looks like a pair of angled lines, to link the value to the arc or circle. *Figure 24* shows a leader line example.

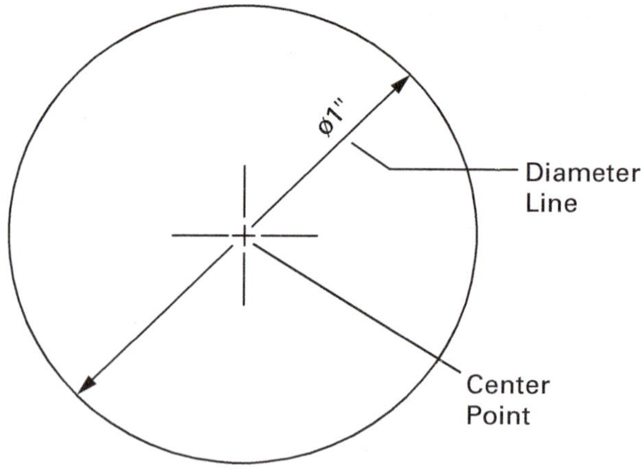

Figure 25 Circle dimension lines.

1.3.8 Sketching Dimensions with a Tablet

Drawing apps designed for mechanical drawing contain dimensioning tools. These allow you to pick a dimension line style and position it. In many cases, the dimension value appears automatically.

Art-oriented drawing apps will not have these tools. To dimension your drawing, create the extension lines and arrows using whatever tools the app provides. Then add a text box containing the dimension value.

1.4.0 Making Orthographic and Pictorial Sketches

Most real objects are three-dimensional (3D). In many situations, however, a single two-dimensional sketch adequately represents them. A flange's bolt circle is a good example. On the other hand, many objects require you to portray their three-dimensional qualities accurately and in detail. While sketches are two-dimensional, they can represent three-dimensional objects in several ways.

1.4.1 Orthographic Sketches

Orthographic sketches portray three-dimensional objects with several two-dimensional sketches. Imagine the object sitting inside a glass box. You walk around the box and use a marker to sketch on the glass what you see. Each sketch is a *view* of the object. Together, the views accurately represent the object (*Figure 26*).

Sometimes you need to show a part concealed by another part. To do this, add *hidden lines* to the sketch. These are dashed lines that show the location of the obscured feature. The right view of *Figure 26* includes a hidden line.

With a little training, you can make good orthographic sketches. But they do not look like the actual object, so some people struggle to interpret them. For this reason, many prefer **pictorial** sketches. They look more like the actual object and fall into two categories: **oblique** and **isometric** (*Figure 27*).

1.4.2 Oblique Sketches

An oblique sketch represents the object with a single view. While it does not look completely realistic, it has a three-dimensional feel. The front face resembles an orthographic sketch, with all lines undistorted and accurate. The other sides appear at an angle to make them look three-dimensional.

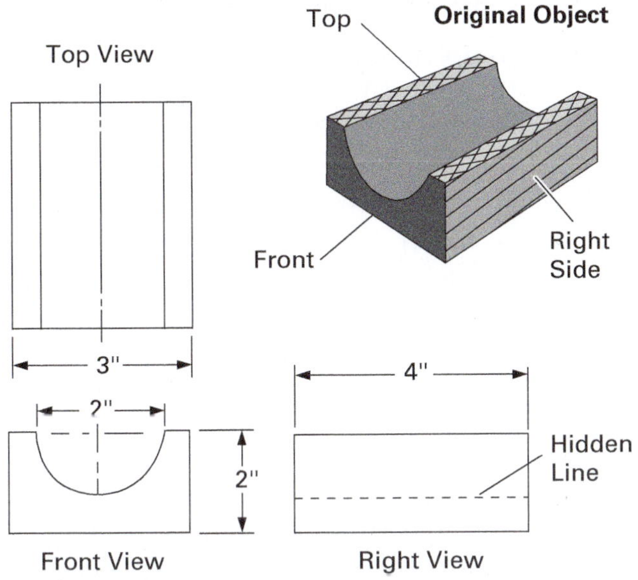

Figure 26 Three-view orthographic sketch.

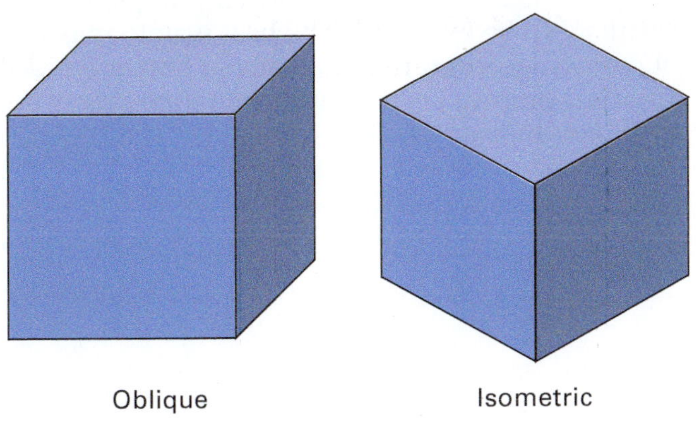

Figure 27 Oblique and isometric representations.

But this drawing method distorts them, so the lines are not their true length.

Follow these steps to make an oblique sketch of a rectangular object:

Step 1 Sketch horizontal and vertical baselines (*Figure 28*).

Step 2 Sketch the object's front view from the baselines (*Figure 29*).

> **NOTE**
> All vertical lines should be parallel to the vertical baseline. All horizontal lines should be parallel to the horizontal baseline. All drawing elements should be their true size and shape.

Step 3 Start at the vertical and horizontal baseline intersection. Sketch a line at a 45-degree angle to create an oblique baseline (*Figure 30*).

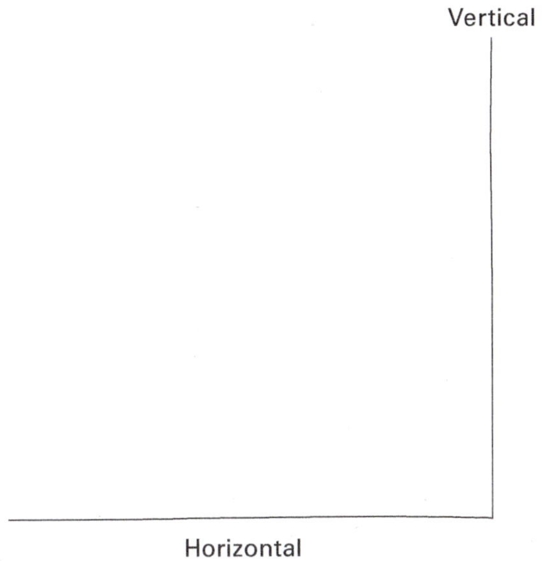

Figure 28 Horizontal and vertical baselines.

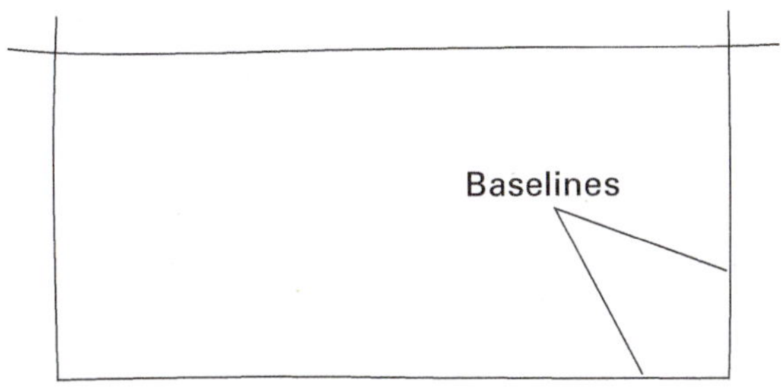

Figure 29 Sketching the front view.

Step 4 Sketch the remaining object lines. Draw the top and right-side view lines parallel to the oblique, horizontal, or vertical baseline as required (*Figure 31*). These lines will not be their true size.

Step 5 Add the dimensions, making them parallel to each axis (*Figure 32*).

Step 6 Erase all unneeded construction elements, leaving behind the completed oblique sketch.

Figure 30 Oblique baseline.

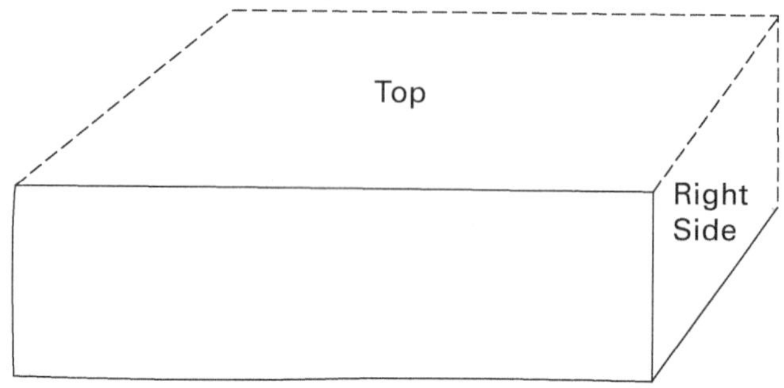

Figure 31 Sketching top and right-side views.

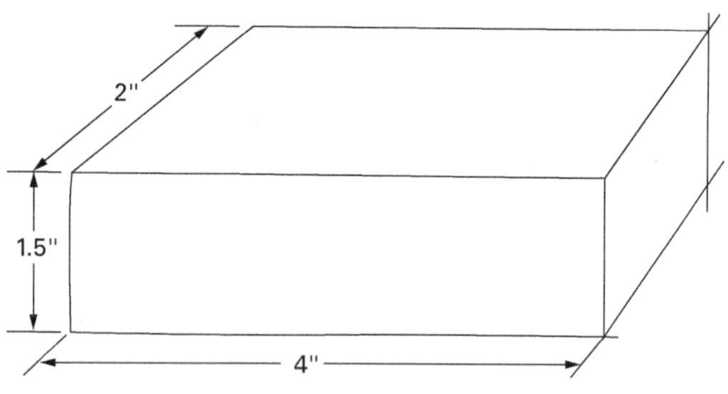

Figure 32 Sketching dimensions.

1.4.3 Isometric Sketches

Isometric sketches resemble oblique ones, but they preserve line length better. They grow from three major lines called *isometric baselines* (or *isometric axes*). The left-side and the right-side isometric baselines form 30-degree angles to the horizontal baseline. The vertical baseline forms a 90-degree angle to the horizontal baseline (*Figure 33*).

Follow these steps to make an isometric sketch of a rectangular object:

Step 1 Sketch the horizontal baseline.

Step 2 Sketch the two 30-degree isometric baselines and the 90-degree vertical baseline.

Step 3 Sketch the object's length on the right-side isometric baseline.

Step 4 Sketch the object's width on the left-side isometric baseline.

Step 5 Sketch the object's height on the vertical isometric baseline (*Figure 34*).

Step 6 Complete the sketch by creating any remaining lines parallel to the three isometric baselines (*Figure 35*).

Step 7 Add the dimensions, making them parallel to each axis (*Figure 36*).

Step 8 Erase all unneeded construction elements, leaving behind the completed isometric sketch.

1.4.4 Sketching 3D Objects with a Tablet

You can create pictorial sketches with any drawing app by using the techniques that you've just learned. The app will make things easier since you'll be able to use the built-in tools to create the key drawing elements.

Some drawing apps can create three-dimensional drawings directly. These tools take a lot longer to learn than basic drawing apps, but the results will be excellent. The best way to master a 3D drawing program is to watch a few free video tutorials online. Then spend some time practicing until you master the app.

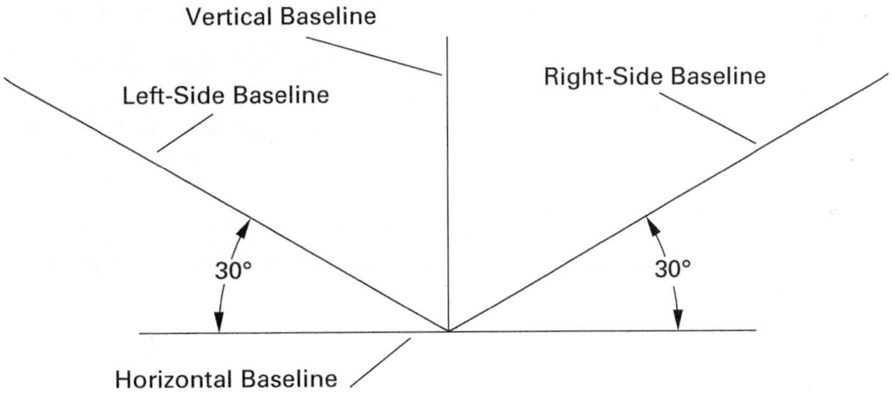

Figure 33 Isometric baselines.

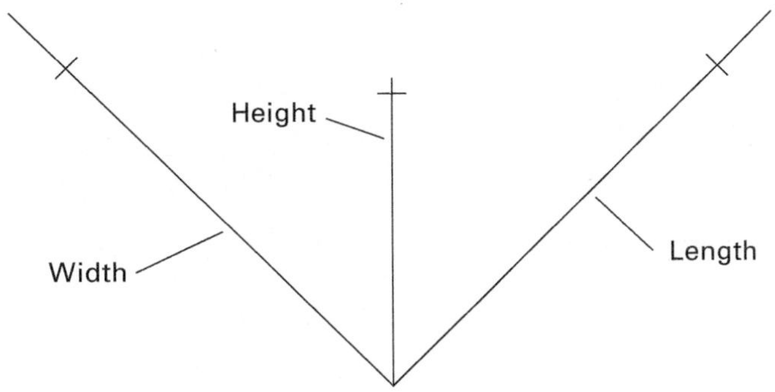

Figure 34 Laying out length, width, and height.

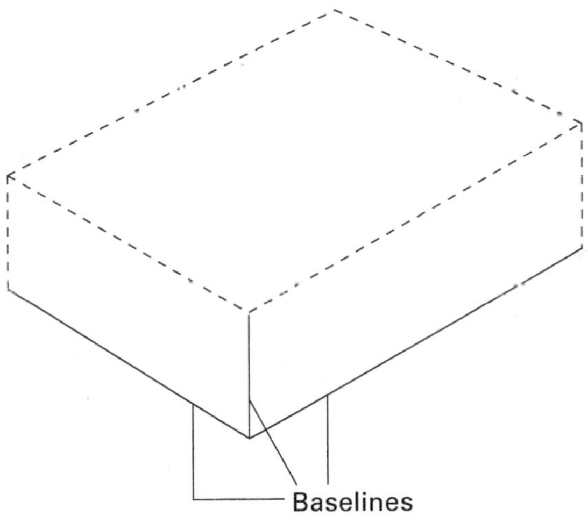

Figure 35 Completing the isometric sketch.

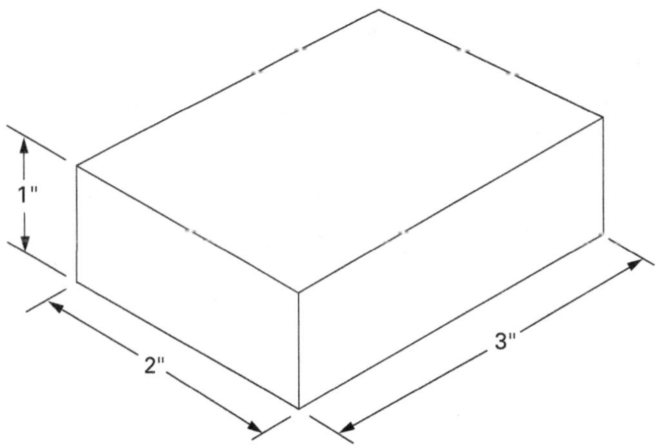

Figure 36 Sketching dimensions.

Did You Know?

Artist, Engineer, and Construction Expert

During the Renaissance, some people held several jobs that rarely go together today. Filippo Brunelleschi (1377–1446) is an amazing example. He started out as a goldsmith, sculptor, and artist. In this capacity, he developed many of the pictorial drawing techniques that we use today.

Next, he became an architect and engineer. His most famous design, the dome of the cathedral in Florence, Italy, occupied the last 26 years of his life.

But Brunelleschi knew that being a good designer was not enough. He also supervised construction, coordinating the work of many crafts. To speed construction, he invented several machines, including a multi-speed hoist and crane for lifting materials.

He even included human factors in his management style. Workers wasted time and energy climbing to and from the dome at lunch. So, he had lunch delivered, allowing workers to enjoy it hundreds of feet off the ground.

Figure Credit: iStock@juriskraulis

1.0.0 Section Review

1. What is the *best* way to sketch a straight line?
   a. Use finger motion.
   b. Use wrist and finger motion.
   c. Use the whole arm.
   d. Use wrist motion.

2. To sketch an arc, start with a pair of _____.
   a. isometric views
   b. perpendicular lines
   c. ellipses
   d. circles

3. Extension lines should be drawn to the part at a _____.
   a. 0-degree angle
   b. 45-degree angle
   c. 60-degree angle
   d. 90-degree angle

4. Which of the following preserves line lengths and requires just one sketch to represent a three-dimensional object?
   a. Orthographic sketch
   b. Isometric sketch
   c. Oblique sketch
   d. Hidden-line sketch

Review Questions

1. Millwrights produce sketches to communicate _____.
   a. an artistically perfect representation of an object
   b. information about wiring specifications
   c. the proper PPE to wear when using the assembly
   d. key ideas about parts and assemblies

2. What is the *best* way to sketch vertical lines?
   a. From bottom to top
   b. From top to bottom
   c. As a series of small, connected lines
   d. Parallel to the paper's bottom edge

3. To sketch a 45-degree angle, _____.
   a. construct two equilateral triangles
   b. divide a 90-degree angle in half
   c. connect three 15-degree angles
   d. divide a circle into eight parts

4. To sketch a circle, construct two perpendicular lines equal to the circle's _____.
   a. radius
   b. circumference
   c. diameter
   d. chord

5. What do ellipses often represent in sketches?
   a. Round objects seen at an angle
   b. Fillets, inside corners, and edges
   c. Orthographic views
   d. Isometric perspectives

6. How should you draw arrowheads on dimension lines?
   a. With gaps between the arrowheads and the line
   b. At 90-degree angles to the dimension line
   c. Just touching the extension lines
   d. As small as possible

7. Centerlines are drawn with _____.
   a. solid lines colored red
   b. a series of short dashes
   c. a series of dots
   d. short and long dashes

8. If you cannot fit a dimension between extension lines, you should _____.
   a. turn it sideways to make it fit
   b. sketch it outside the extension lines
   c. sketch it as small as possible
   d. make it fit by any method that works

9. Any sketch that somewhat realistically represents a three-dimensional object by a single view is _____.
   a. translational
   b. multi-view
   c. orthographic
   d. pictorial

10. An oblique sketch distorts all views but the _____.
    a. left-side view
    b. right-side view
    c. front view
    d. top view

Trade Terms Quiz

Fill in the blank with the correct term learned from your study of this module.

1. A(n) ____________ sketch portrays a three-dimensional object with several two-dimensional views.

2. To represent a three-dimensional object with a single drawing that shows all three faces equally tilted, create a(n) ____________ sketch.

3. ____________ sketches represent three-dimensional objects somewhat realistically.

4. Geometric figures resembling stretched circles are ____________.

5. To represent a three-dimensional object with a single drawing that shows one face parallel to the paper surface, create a(n) ____________ sketch.

6. A freehand drawing that represents an object's key features is a(n) ____________.

Trade Terms

Ellipses
Isometric
Oblique
Orthographic
Pictorial
Sketch

Trade Terms Introduced in This Module

Ellipses: Stretched circles, similar to ovals.

Isometric: A drawing method that represents a three-dimensional object with all three faces equally tilted toward the paper surface.

Oblique: A drawing method that represents a three-dimensional object with one face parallel to the paper surface.

Orthographic: A drawing method that represents a three-dimensional object by several two-dimensional views.

Pictorial: Any drawing method that represents three-dimensional objects somewhat realistically.

Sketch: A rough drawing that represents an object's key features.

Additional Resources

This module presents thorough resources for task training. The following reference material is suggested for further study.

Architectural Drawing and Light Construction, 8th Edition. Philip A. Grau III, Edward J. Muller, James G. Fausett. Upper Saddle River, NJ: Pearson.

Figure Credits

iStock@Nostal6ie, Module Opener

iStock@Prykhodov, Figure 1

Section Review Answer Key

Section 1.0.0

Answer	Section Reference	Objective
1. c	1.1.1	1a
2. b	1.2.1	1b
3. d	1.3.1	1c
4. b	1.4.3	1d

NCCER CURRICULA — USER UPDATE

NCCER makes every effort to keep its textbooks up-to-date and free of technical errors. We appreciate your help in this process. If you find an error, a typographical mistake, or an inaccuracy in NCCER's curricula, please fill out this form (or a photocopy), or complete the online form at **www.nccer.org/olf**. Be sure to include the exact module ID number, page number, a detailed description, and your recommended correction. Your input will be brought to the attention of the Authoring Team. Thank you for your assistance.

Instructors – If you have an idea for improving this textbook, or have found that additional materials were necessary to teach this module effectively, please let us know so that we may present your suggestions to the Authoring Team.

NCCER Product Development and Revision

13614 Progress Blvd., Alachua, FL 32615

Email: curriculum@nccer.org
Online: www.nccer.org/olf

❑ Trainee Guide ❑ Lesson Plans ❑ Exam ❑ PowerPoints Other ____________

Craft / Level: Copyright Date:

Module ID Number / Title:

Section Number(s):

Description:

Recommended Correction:

Your Name:

Address:

Email: Phone:

This page is intentionally left blank.

Glossary

Accredited: In the context of NCCER craft training programs, a status granted to an organization or site recognized as having met the established minimum standards and criteria for the administration of NCCER training and assessment programs.

Apprentice: A person learning a craft through practical experience with skilled workers, as well as through instruction and study, while earning a wage.

Baselines: Key reference lines around which all layout elements develop.

Benchmarks: Metal markers that precisely identify specific locations.

Bill of materials (BOM): The list of parts and materials required to construct a system or assembly.

Bisect: To divide into two equal parts.

Blind hole: A hole that does not go completely through the material and therefore has a bottom.

Blind rivet: A fastener that installs from one side of the work surface and requires a special tool to expand and tighten it.

Bonnet: The removeable cover of a valve body, through which the valve stem usually passes. Valve repair typically requires removal of the bonnet to gain access to the valve components beneath it.

Break-away torque: The torque required to loosen a fastener.

Centerlines: Lines made from alternating long and short dashes that identify the middle of parts or features, such as holes.

Cheater bar: A length of pipe or similar material slipped over a wrench or other tool to substantially increase leverage. A cheater bar creates a hazard, as tools and hardware are not constructed to withstand the added stress.

Chuck: A special type of clamp equipped with jaws in a radial pattern that is typically used to hold rotating tools, such as drill bits and thread taps. Some chucks require a chuck key, which is a small, geared tool used to loosen and tighten the jaws of the chuck.

Clearance hole: A fastener hole large enough for everything but the fastener's head to pass through.

Computer numerical control (CNC) machines: Equipment used in a subtractive manufacturing process based on computerized controls and machine tools that remove material from a workpiece to produce a component. Many materials can be machined in this manner, including metals, plastics, and composites.

Countersunk: A fastener installation method in which the head is flush with or below the surface.

Desiccant: A material such as silicon dioxide (silica gel) that attracts and captures moisture from the surrounding air or nearby moist materials.

Dies: Threading tools typically shaped like a nut, used to cut external (male) threads on properly sized round stock.

Dimension line: A size-, location-, or angle-identifying number combined with an arrow or pair of arrows.

Dress: The process of forming, shaping, and changing the finish of stone to a desired shape or condition.

Drive coupling: A mechanical component used to connect the shaft of a driver, such as an electric motor, to the shaft of a driven load, such as a pump or compressor. The two shafts must be precisely aligned so that they share a common centerline. Drive couplings are typically classified as either rigid or flexible.

Elastic limit: The point after which stretching a material permanently deforms it.

Ellipses: Stretched circles, similar to ovals.

Extension lines: Pairs of short lines perpendicular to a drawing element that identify a dimension's limits.

Ferrous: Containing or consisting of iron.

Fillet: A concave, rounded surface that serves as a transition between two intersecting surfaces. In welding, a fillet is formed when two pieces of metal are joined at an inside corner.

Foot-pound (ft-lb): A US standard torque unit equal to the force of one pound applied to a lever one foot long. One foot-pound equals 12 inch-pounds. Also called the *pound-foot (lb-ft)*.

Friction: A motion-resisting force produced when two surfaces rub together.

Grade: A fastener-ranking system that identifies quality and strength.

Grit: A number indicating the density of particles for grinding stones, honing stones, sandpaper, and similar materials. High numbers indicate smaller particles packed more closely together (fine), while low numbers indicate larger particles spaced farther apart (coarse).

Handedness: The direction that the threads wrap around a fastener, determining how it tightens or loosens.

Hidden lines: Lines made from short dashes that reveal features not visible in a particular view.

Honing: A process that uses abrasive stones to create a smooth, precise surface on a metal workpiece in a controlled, consistent manner.

Inch-pound (in-lb): A small US standard torque unit equal to the force of one pound applied to a lever one inch long. One inch-pound equals $^1/_{12}$ foot-pound. Also called the *pound-inch (lb-in)*.

Intersection: The point where two lines or arcs cross.

Isometric (ISO) drawing: A drawing method that represents a three-dimensional object with all three faces equally tilted toward the drawing surface.

Jig: A device constructed or assembled to maintain a positional relationship between a workpiece and a tool or between components of an assembly.

Lapping plates: Coarse stones used to rub other stones and restore their flat surface.

Lathe: A machine tool designed to shape metal, wood, and similar materials using a rotating drive that turns the workpiece in opposition to handheld or stationary cutting tools.

Micrometers: Precision measuring tools that measure small distances or thicknesses between two faces, adjusted to contact the workpiece(s) by turning an adjusting barrel.

Newton-meter (Nm): A metric torque unit equal to the force of one newton applied to a lever one meter long.

Nominal size: The typical diameter of a bolt or screw.

Oblique: A drawing method that represents a three-dimensional object with one face parallel to the paper surface.

Orthographic projection: A drawing method that represents a three-dimensional object by several two-dimensional views.

Packing: Rope-like material with a consistent shape used to prevent fluid leakage by compressing it around a shaft or valve stem.

Perpendicular: A line that is at a 90-degree angle to another line.

Pictorial: Any drawing method that represents three-dimensional objects somewhat realistically.

Pipe scale: Deposits remaining on new ferrous pipe products left over from the heating and forming process. In operational piping systems, scale refers to the buildup of corrosion and/or mineral deposits on the inside walls of the piping.

Pitch: Relative to threaded hardware and components, refers to the number of threads found over a unit of distance, e.g., a $^1/_4$" bolt with coarse threads has a pitch of 20 threads per inch.

Plasma arc cutters: Equipment used like a cutting torch to cut through electrically conductive materials using a jet of superheated plasma at a temperature of roughly 40,000°F (~22,200°C) generated by a handheld or automated torch head. Plasma cutting offers improved speed and precision compared to traditional oxyfuel cutting processes.

Polyband belts: Drive belts that combine two or more belts on a common band, side by side.

Ratio: A proportional relationship between two or more numbers.

Run-down resistance: Additional torque required to overcome the resistance created by extra hardware, such as washers and locknuts.

Schematics: A diagram-type drawing that shows a system's essential components and interconnections.

Set/seizure: Fastener lockup that occurs when approaching the target torque.

Sheave: A grooved wheel, also called a *pulley*, mounted to a shaft and used to transmit power from a driver to a driven load through drive belts that ride in its grooves.

Shim stock: Thin material, typically metal, placed between two objects to fill space and allow for the precise positioning of one or both objects. Shims are often used in *shaft alignment* to adjust the position of the motor relative to the driven shaft. Shim stock typically refers to the raw material from which individual shims are cut.

Shutdowns: Periods of time when facility production is largely halted. Although the term can be applied to the cessation of production for any reason, industrial crafts often use the term *shutdown* to identify pre-planned production interruptions that may last for days or even weeks to allow for scheduled maintenance and significant changes to production equipment.

Sketch: A rough drawing that represents an object's key features.

Stuffing box: A casing or chamber in which packing material is placed and compressed around a moving shaft or axle to reduce or eliminate gas or liquid leakage.

Tangent: A line that touches a circle's circumference at exactly one point.

Taps: Threading tools used to cut internal (female) threads in properly sized round holes.

Tensile strength: The point at which stretching a material breaks it.

Tension: A pulling force, such as that produced by some springs.

Thread class: A code indicating the amount of looseness between the threads of matching bolts and nuts.

Thread pitch: The distance from the crest (top) of one thread to the next.

Thread series: Combinations of fastener diameter and threads-per-inch associated with specific thread diameters.

Thread standard: An established set of specifications for machining threads.

Thread-locking fluid: A hardening liquid placed on threaded fasteners to make them more resistant to vibration loosening.

Threads-per-inch (tpi): The number of threads in one inch of fastener length.

Tolerance: A term that quantifies the allowable difference or deviation from specified limits.

Torque: A twisting force that produces rotary motion.

Turbine: A machine for producing power in which a wheel or rotor fitted with a series of vanes is forced to rotate by the flow of steam, gas, water, air, or similar fluids.

Ultrasonic: Related to the use of sound waves moving at a frequency above the upper auditory limit of the human ear (about 20,000 hertz).

View: Part of an orthographic projection, a two-dimensional drawing that accurately represents a three-dimensional object from one vantage point.

This page is intentionally left blank.

This page is intentionally left blank.

This page is intentionally left blank.

This page is intentionally left blank.

This page is intentionally left blank.

This page is intentionally left blank.

This page is intentionally left blank.

This page is intentionally left blank.

This page is intentionally left blank.